RESEARCH METHODOLOGY IN FOOD SCIENCES

Integrated Theory and Practice

RESEARCH METHODOLOGY IN FOOD SCIENCES

Integrated Theory and Practice

Edited by

C. O. Mohan, PhD

Elizabeth Carvajal-Millan, PhD

C. N. Ravishankar, PhD

AAP | APPLE ACADEMIC PRESS

Apple Academic Press Inc.
3333 Mistwell Crescent
Oakville, ON L6L 0A2
Canada

Apple Academic Press Inc.
9 Spinnaker Way
Waretown, NJ 08758
USA

International Standard Book Number-13: 978-1-77188-624-6 (Hardcover)
International Standard Book Number-13: 978-1-315-11435-4 (eBook)

Library and Archives Canada Cataloguing in Publication

Research methodology in food sciences : integrated theory and practice /
edited by C.O. Mohan, PhD, Elizabeth Carvajal-Millan, PhD, C.N. Ravishankar, PhD.

Includes bibliographical references and index.
Issued in print and electronic formats.
ISBN 978-1-77188-624-6 (hardcover).--ISBN 978-1-315-11435-4 (PDF)

1. Food industry and trade--Research. I. Carvajal-Millan,
Elizabeth, editor II. Mohan, C. O., editor III. Ravishankar, C. N., editor

TP370.8.R47 2018 664 C2018-902275-2 C2018-902276-0

Library of Congress Cataloging-in-Publication Data

Names: Mohan, C. O., editor. | Carvajal-Millan, Elizabeth, editor. | Ravishankar, C. N., editor.
Title: Research methodology in food sciences : integrated theory and practice / editors,
C.O. Mohan, PhD, Elizabeth Carvajal-Millan, PhD, C.N. Ravishankar, PhD.
Description: Toronto ; Waretown, New Jersey : Apple Academic Press, 2018. | Includes bibliographical references and index.
Identifiers: LCCN 2018018958 (print) | LCCN 2018021104 (ebook) |
ISBN 9781315114354 (ebook) | ISBN 9781771886246 (hardcover : alk. paper)
Subjects: LCSH: Food science--Research. | Food--Analysis. | Food--Safety measures.
Classification: LCC TP370.8 (ebook) | LCC TP370.8 .R47 2018 (print) | DDC 664/.07--dc23
LC record available at https://lccn.loc.gov/2018018958

Apple Academic Press also publishes its books in a variety of electronic formats. Some content that appears in print may not be available in electronic format. For information about Apple Academic Press products, visit our website at **www.appleacademicpress.com** and the CRC Press website at **www.crcpress.com**

ABOUT THE EDITORS

C. O. Mohan, PhD

C. O. Mohan, PhD, is currently a Scientist at the Fish Processing Division of the Central Institute of Fisheries Technology (Indian Council of Agricultural Research [ICAR]), Willingdon Island, Kochi, Kerala, India. Dr. Mohan is from the historic place of Chitradurga, Karnataka, India. He graduated in fisheries sciences from the College of Fisheries, Mangalore, Karnataka. During his master's and PhD studies, he specialized in fish processing technology at the ICAR-Central Institute of Fisheries Education, Deemed University, Mumbai, India. His areas of interest are thermal processing and active and intelligent packaging. He has guided many postdoctoral students and has published in many reputed national and international journals. He has an h-index of 8.0. He has been awarded with the Jawaharlal Nehru Award for his doctoral thesis research from the ICAR, New Delhi.

Elizabeth Carvajal-Millan, PhD

Elizabeth Carvajal-Millan, PhD, is a Research Scientist at the Research Center for Food and Development (CIAD), Hermosillo, Mexico. She obtained her PhD in France at Ecole Nationale Supérieure Agronomique à Montpellier (ENSAM), her MSc degree from CIAD, and her undergraduate degree from the University of Sonora in Mexico. Her research interests are focused on biopolymers, mainly in the extraction and characterization of high-value-added polysaccharides from co-products recovered from the food industry and agriculture, especially ferulated arabinoxylans. In particular, Dr. Carvajal-Millan studies covalent arabinoxylans gels as functional systems for the food and pharmaceutical industries. Globally, Dr. Carvajal-Millan is a pioneer in in vitro and in vivo studies on covalent arabinoxylans gels as carriers for oral insulin focused on the treatment of diabetes type 1. She has published 57 refereed papers, 23 chapters in books, over 80 conference presentations, and has one patent registered, with two more submitted.

C. N. Ravishankar, PhD

C. N. Ravishankar, PhD, is at present the Director of the Indian Council of Agricultural Research (ICAR), Central Institute of Fisheries Technology (CIFT), Cochin, India. He completed his graduate studies in fisheries sciences and specialized in fish processing technology during his masters and PhD degrees from the College of Fisheries, Mangalore, Karnataka, India. He is an expert in the field of fish processing and packaging, and he developed, popularized, and transferred many technologies to the seafood industry. He participated in the First Indian Antarctic Expedition and traveled widely abroad for training and consultancy programs. He has more than 200 international and national publications to his credit, and he has an h-index of 15.0 and has filed 17 patents. He received the Outstanding Team Research Award in the field of fish products technology from the ICAR, New Delhi, the K. Chidambaram Memorial Award from the Fisheries Technocrats Forum, as well as a Gold Medal for his PhD work and a Merit Certificate from the Royal Institute of Public Health & Hygiene, London. He was instrumental in establishing the Business Incubation Centre with an office and pilot plant facility for entrepreneurship development in fish and other food products. In addition to his many other activities, he has delivered numerous invited talks on fish preservation techniques, food packaging, business incubation, and other related areas.

CONTENTS

LIST OF CONTRIBUTORS

Cristóbal N. Aguilar
Food Research Department, Chemistry School, Coahuila Autonomous University, Saltillo Unit 25280, Coahuila, México. E-mail: cristobal.aguilar@uadec.edu.mx

Miguel A. Aguilar-González
Cinvestav, Center for Research and Advanced Studies, IPN Unit Ramos Arizpe, Coahuila, México

Olga B. Alvarez-Pérez
Department of Food Research, School of Chemistry, Universidad Autónoma de Coahuila, Saltillo 25280, Coahuila, México

Teuku Beuna Bardant
Indonesian Institute of Science, Jakarta, Indonesia

Daniel Boone-Villa
School of Health Sciences, University of the Valley of Mexico campus Saltillo, Tezcatlipoca 2301, Saltillo 25204, Fraccionamiento El Portal, Coahuila, Mexico

Gabriela Sanchez Brambila
Russell Research Center-ARS, Quality and Safety Assessment Research Unit USDA, 950 College Station Road, Athens 30605, GA, USA

A. C. Campa-Mada
Research Center for Food and Development, CIAD, AC., Hermosillo 83000, Sonora, Mexico

Antonio Francisco Aguilera Carbo
Department of Animal Nutrition, Universidad Autonoma Agraria Antonio Narro, Calzada Antonio Narro 1923, Colonia Buenavista, Saltillo 25315, Coahuila, Mexico Z

Elizabeth Carvajal-Millan
Biopolymers, CTAOA, Centro de Investigación en Alimentación y Desarrollo, Hermosillo, Sonora, México. E-mail: ecarvajal@ciad.mx

Cecilia Castro-López
Departamento de Investigación en Alimentos, Facultad de Ciencias Químicas, Universidad Autónoma de Coahuila, Venustiano Carranza e Ing, José Cárdenas s/n, Col. República, Saltillo 25280, Coahuila, México

Xochitl Ruelas Chacon
Department of Food Research, Faculty of Chemistry, Universidad Autonoma de Coahuila, Blvd. V. Carranza, Colonia Republica Oriente, Saltillo 25280, Coahuila, Mexico.
E-mail: xochitl.ruelas@uaaan.mx

Juan C. Contreras-Esquivel
Food Research Department, School of Chemistry, University Autonomous of Coahuila, Saltillo CP 25280, Coahuila, Mexico

Sanjoy Das
ICAR–Central Institute of Fisheries Technology, Wellington Island, Matsyapuri P.O., Cochin 682029, India

G. P. Girija
Mumbai Research Centre of Central Institute of Fisheries Technology, CIDCO Admin Building, Sector 1, Vashi, Navi Mumbai 400703, Maharashtra, India

Bindu J.
ICAR–Central Institute of Fisheries Technology, Wellington Island, Matsyapuri P.O., Cochin 682029, India. E-mail: bindujaganath@gmail.com

Pankaj Kishore
ICAR–Central Institute of Fisheries Technology, Wellington Island, Matsyapuri P.O., Cochin 682029, India

Jorge A. Marquez-Escalante
Research Center for Food and Development, CIAD, A. C., Hermosillo 83000, Sonora, Mexico. E-mail: marquez1jorge@gmail.com

José L. Martínez
Food Research Department, Chemistry School, Coahuila Autonomous University, Saltillo Unit 25280, Coahuila, México

Guillermo C. G. Martinez-Avila
Facultad de Agronomía, Universidad Autónoma de Nuevo León, Francisco Villa s/n, Col. Ex-Hacienda El Canadá, Escobedo 66050, Nuevo León, México

Gloria Alicia Martínez-Medina
Food Research Department, Chemistry School, Coahuila Autonomous University, Saltillo Unit 25280, Coahuila, México

Karla Martínez-Robinson
Biopolymers, CTAOA, Centro de Investigación en Alimentación y Desarrollo, Hermosillo, Sonora, México

K. G. Martínez-Robinson
Research Center for Food and Development, CIAD, AC., Hermosillo 83000, Sonora, Mexico

V. A. Minimol
ICAR–Central Institute of Fisheries Technology, Wellington Island, Matsyapuri P.O., Cochin 682029, India

Sumit Mishra
Chemistry Department, Birla Institute of Technology, Mesra, Ranchi 835215, Jharkhand, India

C. O. Mohan
ICAR–Central Institute of Fisheries Technology, Wellington Island, Matsyapuri P.O., Cochin 682029, India

Julio Montañez
Department of Chemical Engineering, Faculty of Chemistry, Universidad Autonoma de Coahuila, Blvd. V. Carranza, Colonia Republica Oriente, Saltillo 25280, Coahuila, Mexico

Diana B. Muñiz-Márquez
Engineering Department, Technological Institute of Ciudad Valles, National Technological of Mexico, Ciudad Valles 79010, San Luis Potosí, Mexico

L. N. Murthy
Mumbai Research Centre of Central Institute of Fisheries Technology, CIDCO Admin Building, Sector 1, Vashi, Navi Mumbai 400703, Maharashtra, India

Pinki Pal
Chemistry Department, Birla Institute of Technology, Mesra, Ranchi 835215, Jharkhand, India.
E-mail: pinkipalhbti@gmail.com

Satyen K. Panda
ICAR–Central Institute of Fisheries Technology, Wellington Island, Matsyapuri P.O., Cochin 682029, India

Jay Prakash Pandey
Chemistry Department, Birla Institute of Technology, Mesra, Ranchi 835215, Jharkhand, India

Ramona Pérez-Leal
Faculty of Agro-Technological Sciences, Autonomous University of Chihuahua, Chihuahua 31125, Mexico

K. Porteen
Department of Veterinary Public Health and Epidemiology, Madras Veterinary College, Chennai 600007, India

Arely Prado-Barragán
Biotechnology Department, Biological and Health Sciences Division, Metropolitan Autonomous University, Iztapalapa Unit 09340, Ciudad de México, México

Nithya Quintoil
Department of Veterinary Public Health, Pondicherry, India

Nancy Ramírez-Chavez
Faculty of Agro-Technological Sciences, Autonomous University of Chihuahua, Chihuahua 31125, Mexico

Priti Rani
Chemistry Department, Birla Institute of Technology, Mesra, Ranchi 835215, Jharkhand, India.
Email: pritijolly2@gmail.com

A. Rascón-Chu
Research Center for Food and Development, CIAD, AC., Hermosillo 83000, Sonora, Mexico

Agustin Rascón-Chu
Biopolymers, CTAOA, Centro de Investigación en Alimentación y Desarrollo, Hermosillo, Sonora, México

C. N. Ravishankar
ICAR-Central Institute of Fisheries Technology, Willingdon Island, Matsyapuri P.O., Cochin 682029, India

Raúl Rodríguez-Herrera
Department of Food Research, School of Chemistry, Universidad Autónoma de Coahuila, Saltillo 25280, Coahuila, México

Rosa M. Rodríguez-Jasso
Department of Food Research, School of Chemistry, Universidad Autónoma de Coahuila, Saltillo 25280, Coahuila, México

Rene Dario Peralta Rodriguez
Research Center for Applied Chemistry, Blvd. Enrique Reyna Hermosillo No. 140, Saltillo 25253, Coahuila, Mexico

Romeo Rojas
Universidad Autónoma de Nuevo León, School of Agronomy, Research Center and Development for Food Industry, General Escobedo 66050, Nuevo León, México

Orlando de la Rosa
Food Research Department, School of Chemistry, University Autonomous of Coahuila, Saltillo CP 25280, Coahuila, Mexico

S. Wilfred Ruban
Department of Livestock Products Technology, Veterinary College, Hassan 573201, India

Héctor Ruiz
Food Research Department, Chemistry School, Coahuila Autonomous University, Saltillo Unit 25280, Coahuila, México

Juan Salmerón-Zamora
Faculty of Agro-Technological Sciences, Autonomous University of Chihuahua, Chihuahua 31125, Mexico

R. Sathishkumar
Department of Botany, PSG College of Arts and Science, Coimbatore, India

Gautam Sen
Chemistry Department, Birla Institute of Technology, Mesra, Ranchi 835215, Jharkhand, India

Heru Susanto
Indonesian Institute of Science, Jakarta, Indonesia
Department of Information Management, College of Management, Tunghai University, Taichung City, Taiwan. E-mail: heru.susanto@lipi.go.id

V. Urias-Orona
Research Center for Food and Development, CIAD, AC., Hermosillo 83000, Sonora, Mexico. E-mail: arascon@ciad.mx

Maria de la Luz Reyes Vega
Department of Food Research, Faculty of Chemistry, Universidad Autonoma de Coahuila, Blvd. V. Carranza, Colonia Republica Oriente, Saltillo 25280, Coahuila, Mexico

Janeth M. Ventura-Sobrevilla
Departamento de Investigación en Alimentos, Facultad de Ciencias Químicas, Universidad Autónoma de Coahuila, Venustiano Carranza e Ing, José Cárdenas s/n, Col. República, Saltillo 25280, Coahuila, México. E-mail: janethventura@uadec.edu.mx

Ina Winarni
Forest Products Research and Development Center, Bogor, Indonesia

Jorge E. Wong-Paz
Engineering Department, Technological Institute of Ciudad Valles, National Technological of Mexico, Ciudad Valles 79010, San Luis Potosí, Mexico

LIST OF ABBREVIATIONS

ACE I	angiotensin I converting enzyme inhibitor
ACE I	angiotensin I converting enzyme inhibitor
ALLs	acute lymphoblastic leukemias
APAF-1	apoptosis activating factor
AR	androgen receptor
ATF	arrow tooth flounder
BCC	basal cell carcinoma
BOD	biochemical oxygen demand
BPH	benign prostatic hyperplasis
CAGR	compound annual growth rate
CCRD	central composite rotatable design
CMHPG	O-carboxymethyl-O-hydroxypropyl guar gum
CMHTPG	carboyxymethyl-O-2-hydroxy-3-(trimethylammonia propyl) guar gum
COD	chemical oxygen demand
CONACyT	Mexican Council for Science and Technology
CVDs	cardiovascular diseases
DBD	DNA-binding domain
DE	degree of esterification
DHFR	dihydrofolate reductase
DSC	differential scanning calorimetry
dTMP	thymidine monophosphate
dUMP	deoxyuridine monophosphate
EBV	Epstein–Barr virus
EFB	empty fruit bunch
EGF	epidermal growth factor
EGFR	epidermal growth factor receptor
EIA	enzyme immunoassay
ELC	essential light chain
ELISAs	enzyme-linked immunosorbent assays
EMEA	European Medicines Agency
ER	estrogen receptors
EVOH	ethyl vinyl alcohol

FDA	Food and Drug Administration
FOS	fructooligosaccharides
FRAP1	rapamycin associated protein 1
GALT	gut-associated lymphoid tissue
GG	guar gum
GOS	galactooligosaccharides
GPER	G protein-coupled receptor GPR30
GR or GCR	glucocorticoid receptor
GSP	general secretory pathway
HER2	human epidermal growth factor receptor 2
HHP	high hydrostatic pressure
HM	high methoxy pectin
HMM	heavy meromyosin
HPLC	high-performance liquid chromatography
HPP	high-pressure processing
HPP	high hydrostatic pressure processing
HTPG	O-2-hydroxy-3-(trimethylammonia propyl) guar gum
IFA	immuno fluorescent antibody
IFT	Institute of Food Technologist
IGF-IR	insulin-like growth factor-IR
INIFAP	Investigation in Forestry, Agriculture and Animal Production in Mexico
LAB	lactic acid bacteria
LBD	ligand-binding domain
LE	leupeptin
LM	low methoxy
LMM	light meromyosin
MC	moisture content
MD	molecular dynamics
Mf-MIP	myofibrillar MIP
MFP	myofibrillar protein
MHC	myosin heavy chain
MIPs	modori-inducing proteinases
MMFFs	molecular mechanics force fields
MS	mass spectrometry
mTOR	mammalian target of rapamycin
MTR	methyltransferase
MTX	methotrexate
MW	microwave

NAIP	National Agricultural Innovation Project
NASBA	nanoprobe-nucleic acid sequence-based amplification
NDOs	non-digestible oligosaccharides
OSCC	oral squamous cell carcinomas
PAMPs	pathogen-associated molecular patterns
PCNSL	primary central nervous system lymphoma
PCR	polymerase chain reaction
PDB	Protein Data Bank
PHA	polyhydroxyalkanoates
PHGG	partially hydrolyzed guar gum
PLA	polylactic acid
PR+	progesterone receptors
PSMA	prostate-specific membrane antigen
RH	relative humidity values
RLC	regulatory light chain
RSM	response surface methodology
RSM-CI	response surface methodology
SB	soybean trypsin inhibitor
SCC	squamous–cell carcinoma
SCCs	silver carbene complexes
SCFA	short-chain fatty acid
SEM	scanning electron microscopy
SERS	surface-enhanced Raman scattering
SmF	submerged fermentation
SPF	supercritical fluids hydrolysis
Sp-MIP	sarcoplasmic MIP
SSF	solid state fermentation
TBA	thiobarbituric acid
TLR	toll-like receptors
TMA	trimethylamine nitrogen
TMAO	trimethylamine oxide
TMR	Transparency Market Research
TNM	tetranitromethane
TRAMP	transgenic adenocarcinoma of the mouse prostate
TS	tensile strength
TVB-N	total volatile base nitrogen
UAAAN	Universidad Autonoma Agraria Antonio Narro
UHP	ultrahigh-pressure processing
VEGF	vascular endothelial growth factor

WEAX	water extractable arabinoxylan
WHC	water holding capacity
WPC	whey protein concentrates
WPI	whey protein isolates
WUAX	water unextractable
WVP	water vapor permeability
WVTR	water vapor transmission rate

PREFACE

As food is one of the basic necessities of our life, utmost importance is given to it across the globe. This is evidenced by the increase in the number of food science courses offered by universities from all the countries. The major emphasis of the food science courses is to inculcate specialization in food technology or food engineering aspects. Introduction of new varieties of foods and ingredients has resulted in the need to understand in detail the entire processing operations. This new book, *Research Methodology in Food Sciences: Integrated Theory and Practice,* helps by providing necessary scientific knowledge on the basic principles of the latest food processing aspects to the readers in simple ways. The book is intended as a resource on the recent research innovations in the field of food processing and food engineering. Chapters are written by eminent researchers in the field of food science and provide in-depth knowledge to readers. Considering the vast variations in readers' disciplines and programs, different topics are included in this book to engage different audiences. This book will help to develop confidence in students and readers to become professionals in food science with the latest innovations in food process engineering.

The topics selected in this book illustrate the application of engineering aspects in food processing, packaging, and ensuring food safety. For better understanding, the chapters have been divided under three different sections: research methodology and practice, food science and technology research, and special topics. Chapter 1 discusses the latest nanotechnology aspects for the detection of foodborne pathogens to ensure safety with respect to these pathogens. The characteristics of edible films prepared from plasticized guar gum are highlighted in Chapter 2. Detailed descriptions on food borne pathogens are given in Chapter 3. Chapter 4 underpins the effect of calcium on the gelation of pectins prepared from chickpea. Detailed notes on the antioxidant activity of oats harvested from draught area are given in Chapter 5.

Under the section on food science and technology research, emphasis is given on production of novel biomolecules and their characterization. In Chapter 6, a detailed note on microstructural properties of arabionoxylan aerogels is given. The effect of quercetin isolated from *Enicostemma littorale* against cancer targets are highlighted in Chapter 7. The latest trends in production of ethanol and fructooligosachharides are detailed in Chapters 8 and 9. A detailed description of functional peptides is provided in Chapter 10. Applications of guar gum for industrial purpose is highlighted in Chapter 11. Selected research topics are given in the last section of this book.

This publication will be a useful guide for students, researchers, academicians, industry, technologists, and entrepreneurs engaged in the area of food processing. Appreciation is given to all the authors for their contribution in this book. The editors would like to acknowledge the effort of their peers for their time to read the drafts and provide us with technical corrections and constructive suggestions, which is invaluable. Special thanks to Prof. A. K. Haghi for his immense support and advice throughout the formulation of this book. Thanks also to the AAP staff for their time and valuable efforts in publishing this book. We thank our entire family for their encouragement, sacrifices, and support throughout our journey, which will enable us to contribute more in the future.

—C. O. Mohan, PhD
Elizabeth Carvajal-Millan, PhD
C. N. Ravishankar, PhD

PART I
Research Methodology and Practice

CHAPTER 1

NANOTECHNOLOGY FOR PATHOGEN DETECTION AND FOOD SAFETY

PANKAJ KISHORE*, SATYEN K. PANDA, V. A. MINIMOL, C. O. MOHAN, and C. N. RAVISHANKAR

ICAR–Central Institute of Fisheries Technology, Wellington Island, Matsyapuri P.O., Cochin 682029, India

Corresponding author. E-mail: pkishore2007@gmail.com

CONTENTS

ABSTRACT

Food safety is an important issue in the globalized world as there are many foodborne outbreaks around the world. These foodborne outbreaks are mainly related to bacteria pathogens like *Escherichia coli*, *Salmonella*, *Listeria*, etc. Hence, pathogens detection needs to be fast, accurate, and easy methods to monitor food quality and control. Nanotechnology is a promising field as a boon for scientific research and application in vast wider range including pathogens detection. The efficient and reliable detection systems for pathogens are very much required particularly in food safety. The traditional ways of detection are cumbersome and exhaustive. Hence, nanotechnology appeared as a better mean for pathogen detection. Detection of microorganisms using nanoparticles includes fluorescence-based detection. This can be used for identification and elimination of bacteria and antibiotic treatment. Nanotechnology in food safety comprises nanosensors and nanofiltration. This can be very useful in risk assessment studies.

1.1 INTRODUCTION

Foodborne diseases are major concern for health as well as economic concerns in both developed and developing countries which encompass a wide spectrum of illnesses. Unsafe food related to harmful bacteria, viruses, parasites, and chemical substances causes more than 200 diseases. Unsafe foods are primarily affecting infants, young children, elderly, and sick. Safe food and nutritious food is important for sustainable life and promoting good health. Diarrhea is the third leading cause in 2010 which results millions childhood deaths annually, predominantly in developing countries.[49] Major considerable factors like *Escherichia coli* O157:H7 and O:121, *Salmonella*, *Clostridium botulinum*, *Listeria monocytogenes*, etc., are responsible for many outbreaks around the globe.

A new era of technology called nanotechnology is an emerging scientific field which has a lot of potential in every field of life including food safety, which involves the design and application of structures, devices, and systems by controlling their shape and size at an extremely small scale, that is, at nanolevel one-billionth of a meter.[22] Nanotechnology products have a substantial impact on the food and feed sector, potentially offering benefits for industry and the consumer. Many are currently working and developing applications in fields like mechanical and sensorial properties of food to achieve changed taste or texture and modified nutritional value along with food safety worldwide. Nanotechnology is also be used in food packaging to ensure better protection, providing a way to have check on fresh food.

A well-known fact is that the primary causes of food spoilage and food-borne illness are microorganisms. The detection of organisms significant for human health is of prime importance to ensure the safety and quality of food. The microbial pathogens detection in foods remains a challenge in spite of potential efforts being made for decades. Numerous technologies have been developed to enumerate microorganisms and to detect and identify specific pathogens present in foods.[28] This is mainly due to several difficulties associated with food analysis, that is, the complexities of food matrices (inhibitors and microflora), the attributes of target analytes in foods (low level, heterogeneous distribution, and cell disruption during processing), including the balance between the amount of food samples and the detection assay volume.[4]

Traditional methods are dependent mostly on suitable culture media. For specific microorganism detection present in foods, which are often present in a small proportion of the total microorganisms, selective media are used to enhance the growth of the target organism(s) by suppressing the growth of the other. Although, there have been many formulations to improved media for selective enrichment and media for isolation of target microorganism(s) by identification on the plate and biochemical tests, are remains lengthy and labor intensive. Hence, there is a need to overcome with methods which has better sensitivity in less time, and which can be a valuable tool in defining the problems and outlining solutions.

The simple and rapid detection of pathogenic microorganisms is of great importance in food, medical, forensic, and environmental sciences are the need of modern era.[9,10] Although advanced technologies such as convenience-based, antibody-based, nucleic acid-based assays, and molecular methods have revolutionized the detection methodology for microbial pathogens in foods. The rapid advancement in the field of nanotechnology may provide unique approach.

1.2 NANOTECHNOLOGY

The word "nano" has been originated from the Greek word "dwarf." The idea of nanotechnology was introduced first time in 1959 by a physicist named Richard Feynman. Norio Taniguchi was first man who used the term "nanotechnology" (Professor of Tokyo Science University) in 1974. Kim Eric Drexler, an American engineer, is known as the father of nanotechnology, who is best known for popularizing the potential of molecular nanotechnology, from the 1970s and 1980s.[24]

A nanometer is one thousand-millionth of a meter which is about 60,000 times smaller than diameter of human hair or even the size of a virus. A typical paper sheet is about 100,000 nm thick, a red blood cell is about 2000–5000 nm in size, and the diameter of DNA ranges from 2.5 nm to 5 µm. Therefore, nanotechnology deals in the range of one-half the diameter of DNA of 20th part of the size of a red blood cell.[11] Further, nanomaterials are so small that it needs electron microscope whereas bacteria would need a microscope to see them.[20] Nanoparticles are those with a particle size below 100 nm which comprise unique potential enable novel applications and benefits.[42]

Nanoparticles may be defined as particles whose sizes range of 1–100 nm is called a nanoparticle, which can be dispersed in gaseous, liquid, or solid medium. These materials are bonded together and aggregated to make bulk material.[7] These properties of nanoparticles can be individual atoms or subdivide bulk materials.[7] The recent advent and rapid upcoming field of nanotechnology has unique approach for detection of pathogen and toxin in foods.[44] A novel method has been reported to incorporate dye molecules into a nanostructure to create dye-doped silica nanoparticles[51] like conjugation with antibodies for *E. coli* O157:H7 make immunofluorescent nanoparticles capable of detecting even 1 CFU/g of the bacterium within 20 min.[51] In addition, the coupling of nanotechnology with microfluidic systems has been successfully applied in molecular biology works in recent times and can be adapted to detecting pathogens and toxins[4] which can significantly reduce the total analysis time. However, the possibility of accumulation in microfluidic system by food particles is a major concern for technologies.[4]

1.3 NANOPARTICLES FORMATION

Nanoparticles are formed by either the breaking down of larger particles or controlled assembly processes which can be natural phenomena and many human industrial and domestic activities, such as cooking, manufacturing, and transportation release nanoparticles into the atmosphere. In recent technological era, nanoparticles intentionally engineered for advanced technologies and various applications.

There are two approaches for the manufacturing of nanomaterials:

1. The "top-down" approach involves the breaking down of large pieces of material to generate the required nanostructures from them. This method is particularly suitable for making interconnected and integrated structures such as in electronic circuitry via attrition/milling which involves mechanical thermal cycles. This process yields a broad size distribution of 10–1000 nm particles with varied particle shape or geometry. These nanoparticles can be applied for nanocomposites, nano-grained, and bulk materials.[37]
2. In the "bottom-up" approach via pyrolysis, inert gas condensation, solvothermal reaction, sol–gel fabrication, and structured media,[37] where single atoms and molecules are assembled into larger nanostructures. This is a very powerful method of creating identical structures with atomic precision, although to date, the manufactured materials generated in this way are still much simpler than nature's complex structures.[19]

1.4 NANOTECHNOLOGY APPLICATIONS

Nanotechnology has a lot of exciting potential benefits along with quality and safety of human foods. This technology has wide application in paints and coatings, textiles and clothing, cosmetics, nanotechnology in catalysis and in food science.[2] Nanotechnology may be used in agriculture and food production in the form of nanosensors which can be used for monitoring crop growth and pest control by early identification of plant diseases. These nanosensors can help enhance productivity and improve food safety.[36] Bacteria identification and monitoring of food quality using biosensors; intelligent, active, and smart food packaging systems; nanocapsulation of bioactive food compounds are few examples. A nanocomposite coating process may help to improve food packaging by placing anti-microbial agents directly on the surface of the coated film. This can also improve the mechanical and heat-resistance properties along with lower the oxygen transmission rate.

There are various applications comprising:

1. Edible food films made with cinnamon or oregano oil or nanoparticles of zinc, calcium, and other materials that kill bacteria which can be called as antimicrobial packaging.

2. Nanoenhanced barrier keeps oxygen sensitive foods fresher means improved storage of foods.
3. Nanoencapsulation improves solubility of vitamins, antioxidants, omega-3 oils, and other "nutraceuticals."
4. Nanofibers made from lobster shells or organic corn is both antimicrobial and biodegradable which is named as green packaging.
5. Cloth saturation technique with nanofibers providing slow releases pesticides, eliminating the need of additional spraying and reducing chemical leakage into the water supply-pesticide reduction.
6. Nanobarcodes may be created to tag individual products and trace outbreaks and recalls.
7. Food spreadability and stability improve with nano-sized crystals and lipids for better low-fat foods.
8. The detection of *E. coli* bacteria using immunofluorescent nanoparticles.
9. Nano carbohydrate particles binded with bacteria can be detected and eliminated—bacteria identification and elimination.
10. Biofilm control to reduce pathogens occurrence.[15]
11. Combined detection system of PCR and oligonucleotide-labeled nanoparticle (Luminex, Austin, TX) for molecular serotyping of *Salmonella*.[5,16]
12. Isothermal amplification technology and transcription-mediated amplification assay (TMA) for the rapid detection of human health significant pathogens like *Listeria* spp., *Salmonella* spp., and *Campylobacter* spp. can be used, which has been developed by Gen-Probe by San Diego, CA, USA.[38]
13. A liposome-PCR assay can be used for the ultrasensitive detection of biological toxins.[29]

1.5 NANOTECHNOLOGICAL APPLICATIONS IN MICROBIOLOGY AND SAFETY OF FOOD

Potential progression in the nanotechnology and its applications to the field of food quality and their safety has lead intensive research in this 21st century for consumer protection from foodborne illness. Chen et al.[8] and Weiss et al.[47,48] reported four principal applications of nanotechnology comprising packaging, process technology, microbiology, and ingredients where intensive research is currently in progress in the food industry. In food microbiology, this technique has led to improve the effectiveness of preservatives, that is, materials which inhibit the growth of or kill microorganisms.[14,17,25,33,46] As foods are mostly affected by microbes, nanotechnology made its application for prevention of growth of spoilage and pathogenic microbes along with not letting them to attach and make biofilms. Other tools provide application in investigating microbes by detecting their attachment to food contact surfaces or specifying microbes and their growth in foods.[3] The main applications of nanotechnology in food safety are antimicrobial effect of nanoparticles and nanosensors for detection of pathogens and spoilage-causing microorganisms.[32]

1.5.1 DETECTION OF MICROORGANISMS USING NANOPARTICLES

Microorganisms which are concerns for bio-security such as *Bacillus anthracis*[39] and *Yersinia pestis*,[52] including foodborne pathogens like *E. coli, Salmonella, Bacillus cereus, Campylobacter jejuni, Clostridium botulinum,* etc., are serious threats for food safety and security. Early detection of these life-threatening pathogens/microbes is critical to prevent disease their outbreaks and therefore protect public health.[35] Nanotechnology provides great opportunity to develop fast, accurate, and cost-effective diagnostics for the detection of pathogens.[21,40] Due to the presence of unique properties in nanoscale materials, devices were able to detect the presence of a pathogenic agent in clinical or environmental samples. The properties observed in nanomaterials were different from those observed in the bulk (micron-size) material due to their small size of 1–100 nm have large surface area, resulting into enhanced surface reactivity, quantum confinement effects enhanced electrical conductivity, and high magnetic properties.[40] Most importantly, modifications in the nanostructures surface may dramatically alter some of their properties to have their wider applications[13,30] Because of such properties, nanotechnology has been engineered to make multiple nanostructures been to detect particular molecular targets in biodiagnostic applications, including pathogen detection.[23]

1.5.2 IDENTIFICATION AND ELIMINATION OF BACTERIA

A chicken feed developed by Clemson University, South Carolina, to remove *Campylobacter*, a bacteria common and benign in poultry that provokes cramps and diarrhea in people ingesting the contaminated and undercooked meat. The feed enriched by nanocarbohydrate particles binds with the bacterium's surface to remove it through the bird's feces. These nanoparticles might one day be combined with sensors in order to identify and remove other bacteria. Now, use in chickens might reduce the one million annual outbreaks of campylobacteriosis.

1.5.3 FLUORESCENCE-BASED DETECTION

The immediate detection of *E. coli* bacteria in a food sample can be possible by using a digital camera and a laser (University of Rochester Medical Center). In this measuring and detecting light scattering by cell mitochondria evolved toward development of a system detecting the presence of *E. coli*. Bacterium proteins are impregnated on a silicon chip for detection of *E. coli* bacteria present in the food sample. A biosensor can detect *Salmonella* bacteria are present in the food being tested using fluorescent dye particles attached to bacteria antibodies in which the nano-sized dye particles become visible (ARS Scientist, Athens, Georgia).

AuNP-conjugated polymer systems have been reported by Phillips et al.[34]for detection of pathogens. There cationic AuNPs and anionic PPE (3:1) carrying carboxylate with oligo (ethylene glycol) arms were combined to generate non-covalent. The sensor array was used to identify 12 microorganisms comprising both Gram complexes. The initial quenched fluorescent polymers recover their fluorescence positive (e.g., *Anchusa azurea* and *Bacillus subtilis*) and Gram-negative (e.g., *E. coli* and *Pseudomonas putida*) species in the presence of bacteria.

The use of quantum dots as a fluorescence labeling system has been successfully used in microorganism detection by Liu.[27] Fluorescent CdSe NPs conjugated to respond to Gram-positive bacteria and the QD-WGA conjugate with sialic acid and N-acetylglucosamine on the bacterial cell walls for detection. QDs can also be conjugated with antibodies for specific pathogens such as *E. coli* and *Salmonella typhimurium,* including parasites as well as oral bacteria detection.[6,45,50,53]

1.5.4 ANTIBIOTIC TREATMENT

Recent advancement in nanotechnology has led to formulate nanoparticles with desired physicochemical properties to defense against multi-drug resistance microorganism like methicillin-resistant *Staphylococcus aureus*, vancomycin resistant enterococci, etc. Such nanoparticles have effective antimicrobial activity against *Acinetobacter baumanii*, *Pseudomonas aeruginosa*, *Klebsiella pneumoniae*, *Mycobacterium tuberculosis*, and others.[43] Seil and Webster[41] have illustrated that zinc and silver possess greater antibacterial properties as particle size when reduced into the nanometer regime. The recent technology may improve the effectiveness of antibiotics by allowing the medicine to be put into an aerosol form. A spray of antibiotics encapsulated in microscopic antimicrobial silver–carbene complexes (SCCs) is effective against *P. aeruginosa* which is a common food spoiler and a causative agent of pneumonia-like respiratory illness in people.

1.5.5 NANO SENSORS

Biosensors may be an effective tool in the food chain with tremendous potential for monitoring of food safety. These sensors might detect pathogenic microorganisms which can give processors an accurate real-time indication of a product's relative quality or safety from farm to fork. Color-coded pathogen indicators using nanotechnology can be used to test the quality and safety of milk where pathogens bind with their specific antibodies on the indicator. Different color levels are used now to indicate the concentration of the pathogen which correspondingly concludes safety and quality of the milk.

The value-added products at any point in the supply chain can be monitored using Radio frequency identification technologies used bio-chips or bio-sensors which can enable processors and reduce the food loss. Chemical produced in spoiled seafood can be detected using enzymes incorporated into sensors. Personnel can determine product safety or quality issues from almost anywhere using radio frequency signals passing through a central control system. Nanosensors are introduced for detection of bioterrorism agents.[1]

Synthetic tree-shaped DNA tagged with color-coded probes has been developed as a nanobarcode device to enables the food pathogens identification.[26] Horner et al.[18] have developed a reflective interferometry, using nanotechnology which is specific, rapid, and label-free optical detection of biomolecules in complex mixtures. This analytical technology may provide food quality assurance by detecting *E. coli* in food samples. Such sensor works as a protein of a known and characterized bacterium set on a silicon chip can bind with any other *E. coli* bacteria present in the food sample may be detectable by analysis of digital images.

Nano-bio-sensing could be an ideal molecular detection approaches for foodborne pathogens. This novel detection technique was developed using 16S rRNA gold nanoprobe-nucleic acid sequence-based amplification (NASBA), which can determine around 5 CFUs *Salmonella* per amplification tube.[31] The surface-enhanced Raman scattering (SERS) nanoparticles detect pathogens in complex matrices using combination with a novel homogeneous, eliminating the wash steps required or extensive sample preparation.[32]

1.5.6 NANO FILTRATION

Food safety is prime concern in respect of biotoxins and synthetic chemical. Nanofilters can also be used to remove toxins such as pesticides. Dutch company Aquamarijn has produced microsieves with fine-tuned nanopores produced by that act as filters for a variety of applications. In food preparation areas, nanofilters are used to clean the environment, and nano-enhanced antibacterial surfaces are existing, and nanocoatings on tools and equipment make them sharper, longer lasting, and easier to clean. A nanofiltration membrane, also known as a molecular sieve can filter out pathogens and spoilage organisms of paramount importance. The nanocoatings on membranes can help to reduce both damage and build-up to the membranes.

1.6 FUTURE PERSPECTIVES

The detection systems with high specificity and sensitivity for microorganism along with fast response time along with reproducibility are the needed. Culture enrichment step that takes most of the testing time has not been eliminated and is likely to stay due to the several advantages. Formulation of improved selective media and their enrichment should facilitate to decrease the time-to-test results. Alternative concentration technologies other than enrichment need to be developed. This requires collaborative efforts among chemists, engineers, molecular biologists, microbiologists, as well as food scientists.

An ideal detection system should comprise high specificity and sensitivity with fast response time. Advance progresses were taken during past decades, including automation and high throughput in sample processing and testing.

There is a lack of "real-time" procedures from sampling to results. Culture enrichment step that takes most of the testing time has not been eliminated due to lack of specific selective media, where a serious effort is required. To achieve the goal of real-time testing, alternative technologies are must to comply with needs. Developing improved universal sample preparation method is also critical toward the next generation of pathogen and toxin detection methodologies in foods.

In addition, desirable features such as quantification and multiplex detection are becoming increasingly important. Technologies that are routinely used in the chemical and physical testing have adopted for microbial testing in foods, such as mass spectrometry (MS) and optical scanning technology. Again, collaborative efforts among scientists with expertise in multiple disciplines should lead to major technological advancements and result in the next generation of pathogen and toxin detection methodologies in foods. Nanotechnology has emerged as a growing and rapidly changing field. New generations of nanomaterial will evolve, and with the new and possibly unforeseen issues. Nanotechnology is the future of advanced development. It is everything today from clothes to foods, there are every sector in its range we should promote it more for our future and for more developments in our current life. Nanotechnology is slowly creeping into popular culture. There is a possibility that the future of nanotechnology is very bright with very wide application in all parts of life.

Risks

Nanotechnology is a field gaining a great interest among scientist as well as multinational companies due to the necessity and applications of nanomaterials in wider areas of human endeavors including industry, agriculture, business, medicine, and public health. As exposure to nanomaterials is inevitable as nanomaterials become part of our daily life, the specific properties and characteristics of nanomaterials need to be considered for any potential health risks.[12] There are effects of the various nanoparticles on the human body and negative consequences on the human health. Nanoparticles are firmly embedded in the matrix which can release and are harmful. Concentrations of nanoparticles at the workplace and in the environment are dangerous. Its waste is very dangerous and can cause the diseases like cancer, and other human-related dangerous diseases.

KEYWORDS

- **foodborne diseases**
- **nanotechnology**
- **foodborne pathogens**
- **nanoparticles**

REFERENCES

1. Abraham, A. M.; Kannangai, R.; Sridharan, G. Nanotechnology: A New Frontier in Virus Detection in Clinical Practice. *Indian J. Med. Microbiol.* **2008,** *26*, 297–301.
2. Arora, S.; Jyutika, M.; Rajwade, K.; Paknikar, M. Nanotoxicology and In Vitro Studies: The Need of the Hour. *Toxicol. Appl. Pharmacol.* **2012,** *258*, 151–165.
3. Bata-Vidács, I.; Adányi, N.; Beczner, J.; Farkas, J.; Székács, A. *Microbial Pathogens and Strategies for Combating Them: Science, Technology and Education.* Méndez-Vilas, A., Ed.; Formatex Research Center: *Badajo,* Spain, 2003.
4. Beilei, G.; Meng, J. Advanced Technologies for Pathogen and Toxin Detection in Foods: Current Applications and Future Directions. *J. Lab. Autom.* **2009,** *14*, 235.
5. CDC Standard Protocol, *Molecular Determination of Serotype in Salmonella;* CDC: Atlanta, GA, 2007.
6. Chalmers, N. I.; Palmer, R. J.; Du-Thumm, L.; Sullivan, R.; Shi, W. Y.; Kolenbrander, P. E. Use of Quantum Dot Luminescent Probes to Achieve Single-cell Resolution of Human Oral Bacteria in Biofilms. *Appl. Environ. Microbiol.* **2007,** *73*, 630–636.
7. Charles Poole, P. Jr.; Owens Frank J. *Introduction to Nanotechnology;* Wiley, John & Sons: Hoboken, NJ, 2003.
8. Chen, H.; Weiss, J.; Shahidi, F. Nanotechnology in Nutraceuticals and Functional Foods. *Food Technol.* **2006,** *03*, 30–36.
9. Kelty, C.; Hutchinson, J. S.The Early History of Nanotechnology (Introduction). In *Nanotechnology: Content and Context*; Connexions: Rice University, Houston, Texas, 2012; 1–24
10. Deisingh, A. K.; Thompson, M. Detection of Infectious and Toxigenic Bacteria. *Analyst* **2002,** *127*, 567–581.
11. Dingman, J. Nanotechnology: Its Impact on Food Safety. (Guest Commentary). *J. Environ. Health.* **2008,** *70* (6), 47–50.
12. EFSA Scientific Committee. Guidance on the Risk Assessment of the Application of Nanoscience and Nanotechnologies in the Food and Feed Chain, EFSA Journal; European Food Safety Authority (EFSA), Parma, Italy, 2011, 9 (5):2140, 1–36.
13. Elghanian, R.; Storhoff, J. J.; Mucic, R. C.; Letsinger, R. L.; Mirkin, C. A. Selective Colorimetric Detection of Polynucleotides Based on the Distance-dependent Optical Properties of Gold Nanoparticles. *Science* **1997,** *277*, 1078–1081.

14. Farhang, B. Nanotechnology and Applications in Food Safety. In *Global Issues in Food Science and Technology;* Gustavo, B. C., Alan, M., David, L. et al., Eds.; Academic Press: San Diego, CA, 2009; pp 401–410.
15. Ferreira, C.; Pereira, A. M.; Melo, L. F.; Simões, M. Advances in Industrial Biofilm Control with Micro-nanotechnology. In *Current Research, Technology and Education Topics in Applied Microbiology and Microbial Biotechnology;* Méndez-Vilas, A. Ed.; Formatex Research Center: Badajoz, Spain, 2010; pp 845–854.
16. Fitzgerald, C.; Collins, M.; van Duyne, S.; Mikoleit, M.; Brown, T.; Fields, P. Multiplex, Bead-based Suspension Array for Molecular Determination of Common Salmonella Serogroups. *J. Clin. Microbiol.* **2007,** *45* (10), 3323–3334.
17. Gaysinsky, S.; Davidson, P. M.; McClements, D. J.; Weiss, J. Formulation and Characterization of Phytophenol-Carrying Microemulsions. *Food Biophys.* **2008,** *3*, 54–65.
18. Horner, S. R.; Mace, C. R.; Rothberg, L. J.; Miller, B. L. A Proteomic Biosensor for Enteropathogenic *E. coli*. Biosens. *Bioelectron.* **2006,** *21*, 1659–1663.
19. Broomfield, M.; Hansen, S. F.; Pelsy, F. *Support for 3rd Regulatory Review on Nanomaterials: Environmental Legislation*; Ricardo Energy & Environment, Milieu Consulting and Danish Technical University, European Union, 2016, 110–130.
20. IOM (Institute of Medicine). *Nanotechnology in Food Products: Workshop Summary;* The National Academies Press: Washington, DC, 2009.
21. Jain, K. K. Nanotechnology in Clinical Laboratory Diagnostics. *Clin. Chim. Acta.* **2005,** *358*, 37–54.
22. Joseph, T.; Morrison, M. *Nanotechnology in Agriculture and Food;* Nanoforum Report 2; The Institute of Nanotechnology: UK, 2006, 2–3.
23. Kaittanis, C.; Santra, S.; Perez, J. Manuel. Emerging Nanotechnology-based Strategies for the Identification of Microbial Pathogenesis. *Adv. Drug Deliv. Rev.* **2010,** *62*, 408–423.
24. Kaur, G.; Singh, T.; Kumar, A. Nanotechnology: A Review. *Int. J. Educ. Appl. Res.* **2012,** *2*, 50–53.
25. Kriegel, C.; Kit, K.; McClements, D. J.; Weiss, J. Nanofibers as Carrier Systems for Antimicrobial Microemulsions. Part I. Fabrication and Characterization. *Langmuir* **2009,** *25*, 1154–1161.
26. Li, N.; Sioutas, C.; Cho, A.; Schmitz, D.; Misra, C.; Sempf, J.; Wang, M.; Oberley, T.; Froines, J.; Nel, A. Ultrafine Particulate Pollutants Induce Oxidative Stress and Mitochondrial Damage. *Environ. Health Perspect.* **2003,** *111*, 455–460.
27. Liu, W. T. Nanoparticles and Their Biological and Environmental Applications. *J. Biosci. Bioeng.* **2006,** *102*, 1–7.
28. Lund, B. M.; Baird-Parker, T. C.; Gould, G. W. *The Microbiological Safety and Quality of Foods;* Aspen Publishers: Gaithersburg, MD, 2000.
29. Mason, J. T.; Xu, L.; Sheng, Z. M.; O'Leary, T. J. A Liposome-PCR Assay for the Ultrasensitive Detection of Biological Toxins. *Nat. Biotechnol.* **2006,** *24* (5), 555–557.
30. Mirkin, C. A.; Letsinger, R. L.; Mucic, R. C.; Storhoff, J. J. A DNA-based Method for Rationally Assembling Nanoparticles into Macroscopic Materials. *Nature* **1996,** *382*, 607–609.
31. Mollasalehi, H.; Yazdanparast, R. An Improved Non-crosslinking Goldnanoprobe-NASBA Based on 16S rRNA for Rapid Discriminative Biosensing of Major Salmonellosis Pathogens. *Biosens Bioelectron.* **2013,** *47*, 231–236.
32. Nasr, N. F. Applications of Nanotechnology in Food Microbiology. *Int. J. Curr. Microbiol. App. Sci.* **2015,** *4* (4), 846–853.

33. Pérez-Conesa, D.; Cao, J.; Chen, L.; Inactivation of *Listeria monocytogenes* and *Escherichia coli* O157:H7 Biofilms by Micelle-Encapsulated Eugenol and Carvacrol. *J. Food Protect.* **2011,** *74* (1), 55–62.
34. Phillips, R. L.; Miranda, O. R.; You, C. C.; Rotello, V. M.; Bunz, U. H. F. Rapid and Efficient Identification of Bacteria Using Gold-nanoparticle-Poly (Para-phenylene Ethynylene) Constructs. *Angew. Chem. Int. Ed.* **2008,** *47*, 2590–2594.
35. Priyanka, S.; Shashank, P.; Prashant, S.; Krishan Pal Singh. Nanotechnology and Its Role in Pathogen Detection: A Short Review. *Int. J. Curr. Sci.* **2014,** *13*, E 9–15.
36. Raliya, R.; Tarafdar, J. C.; Gulecha, K.; Choudhary, K.; Rameshwar Ram; Prakash Mal.; Saran, R. P. Review Article: Scope of Nanoscience and Nanotechnology in Agriculture. *J. Appl. Biol. Biotechnol.* **2013,** *1* (03), 041–044.
37. Overney, R. Nanothermodynamics and Nanoparticle Synthesis, NME 498A/A.ppt. http://courses.washington.edu/overney/NME498_Material/NME498_Lectures/Lecture4-Overney-NP-Synthesis.pdf.
38. Reshatoff, M.; Ong, E.; Ritter, J.; Garcia, J.; Motta, C.; Lewis, C.; Fullerton, M.; Deras, M.; Eusebio, A.; Pekny, K.; Hedrick, N.; Lee, S.; McDonough, D.; Hogan, J. In *Rapid Method of Detecting Listeria Genus, Salmonella Genus, and Campylobacter Using Real Time Transcription-mediated Amplification Assays Targeted to Ribosomal RNA (P-000);* American Society for Microbiology Annual Meeting; Boston, MA, 2008.
39. Roffey, R.; Lantorp, K.; Tegnell, A.; Elgh, F. Update on Biological Weapons and Bioterrorism. Important That Health Services Pay Attention to Unusual Events. *Lakartidningen* **2001,** *98* (50), 5746–5748.
40. Rosi, N. L.; Mirkin, C. A. Nanostructures in Biodiagnostics. *Chem. Rev.* **2005,** *105*, 1547–1562.
41. Seil, J. T.; Webster, T. J. Antimicrobial Applications of Nanotechnology: Methods and Literature. *Int. J. Nanomed.* **2012,** *7*, 2767–2781.
42. Sekhon, B. S. Food Nanotechnology–An Overview. *Nanotechnol. Sci. Appl.* **2010,** *3*, 1–15.
43. Singh, R.; Smitha, M. S.; Singh, S. P. The Role of Nanotechnology in Combating Multidrug Resistant Bacteria. *J. Nanosci. Nanotechnol.* **2014,** *14* (7), 4745–4756.
44. Stevens, K. A.; Jaykus, L. A. Bacterial Separation and Concentration from Complex Sample Matrices: A Review. *Crit. Rev. Microbiol.* **2004,** *30* (1), 7–24.
45. Su, X. L.; Li, Y. B. Quantum Dot Biolabeling Coupled with Immunomagnetic Separation for Detection of *Escherichia coli* O157:H7. *Anal. Chem.* **2004,** *76*, 4806–4810.
46. Taylor, T. M.; Gaysinksy, S.; Davidson, P. M.; Barry D. Bruce.; Weiss, J. Characterization of Antimicrobial Bearing Liposomes by Zeta-Potential, Vesicle Size and Encapsulation Efficiency. *Food Biophys.* **2007,** *2*, 1–9.
47. Weiss, J.; Gibis, M.; Stuttgart, H. Nanotechnology in the Food Industry. *Ernaehrungs Umschau Int.* **2013,** *60* (4), 44–51.
48. Weiss, J.; Takhistov, P.; McClements, D. J. IFT Status Summary: Nanotechnology–Applications in Food Processing and Product Development. *J. Food Sci.* **2006,** *71*, R107–R116.
49. WHO, *World Health Statistics 2015. World Health Organization;* WHO Press: Geneva, Switzerland, 2015.
50. Yang, L. J.; Li, Y. B. Simultaneous Detection of Escherichia coli O157:H7 and Salmonella Typhimurium Using Quantum Dots as Fluorescence Labels. *Analyst* **2006,** *131*, 394–401.

51. Zhao, X.; Hilliard, L. R.; Mechery, S. J.; Wang, Y.; Bagwe, R. P.; Jin, S.; Tan, W. A Rapid Bioassay for Single Bacterial Cell Quantitation Using Bioconjugated Nanoparticles. *Proc. Natl. Acad. Sci. USA* **2004,** *101* (42), 15027–15032.
52. Zhou, D.; Han, Y.; Yang, R. Molecular and Physiological Insights into Plague Transmission, Virulence and Etiology. *Microbes Infect.* **2006,** *8* (1), 273–284.
53. Zhu, L.; Ang, S.; Liu, W. T. Quantum Dots as a Novel Immunofluorescent Detection System for *Cryptosporidium parvum* and *Giardia lamblia. Appl. Environ. Microbiol.* **2004,** *70*, 597–598.

CHAPTER 2

WATER VAPOR PERMEABILITY, MECHANICAL, OPTICAL, AND SENSORIAL PROPERTIES OF PLASTICIZED GUAR GUM EDIBLE FILMS

XOCHITL RUELAS CHACON[1,2*],
JUAN C. CONTRERAS-ESQUIVEL[1], JULIO MONTAÑEZ[3], ANTONIO FRANCISCO AGUILERA CARBO[4],
MARIA DE LA LUZ REYES VEGA[1*],
RENE DARIO PERALTA RODRIGUEZ[5], and
GABRIELA SANCHEZ BRAMBILA[6]

[1]*Department of Food Research, Faculty of Chemistry, Universidad Autonoma de Coahuila, Blvd. V. Carranza, Colonia Republica Oriente, Saltillo 25280, Coahuila, Mexico*

[2]*Department of Food Science and Technology, Universidad Autonoma Agraria Antonio Narro, Calzada Antonio Narro 1923, Colonia Buenavista, Saltillo 25315, Coahuila, Mexico*

[3]*Department of Chemical Engineering, Faculty of Chemistry, Universidad Autonoma de Coahuila, Blvd. V. Carranza, Colonia Republica Oriente, Saltillo 25280, Coahuila, Mexico*

[4]*Department of Animal Nutrition, Universidad Autonoma Agraria Antonio Narro, Calzada Antonio Narro 1923, Colonia Buenavista, Saltillo 25315, Coahuila, Mexico*

[5]*Research Center for Applied Chemistry, Blvd. Enrique Reyna Hermosillo No. 140, Saltillo 25253, Coahuila, Mexico*

[6]*Russell Research Center-ARS, Quality and Safety Assessment Research Unit USDA, 950 College Station Road, Athens 30605, GA, USA*

[*]*Corresponding authors. E-mail: xochitl.ruelas@uaaan.mx; xruelas@yahoo.com; mlrv20@yahoo.com*

CONTENTS

ABSTRACT

Edible films were prepared by casting method using guar gum (1.0%, 1.5%, and 2.0%) and glycerol (20%, 30%, and 40%, w/v) in different ratios. The water vapor permeability (WVP), mechanical properties, thickness, optical properties, solubility, moisture content (MC), and sensory acceptability were investigated. As the plasticizer concentration increased, the MC, solubility, and WVP of the films increased significantly ($p < 0.05$). The tensile strength (TS) decreased as levels of glycerol increased, and the elongation at break increased as polyol and guar gum levels increased. Thickness and optical properties were affected significantly by guar gum and glycerol concentrations ($p < 0.05$). Sensory properties of films showed that taste and overall acceptability were not significantly different, while color and stickiness were significantly different ($p < 0.05$). This study provides basic information on the properties of these biodegradable and flexible films, which are made with a natural biopolymer and represent an attractive option for future trends in food applications.

2.1 INTRODUCTION

The deterioration of packaged foodstuffs largely depends on the mass and heat transfer that may occur between the internal and the external environments of the packaged food.[1,2] The use of natural biopolymers as edible films has increased the shelf life of food by modifying these environment exchanges[3–5] and offered better means of availability by allowing to preserve food for longer time. The properties of these films and their applications will depend on the composition as well as their conditions during their preparation.[6] Biopolymers used to prepare packaging materials include polysaccharides, lipids, proteins, and their derivatives.[7] The polysaccharides, such as cellulose derivatives, chitosan, starch, alginate, carrageenan, and pectin, are the most preferred because of their high film-forming ability and mechanical properties.[7–9] On the other hand, the fact that they come from a natural resource meets the trends in consumer preferences.

In order to apply edible films to foods, it is necessary to study their transport properties (water vapor permeability, WVP), mechanical properties, and sensory characteristics. WVP and the mechanical properties are the most important characteristics of edible films. The WVP is related to the property that films have as barriers against gases (O_2 and CO_2), water vapor, or oil, thus influencing shelf life and improving quality and handling management of food products.[10] Elongation (stretchability) and toughness (film strength × elongation at break) of the film determine their application as food wrap or coating.[11] The edible films should stand mechanical stress and strain to such an extent that they do not break easily under various mechanical forces. Other important elements for edible films and coatings are those related to sensory evaluation such as acceptable color, odor, taste, flavor, and texture (stickiness), which are of main importance to the acceptability for consumers.[12–14]

Edible films can be plasticized by low molecular weight carbohydrates, for example, polyols.[10] Plasticizers such as glycerol, polyethylene glycol, and sorbitol increase the flexibility of the films due to their capacity to reduce the internal hydrogen links among the polymer chains as they increase the molecular space.[15] The plasticizers most often used and recommended in film formulations are sorbitol and glycerol.[15–17] The incorporation of plasticizers is necessary to reduce polymer intermolecular forces, increasing the mobility of the polymeric chains, and improving the mechanical characteristics of the film, for example, extensibility.[18,19] Also, plasticizers affect the water barrier property of the films, since they have a great affinity for water.[16]

Guar gum, is a galactomannan-rich flour, water-soluble polysaccharide obtained from the leguminous Indian cluster bean *Cyamopsis tetragonoloba* (L.) *Taub.* The backbone of this hydrocolloid is a linear chain of D-mannopyranosyl units connected to each other by β-1,4-bonds linked to galactose residues by 1,6-bonds forming short side-branches.[20–22] Guar gum is one of the most important thickeners and a versatile material used for many food applications due to its different physicochemical properties as well as its high availability, low cost, and biodegradability. The guar gum is an excellent non-toxic stiffener used in the textile, pharmaceutical, biomedical, cosmetic, and food industries.[23,24] In addition, this galactomannan exhibits surface, interfacial, and emulsification activities.[20,24,25] Films based on galactomannans can be used to reduce water vapor, oxygen, lipid, and flavor migration between components of multi-component food products, and between food and its surroundings.[8] There is a little published information about the WVP, mechanical, optical, and sensorial properties of plasticized guar gum films. Therefore the aim of this research was to develop edible films based on guar gum biopolymer and to evaluate the effect of glycerol as plasticizer on the WVP, mechanical, optical, and sensory acceptability properties of these films.

2.2 MATERIALS AND METHODS

2.2.1 REAGENTS

Guar gum (G4129-500G) was purchased from Sigma-Aldrich (St. Louis, MO, USA). Glycerol, anhydrous $CaCl_2$, NaCl, and KCl were purchased from Jalmek Co. (Monterrey, Nuevo Leon, Mexico).

2.2.2 FILM PREPARATION

A total of nine different formulations of films were prepared. Guar gum and glycerol were used; the two components were dissolved in distilled water at 60°C with constant agitation (500 rpm) for 40 min. A series of blends were prepared with varying concentration (%) of guar gum and plasticizer (1.0/20, 1.0/30, 1.0/40, 1.5/20, 1.5/30, 1.5/40, 2.0/20, 2.0/30, and 2.0/40). Film-forming solutions were homogenized using a hot plate/stirrer (Talboys, Thorofare, NJ, USA) at 700 rpm. The polymer films were prepared by a casting method, 20 mL of recently prepared suspensions were immediately poured on polyethylene Petri dishes of 8 cm in diameter, resting on a level surface. The suspensions in the Petri dishes were then dried at 50°C in a ventilated oven (Quincy Lab Inc., Chicago, IL, USA) during 10 h. The dried films were peeled from the casting surface and stored in desiccators at 0% RH with silica gel at 25°C.

2.2.3 THICKNESS MEASUREMENT

Thickness was determined using a dial thickness gauge micrometer (Mitutoyo Manufacturing Co. Ltd., Tokyo, Japan) by generating the mean of five measurements at randomized positions on the film. Samples with air bubbles and nicks or tears were excluded from analysis.

2.2.4 OPTICAL PROPERTIES

Optical properties of light transmittance and opacity of the films were obtained using a Genesis 10 UV spectrophotometer (Thermo Electron Corporation, Madison, WI, USA). For each film specimen, a sample of a rectangular piece (1 × 3 cm) was placed directly in a spectrophotometer test cell, and measurements were performed using air as reference. The light transmittance of the films was scanned from wavelength of 400–800 nm using the spectrophotometer. The determination was done by triplicate and, from these spectra; the average transparency at 600 nm (T600) was calculated. The T600 was obtained from the following eq 2.1:[26]

$$\text{T600} = \log \frac{\%T}{b} \tag{2.1}$$

where $\%T$ is the percentage transmittance and b is the film thickness (mm).

The opacity of the films was calculated by the following eq 2.2 according to the method described by Gontard et al.:[27]

$$\text{Opacity} = \text{AU}_{500\text{nm}} * b \tag{2.2}$$

where $\text{AU}_{500\text{nm}}$ is absorbance units at 500 nm and b is the film thickness (mm).

2.2.5 COLOR MEASUREMENT

Color of the films was assessed using a colorimeter (Minolta CR-400, Tokyo, Japan). A white standard color plate (L = 97.75, a = 0.49, b = 1.96, supplied by Minolta Co.) for instrument calibration was used as a background for color measurements of the films. The system provides the values of three-color components; L* (black-white component, luminosity) and chromaticness coordinates, a* (+red to −green component) and b* (+yellow to −blue component) were recorded by triplicates for each sample. Total color change from standard (ΔE), the yellowness (YI), and whiteness (WI) indexes of samples were calculated following Ahmadi et al.[15] and Bolin et al.[28] recommended eqs 2.3–2.5. The samples were analyzed at five random positions.

$$\Delta E = \sqrt{(L' - L)^2 + (a' - a)^2 + (b' - b)^2} \tag{2.3}$$

$$YI = \frac{(142.86 * b)}{L} \tag{2.4}$$

$$WI = 100 - \sqrt{(100 - L)^2 + a^2 + b^2} \tag{2.5}$$

2.2.6 MOISTURE DETERMINATION

Percent MC of each film was determined according to the method reported by Mei et al.[29] The films were cut in squares of 2 cm × 2 cm and were placed into a previously dried and cooled aluminum dishes. The films along with aluminum dishes were dried inside a laboratory oven (Quincy Lab Inc., Chicago, IL, USA) at 100°C for 24 h. Weights of the film samples were taken before and after drying using a digital balance (Adventurer Ohaus Corp., Pine Brook, NJ, USA) with an accuracy of 0.0001 g. The MC was determined according to the eq 2.6. Three replications of each film were measured for MC values.

$$\text{Moisture content}(\%) = \frac{(\text{Initial dry weight - Final dry weight}) * 100}{\text{Initial dry weight}} \tag{2.6}$$

2.2.7 SOLUBILITY MEASUREMENT

The solubility of the films was carried out following the methodology reported by Romero-Bastida et al.[30] Dried film samples were placed in a glass beaker with 80 mL of distilled water under constant agitation of 300 rpm at 25°C during 60 min. After this time, the samples were removed and dried for 24 h at 60°C to achieve constant weight. Samples weight was determined by using a digital balance (Adventurer Ohaus Corp., Pine Brook, NJ, USA) with an accuracy of 0.0001 g. Films solubility was calculated using the eq 2.7. The samples were analyzed and the average values were reported.

$$\text{Solubility}(\%) = \frac{(\text{Initial dry weight - Final dry weight}) * 100}{\text{Initial dry weight}} \tag{2.7}$$

2.2.8 WATER VAPOR PERMEABILITY

The WVP of the prepared films was measured following the methodology reported by ASTMA E 96.[31] Granular anhydrous $CaCl_2$ (approximately 3.0 g) was used as a desiccant in the acrylic permeability cell covered with the studied film. Distance between the surface of desiccant and film was less than 6 mm as suggested by ASTM E 96.[31] Thickness of each film was measured with a micrometer (Mitutoyo, Japan with an accuracy of 0.01 mm) at five randomly selected points. The cells were placed in desiccators at different relative humidity values (RH) 75%, 85%, and 100% at 25 ± 1.5°C.[32] Weight gain due to water vapor permeation was determined gravimetrically as a function of time during 10 h. When the relationship between weight gain (Δw) and time (Δt) is linear, the slope of the plot is used to calculate the water vapor transmission rate (WVTR) and WVP.[1] The slope is obtained by linear regression. WVTR was calculated from the slope ($\Delta w/\Delta t$) divided by the test area (A) ($g{*}m^{-2}{*}d^{-1}$), with eq 2.8:

$$\text{WVTR} = \left(\frac{\Delta w}{\Delta t}\right) \div A \tag{2.8}$$

2.2.9 MECHANICAL PROPERTIES

Where $\Delta w/\Delta t$ = transfer rate, amount of moisture loss per unit time ($g\ d^{-1}$); A = area exposed to moisture transfer (m^2).

WVP is calculated as following eq 2.9:

$$\text{WVP} = \frac{\text{WVTR}^{*}L}{\Delta p} \tag{2.9}$$

where WVP is ($g\ mm\ m^{-2}\ d^{-1}\ kPa^{-1}$), WVTR is WVTR ($gm^{-2}d^{-1}$), L is thickness of film (mm), and Δp is water vapor pressure difference at both sides of the film (kPa).[1] All determinations were evaluated in triplicate.

The films were cut in 1 cm × 9 cm strips and conditioned in a desiccator for 5 days at 57% RH.[33] Tensile strength (TS) and elongation-at-break (E) were determined from a stress–strain curve using a texture analyzer instrument (TA-XT2; Texture Technologies Corp., Scarsdale, NY, USA) following the procedure outlined in ASTM method D 82-91.[33] Initial grip distance and crosshead speed were 5 cm and 1.00 $mm{*}min^{-1}$, respectively. TS was calculated by dividing the peak load by the cross-section area (thickness × 1 cm) of the initial specimen. E was expressed as the percentage of change in the length of the specimen to the original length between the grips (5 cm). TS and E values were obtained from 10 replications.

2.2.10 SENSORY EVALUATION

Sensory tests were carried out with a sensory panel of 30-trained members who were instructed to give a subjective evaluation of the films. The sensory characteristics such as color, taste, stickiness and overall acceptability of each of the samples were rated on a 9-point hedonic scale (where 1 = disliked extremely and 9 = liked extremely).[14] The samples consisted of pieces of 2.0×2.0 cm^2 of each film samples which were presented to the panelists in coded disposable plates and in randomized order presentation. Panelists were asked to evaluate color followed by tasting and chewing the samples to score the taste, stickiness, and overall acceptability, respectively. Cups with water and unsalted crackers were provided to the panelists to clean palate during tasting.[14]

2.2.11 STATISTICAL ANALYSIS

Differences between the variables were tested for significance by factorial analysis of variance ANOVA with Tukey's post-test using JMP software (Version 5.01; SAS Institute Inc., Cary, NC, USA). Differences at $p < 0.05$ were considered to be significant.

2.3 RESULTS AND DISCUSSION

2.3.1 FILM APPEARANCE

Homogeneous, thin, and flexible films were obtained from guar gum and glycerol solutions. Films could be easily removed from the acrylic plates. The films did not roll over or break. Visually, all the films were colorless and translucent similar to findings reported by Pereda et al.[34] with chitosan–gelatin films and by Matta Fakhoury et al.[35] with blends of manioc starch and gelatin films.

2.3.2 FILM THICKNESS

Thickness of the films showed significant differences ($p < 0.05$) depending on the percentage of guar gum and glycerol used in the formulation (Table 2.1). This has to be done with the plasticizer capacity to reduce the internal hydrogen links among the polymer chains as they increase the molecular space as reported by Farahnaky et al.[16] and Mali et al.[17] and with the amount of galactomannan used as reported by Cerqueira et al.[36]

TABLE 2.1 Effect of Guar Gum and Glycerol Concentration on the Film Thickness, Opacity, Percentage of Solubility, and Moisture Content.

Guar gum (%)	Glycerol (%, v/v)	Thickness (mm)	Opacity$^{500\ nm}$ (UA*mm)	Solubility (%)	Moisture content (%)
1.0	20	0.038 ± 0.003^{e}	0.0032 ± 0.0002^{a}	85.53 ± 1.05^{bc}	16.99 ± 0.99^{ab}
1.5	20	0.087 ± 0.009^{a}	0.0111 ± 0.0029^{ab}	78.85 ± 1.05^{e}	13.32 ± 1.09^{c}
2.0	20	0.088 ± 0.001^{a}	0.0121 ± 0.0020^{ab}	65.41 ± 0.79^{g}	6.90 ± 0.91^{e}
1.0	30	0.045 ± 0.002^{de}	0.0038 ± 0.0018^{a}	88.41 ± 1.03^{ab}	18.16 ± 0.86^{a}
1.5	30	0.071 ± 0.002^{bc}	0.0090 ± 0.0003^{ab}	81.04 ± 0.94^{de}	12.90 ± 0.50^{c}
2.0	30	0.079 ± 0.005^{ab}	0.0162 ± 0.0049^{bc}	66.44 ± 0.88^{g}	8.57 ± 0.50^{de}
1.0	40	0.058 ± 0.003^{cd}	0.0059 ± 0.0004^{a}	90.34 ± 1.84^{a}	19.41 ± 1.01^{a}
1.5	40	0.063 ± 0.007^{c}	0.0062 ± 0.0010^{a}	82.70 ± 1.01^{cd}	14.54 ± 0.98^{bc}
2.0	40	0.085 ± 0.007^{ab}	0.0251 ± 0.0017^{c}	69.86 ± 0.98^{f}	9.36 ± 0.83^{d}

Means and standard deviation values followed by different superscripts letters within the same column were significantly different ($p < 0.05$). Three replications of each film were analyzed.

2.3.3 *TRANSPARENCY AND OPACITY*

Addition of guar gum and glycerol in various levels of the formulation led to changes in transparency and opacity of films (Table 2.1 and Fig. 2.1). It can be seen that, opacity and transparency are inversely correlated. There is a significant difference ($p < 0.05$) on transparency of guar gum films since 1% (w/v) guar gum film showed highest transparency followed by films with 1.5% and 2% (w/v) as the concentration of this galactomannan increased and as the glycerol content increased the films were least transparent (Fig. 2.1). The interaction of water molecules with glycerol modifies the refractive index of guar gum affecting the film's transparency (Fig. 2.1). The opacities of the films also varied with guar gum concentration. Guar gum films with 2% (w/v) of this galactomannan were most opaque, followed by guar gum films with 1.5%and 1% (w/v) and the influence of glycerol concentrations at 20%, 30%, and 40% (v/v) (Table 2.1) affected opacity. There were no significant differences ($p < 0.05$) on films with 20% of glycerol, meanwhile there were significant differences ($p < 0.05$) on films with 30% and 40% of glycerol (Table 2.1) due to the nature and structure of the edible film as Mu et al.[37] and Zhang et al.[38] mentioned on their investigations. These findings are important since film transparency and opacity are critical properties in various film applications, particularly if the films will be used as food coatings or to improve the product's appearance.[39]

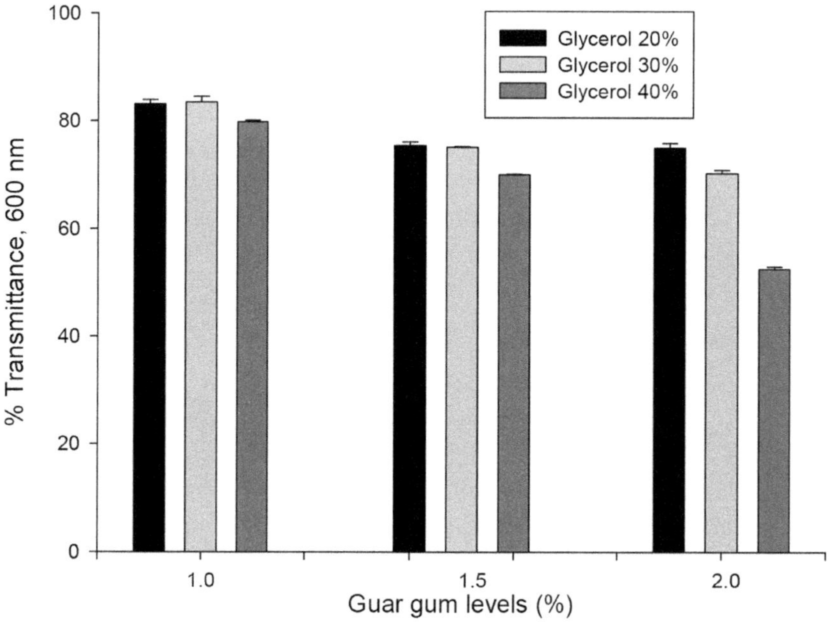

FIGURE 2.1 Effect of guar gum films with different levels of glycerol on transmittance at 600 nm.

2.3.4 SOLUBILITY OF FILMS

The solubility of the films increased basically ($p < 0.05$) at high concentrations of glycerol and this effect is shown in Table 2.1. This can be attributed to the flexibility of the polymer structure due to glycerol and that guar gum, as a polysaccharide, is hydrophilic so the films made with it are highly soluble in water; increasing the potential elasticity (E) and decreasing guar gum resistance (TS) at all dry material concentrations (Table 2.3). Since solubility is an important property for films because it strongly affects the shelf life of packed foods,[40,41] an alternative to decrease the rate of solubility in water is to modify the dry material or the plasticizer concentration used on edible films.[39] Similar investigations agree with these results, for example, Farahnaky et al.[16] and Ramos et al.[42] mentioned solubility (%) values ranging from 6.51% to 27.66% and from 63.91% to 84.22%, respectively. Farahnaky et al.[16] worked with wheat starch edible films plasticized with glycerol at 20% reporting a solubility of 6.51%. Mali et al.[43] worked with yam starch edible films plasticized with glycerol and the solubilities observed ranged from 23.63% to 27.72%. Kurt et al.[40] reported a solubility of 75.94% for an edible film made with 1.5% of guar gum and 10% of glycerol content. A higher solubility is directly related to a high level of glycerol[44] and a low content of biopolymer.[45] However, in this study, using guar gum concentrations of 1.5% and 2.0% with 40% of glycerol content, the solubility did not show the effect mentioned earlier, this could be attributed to the high concentration of dry material that probably causes an inappropriate spreading of the film solution and weakening of solubility of the film in water (Table 2.1).

2.3.5 *MOISTURE CONTENT*

The MC of guar gum films prepared with different glycerol concentrations is shown in Table 2.1. The MC of the films increased significantly ($p < 0.05$) at high levels of plasticizer content; however, the MC decreased with high levels of guar gum content and the values ranged from 6.90% to 19.41%. These results may be attributed to the fact that water acts as a plasticizer in most hydrophilic films and is not only associated with the galactomannan film's structure, but also with the glycerol hydrophilic nature that retains water in the matrix.[46] Similar findings were mentioned by Osés et al.[47] whose composite films based on whey protein isolate and mesquite gum tended to become more hydrophilic with high levels of plasticizer content and the MC values ranged from 9.20% to 11.00% at 50% RH and from 15.00% to 16.80% at 75% RH. Our findings reveal that glycerol-plasticized guar gum films had greater MC when glycerol was at 40%, the values ranged from 9.36% to 19.41%. These results may be interpreted as follows: films with high concentration of glycerol have higher MC; this fact is probably due to the hydrophilic properties of guar gum and glycerol which has higher affinity for binding water thus increases the plasticizing activity and greatly influences mechanical properties of films.[48]

2.3.6 *COLOR ATTRIBUTES*

Color attributes are of prime importance because they directly influence consumer acceptability. Table 2.2 shows the measured color parameters including L (black-white), a* (green-red), and b* (blue-yellow) of guar gum films. Several experiments showed that all of the color parameters of guar gum films were significantly ($p < 0.05$) altered when glycerol and guar gum concentrations increased. Increasing of glycerol concentration provoke an

TABLE 2.2 Effect of Guar Gum and Glycerol Concentration on L, a*, b* Parameters, ΔE, Yellowness (YI), and Whiteness (WI) Indexes.

Guar gum (%)	Glycerol (%)	L	a*	b*	ΔE	YI	WI
1.0	20	90.91 ± 0.06^{abc}	0.39 ± 0.06^{e}	3.03 ± 0.11^{bc}	8.01 ± 0.21^{abc}	4.77 ± 0.16^{b}	90.41 ± 0.05^{bc}
1.0	30	90.68 ± 0.29^{abc}	0.23 ± 0.05^{e}	2.68 ± 0.04^{b}	8.99 ± 0.17^{a}	5.24 ± 0.06^{b}	90.11 ± 0.15^{cd}
1.0	40	91.33 ± 0.18^{ab}	0.92 ± 0.02^{cd}	1.05 ± 0.06^{e}	7.11 ± 0.19^{cd}	2.03 ± 0.48^{d}	91.12 ± 0.22^{ab}
1.5	20	91.03 ± 0.66^{a}	1.04 ± 0.09^{bc}	0.66 ± 0.16^{f}	8.91 ± 0.53^{a}	0.72 ± 0.24^{e}	91.35 ± 0.28^{bcd}
1.5	30	90.43 ± 0.53^{bc}	0.90 ± 0.06^{cd}	3.48 ± 0.33^{e}	7.53 ± 0.59^{bcd}	2.54 ± 0.51^{d}	90.40 ± 0.26^{bcd}
1.5	40	91.38 ± 0.49^{d}	0.89 ± 0.07^{f}	2.17 ± 0.26^{a}	6.62 ± 0.54^{d}	11.41 ± 0.43^{a}	91.07 ± 0.52^{ab}
2.0	20	90.50 ± 0.41^{cd}	1.18 ± 0.09^{ab}	2.19 ± 0.32^{cd}	8.47 ± 0.65^{ab}	4.10 ± 0.52^{bc}	89.83 ± 0.19^{d}
2.0	30	90.86 ± 0.06^{abc}	1.39 ± 0.13^{a}	1.88 ± 0.19^{e}	7.70 ± 0.30^{abcd}	2.81 ± 0.29^{d}	90.64 ± 0.07^{abc}
2.0	40	91.01 ± 0.72^{abc}	0.74 ± 0.03^{d}	1.87 ± 0.48^{de}	7.03 ± 0.61^{cd}	3.02 ± 0.74^{cd}	90.70 ± 0.42^{abc}

Means and standard deviation values followed by different superscripts letters within the same column were significantly different ($p < 0.05$). Three replications of each film were analyzed.

increase in luminosity (L) while b* (blue-yellow) and a* (green-red) parameters decreased.

The ΔE indicates the degree of total color difference from the standard color plate, YI indicates the degree of yellowness, and WI indicates the degree of whiteness which can be more described as a result of increasing glycerol concentration in guar gum films (Table 2.2).

While WI increased significantly ($p < 0.05$) as a result of greater glycerol content, YI and ΔE decreased considerably. Hence, guar gum films became very slightly greenish and yellowish, but there were still transparent. Furthermore, visual observation confirmed this fact.

Comparing the results with Rao et al.[2] values for L* (43.83–49.18) are higher probably because the chitosan they used makes the films less luminescent. And for a* (−2.37 to −4.72) and b* (4.91 to 7.62) the values for the films of chitosan-guar gum are negative and positive but leading to the greenish zone, and the results obtained by Ekrami et al.[39] working with a salep based edible film mentioned that increasing solid material content did not cause changes in L (62.07–69.37); however, b* (−0.36 to −2.73) increased, a* (0.43 to 0.85) decreased significantly ($p < 0.05$), and increasing the glycerol content results in a new bond formation, which altered the color index. The film color tends toward yellow, and an increase in YI confirms this apparent change.

Kurt et al.[40] report L values for guar gum films such as 81.82, a* values 5.69 and b* values 1.28, this values are somewhat different as the ones found in this research (Table 2.2). The difference could be due to several factors such as the concentration of glycerol used, the amount of solution cast in acrylic plates and the kind of guar gum used. Even though the differences among the results the guar gum films formulated are between the range of the values reported by others.[2,39–41]

2.3.7 MECHANICAL PROPERTIES

The mechanical nature of edible films is influenced by the formulation and concentration of ingredients and by changes in plasticizer and biopolymers. An important factor that influences the evaluation of the mechanical properties is the RH because water acts as softener; when RH increases in the film, mechanical resistance decreases and elasticity increases.[39] In this study, prior to assay, the samples were conditioned at 57% RH for 5 days (120 h). The TS and elongation at break (E) of plasticized guar gum films are shown in Table 2.3. There were significant differences ($p < 0.05$) between the TS and

E of the film, the elongation at break of the guar gum films increased as the guar gum content increased. By increasing the glycerol content, small molecules of this plasticizer enter the polymer structure, so film solubility and chain mobility increases, which can led to improve the film elasticity (Table 2.3), similar findings are reported by other researchers.[44,49] The incorporation of plasticizers is necessary to reduce polymer intermolecular forces, increasing the mobility of the polymeric chains, and improving the mechanical characteristics of the film, such as film extensibility, since they have a great affinity for water.[11,19,50] Similar results have been reported by Leceta et al.,[52] who worked with chitosan and glycerol films, the TS values ranged from 31.89 to 61.82 MPa while the E values ranged from 4.59% to 30.51%; Kurt et al.[40] prepared films with salep gum, locust gum, and guar gum using 10% of glycerol and the range values for TS were 42.89–71.41 MPa and for E the range values were 16.1–37.2%; and Banegas et al.[51] have characterized cross-linked guar gum films, the TS range values were 16.00–43.80 MPa and for E the values were 2.00–2.90%. Other researchers using different bases for the film and using glycerol as a plasticizer obtained range values for TS from 0.57 to 9.87 MPa and for E from 17.11% to 250.00%.[19,44,52,53]

TABLE 2.3 Effect of Guar Gum and Glycerol Concentration on Tensile Strength (TS) and Elongation at Break (E) of Edible Films.

Guar gum (%)	Glycerol (%)	TS (MPa)	E (%)
1.0	20	32.68 ± 0.42[bcd]	8.97 ± 2.64[e]
1.0	30	23.97 ± 3.89[d]	18.60 ± 2.55[b]
1.0	40	30.63 ± 3.12[cd]	14.51 ± 0.14[cd]
1.5	20	32.41 ± 2.03[bcd]	12.40 ± 0.35[de]
1.5	30	23.11 ± 1.57[d]	19.30 ± 0.72[c]
1.5	40	27.88 ± 2.87[d]	17.43 ± 1.33[cd]
2.0	20	48.26 ± 3.49[a]	18.50 ± 0.85[b]
2.0	30	38.66 ± 0.19[abc]	21.15 ± 2.05[c]
2.0	40	41.08 ± 1.04[ab]	39.60 ± 1.94[a]

Means and standard deviation values followed by different superscripts letters within the same column were significantly different ($p < 0.05$). Ten replications of each film were analyzed.

2.3.8 WATER VAPOR PERMEABILITY

WVP of a film is an important property that greatly influences the usefulness of the film in foods.[45] Our results indicate that low WVP on films was related to the concentration of guar gum, the levels of glycerol used and the RH where the films were kept in a conditioning desiccator. WVP of guar gum films increased as glycerol and guar gum concentration increased ($p < 0.05$). WVP of films is dependent on both solubility and diffusion rate of water vapor.[44] Figure 2.2A–C illustrates that WVP increased as plasticizer levels increased. Hydrogen bonding between guar gum and glycerol is disrupted by water sorption therefore the intensity of the guar gum–water and the polyol–water interactions increases resulting in swelling of the film and an increased in water diffusion through the film.[44] This is due to increases in free volume and chain movement, reducing the rigidity and increasing the molecular mobility of films, thus allowing higher water vapor diffusion through their structure and therefore increasing WVP as stated by Cerqueira et al.[36] Plasticizers may also promote water vapor diffusivity through the polymeric structure by increasing interchain spacing between polymer chains; consequently, accelerating the water vapor transmission.[19] These results of the permeability of the films are in agreement with other works where the increase of plasticizer concentrations has increased the values of WVP.[19,36,40,41] WVP of films with good water vapor barrier properties (low or no water permeation and diffusion through the film) should not increase or increase very little with increasing relative vapor pressure.[10]

2.3.9 SENSORY EVALUATION

Table 2.4 illustrates results scored by the panelists of the sensory acceptability of the guar gum and glycerol edible films. Color was rated by the trained sensory panelists in a range from 5.13 to 6.17 of hedonic scale. The film with the highest color acceptability was the one formulated using 1.5% of guar gum and 30% of glycerol. There was a significantly difference ($p < 0.05$) in sensory panelists rating on the color attribute of the edible films which is an important factor that determines the general appearance and the consumer acceptability of the biofilms.[14] Other important sensory attributes are stickiness and taste because they provide fundamental information on the applicability of edible films and coatings on food surfaces as protective layers.[54] The rating given by the sensory panelists for the stickiness of each film ranged from 5.50 to 6.86 of hedonic scale. The increase in stickiness may be attributed to the water holding capacity of glycerol and likewise depends on its high water absorption capacity.[14,40,41,55] The sensory scores of taste from guar gum films ranged from 5.61 to 6.90 and were not significantly different ($p < 0.05$). The overall acceptability of edible films

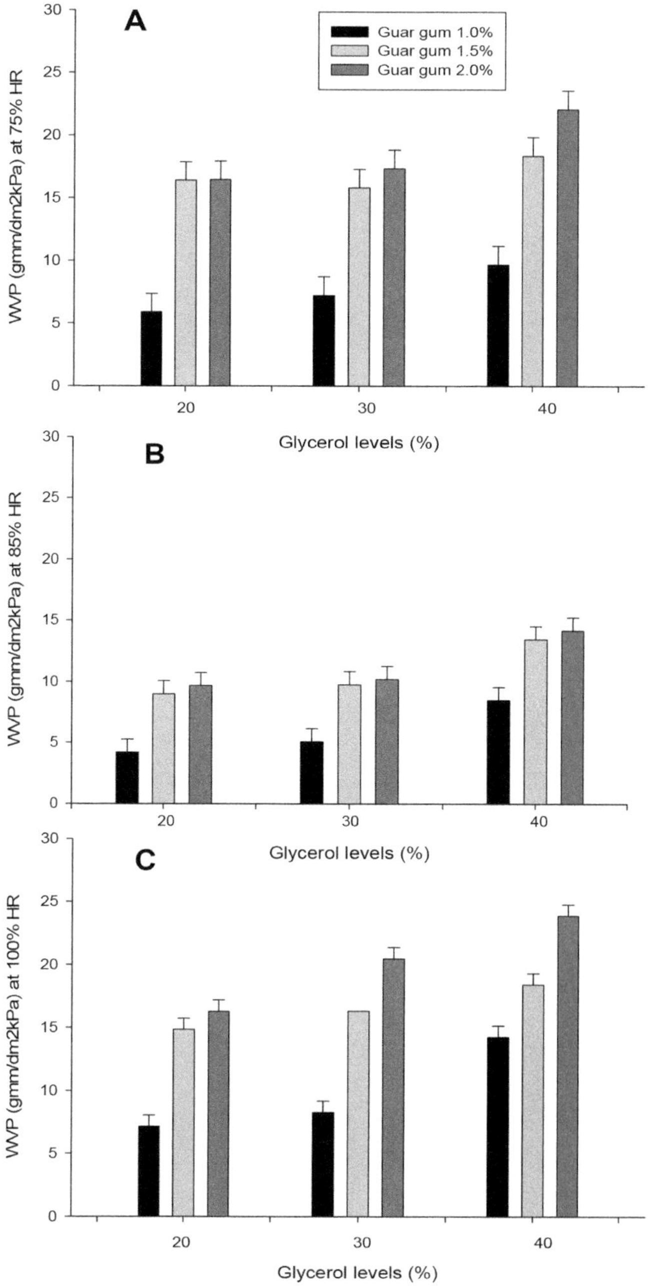

FIGURE 2.2 Effects of guar gum and glycerol content on WVP at 75% (A), 85% (B), and 100% (C) of relative humidity.

ranged from 5.42 to 7.16. The film samples were not significantly different ($p < 0.05$) in terms of overall acceptability even though other authors report that some factors like transparency, flexibility, brittleness, and elasticity of films influence on this attribute.[14,56,57] This is an important result since all formulations of the films could be eventually accepted by consumers.

TABLE 2.4 Sensory Acceptability of Guar Gum Films.

Guar gum (%)	Glycerol (%, v/v)	Color	Taste	Stickiness	Overall acceptability
1.0	20	5.88 ± 1.03^{ab}	6.45 ± 1.12^{a}	6.03 ± 1.05^{abc}	6.21 ± 0.88^{a}
1.5	20	5.99 ± 1.13^{ab}	6.65 ± 0.89^{a}	6.16 ± 1.05^{abc}	6.35 ± 1.17^{a}
2.0	20	5.54 ± 0.91^{b}	6.90 ± 0.72^{a}	6.21 ± 0.79^{abc}	5.42 ± 1.22^{a}
1.0	30	6.05 ± 0.82^{ab}	6.35 ± 0.68^{a}	6.11 ± 1.03^{abc}	6.19 ± 0.86^{a}
1.5	30	6.17 ± 1.02^{ab}	6.25 ± 1.18^{a}	6.65 ± 0.94^{ab}	7.16 ± 1.24^{a}
2.0	30	5.41 ± 1.05^{b}	5.90 ± 0.83^{a}	6.74 ± 0.88^{ab}	5.57 ± 1.16^{a}
1.0	40	5.86 ± 0.79^{ab}	5.80 ± 0.74^{a}	5.50 ± 1.64^{c}	5.85 ± 1.01^{a}
1.5	40	6.13 ± 0.97^{ab}	5.84 ± 1.23^{a}	6.24 ± 1.01^{abc}	5.54 ± 0.98^{a}
2.0	40	5.13 ± 1.07^{a}	5.61 ± 0.78^{a}	6.86 ± 0.98^{a}	5.51 ± 0.83^{a}

Means and standard deviation of 30 panelists score. Means values followed by different superscripts letters within the same column were significantly different ($p < 0.05$).

2.4 CONCLUSIONS

Elastic and flexible guar gum based films plasticized with different levels of glycerol were prepared successfully. The concentration of the components in the film affected the optical, barrier, mechanical and sensorial properties of the films to various extents. Optical properties such as opacity and color (L, a*, and b*) were affected by the concentration of guar gum and glycerol used. Solubility of films depended on the interaction of guar gum and glycerol content. WVP and elongation at break increased as the concentration of glycerol increased, on the other hand TS decreased as the level of glycerol and guar gum increased. Overall acceptability of films was independent of formulation composition. Sensory acceptability of color and stickiness was affected by the concentration of guar gum and glycerol. Further studies should consider films prepared with guar gum at 1.5% and 30% of glycerol to determine their application in real food systems.

ACKNOWLEDGMENTS

The authors are grateful to the National Council of Science and Technology (CONACyT) for the financial support through a Ph.D. Scholarship for M.Sc. Ruelas-Chacon and to the Universidad Autonoma Agraria Antonio Narro (UAAAN) for the permission granted to M.Sc. Ruelas-Chacon to pursue Ph.D. studies.

KEYWORDS

- *Cyamopsis tetragonoloba* **(L.)** *Taub.*
- **edible gum**
- **optical characteristics**
- **water vapor permeability**
- **mechanical characteristics**
- **glycerol**

REFERENCES

1. Arevalo-Niño, K.; Aleman-Huerta, M. E.; Rojas-Verde, M. G.; Morales-Rodriguez, L. A. Peliculas Biodegradables a Partir de Residuos de Cítricos: Propuesta de Empaques Activos. *Rev. Latinoam. Biotecnol. Amb. Algal.* **2010,** *1*, 124–134.
2. Rao, M. S.; Kanatt, S. R.; Chala, S. P.; Sharma, A. Chitosan and Guar Gum Composite Films: Preparation, Physical, Mechanical and Antimicrobial Properties. *Carbohydr. Polym.* **2010,** *82*, 1243–1247.
3. Tefera, A.; Seyoum, T.; Woldetsadik, K. Effect of Disinfection, Packaging and Storage Environment on the Shelf Life of Mango. *Biosyst. Eng.* **2007,** *96*, 201–212.
4. Saucedo-Pompa, S.; Rojas-Molina, R.; Aguilera-Carbó, A. F.; Sáenz-Galindo, A.; De la Garza, H.; Jasso-Cantú, D.; Aguilar, C. N. Edible Film Based on Candelilla Wax to Improve the Shelf Life and Quality of Avocado. *Food Res. Int.* **2009,** *42*, 511–515.
5. Pushkala, R.; Raghuram, P. K.; Srividya, N. Chitosan Based Powder Coating Technique to Enhance Phytochemicals and Shelf Life Quality of Radish Shreds. *Postharvest Biol. Technol.* **2013,** *86*, 402–408.
6. Ettelaie, R.; Tasker, A.; Chen, J.; Alevisopoulos, S. Kinetics of Food Biopolymer Film Dehydration: Experimental Studies and Mathematical Modeling. *Ind. Eng. Chem. Res.* **2013,** *52*, 7391–7402.
7. Adeodato Vieria, M. G.; Altenhofen da Silva, M.; Oliviera dos Santos, L.; Beppu, M. M. Natural-based Plasticizers and Biopolymer Films: A Review. *Eur. Polym. J.* **2011,** *47*, 254–263.

8. Hendrix, K. M.; Morra, M. J.; Lee, H. B. Min, S. C. Defatted Mustard Seed Meal-based Biopolymer Film Development. *Food Hydrocoll.* **2012,** *26*, 118–125.
9. Wang, Y.; Dong, L.; Wang, L.; Yang, L.; Özkan, N. Dynamic Mechanical Properties of Flaxseed Gum Based Edible Films. *Carbohydr. Polym.* **2011,** *86*, 499–504.
10. Talja, R. A.; Helen, H.; Roos, Y. H.; Jouppila, K. Effect of Various Polyols and Polyol Contents on Physical and Mechanical Properties of Potato Starch-based Films. *Carbohydr. Polym.* **2007,** *67*, 288–295.
11. Sothornvit, R.; Krochta, J. M. Plasticizer Effect on Mechanical Properties of β-lactoglobulin Films. *J. Food Eng.* **2001,** *50*, 149–155.
12. Krogars, K.; Heinamaki, J.; Karjalainen, M.; Niskanen, A.; Leskela, M.; Yliruusi, J. Enhanced Stability of Rubbery Amylose-rich Maize Starch Films Plasticized with a Combination of Sorbitol and Glycerol. *Int. J. Pharm.* **2003,** *251*, 205–208.
13. Mali, S.; Grossmann, M. V. E.; García, M. A.; Martino, M. N.; Zaritzky, N. E. Effects of Controlled Storage on Thermal, Mechanical and Barrier Properties of Plasticized Films from Different Starch Sources. *J. Food Eng.* **2006,** *75*, 453–460.
14. Chinma, C. E.; Ariahu, C. C.; Abu, J. O. Development and Characterization of Cassava Starch and Soy Protein Concentrate Based Edible Films. *Int. J. Food Sci. Technol.* **2012,** *47*, 383–389.
15. Ahmadi, R.; Kalbasi-Ashtari, A.; Oromiehie, A.; Yarmand, M. S.; Jahandideh, F. Development and Characterization of a Novel Biodegradable Edible Film Obtained from Psyllium Seed (*Plantago ovata* Forsk). *J. Food Eng.* **2012,** *109*, 745–751.
16. Farahnaky, A.; Saberi, B.; Majzoobi, M. Effect of Glycerol on Physical and Mechanical Properties of Wheat Starch Edible Films. *J. Texture Stud.* **2013,** *44*, 176–186.
17. Mali, S.; Grossmann, M. V. E.; Garcia, M. A.; Martino, M. N.; Zaritzky, N. E. Mechanical and Thermal Properties of Yam Starch Films. *Food Hydrocoll.* **2005,** *19*, 157–164.
18. Krochta, J. M. Proteins as Raw Materials for Films and Coatings: Definitions, Current Status, and Opportunities. In *Protein-based Films and Coatings;* Gennadios, A., Ed.; CRC Press LCC: Boca Raton, FL, 2002; pp 1–41.
19. Rezvani, E.; Schleining, G.; Sümen, G.; Taherian, A. R. Assessment of Physical and Mechanical Properties of Sodium Caseinate and Stearic Acid Based Film-Forming Emulsions and Edible Films. *J. Food Eng.* **2013,** *116*, 598–605.
20. Heyman, B.; De Vos, W. H.; Depypere, F.; Van der Meeren, P.; Dewettinck, K. Guar and Xanthan Gum Differentially Affect Shear Induced Breakdown of Native Waxy Maize Starch. *Food Hydrocoll.* **2014,** *35*, 546–556.
21. Moser, P.; Lopes Cornelio, M.; Nicoletti Telis, V. R. Influence of the Concentration of Polyols on the Rheological and Spectral Characteristics of Guar Gum. *LWT Food Sci. Technol.* **2013,** *53*, 29–36.
22. Roberts, K. T. The Physiological and Rheological Effects of Foods Supplemented with Guar Gum. *Food Res. Int.* **2011,** *44*, 1109–1114.
23. Srivastava, M.; Kapoor, V. P. Seed Galactomannans: An Overview. *Chem. Biodivers.* **2005,** *2*, 295–317.
24. Cerqueira, M. A.; Souza, B. W. S.; Simões, J.; Teixeira, J. A.; Domigues, M. R. M.; Coimbra, M. A.; Vicente, A. A. Structural and Thermal Characterization of Galactomannans from Non-conventional Sources. *Carbohydr. Polym.* **2011,** *83*, 179–185.
25. Cui, W.; Eskin, M. A. M.; Wu, Y.; Ding, S. Synergisms between Yellow Mustard Mucilage and Galactomannans and Applications in Food Products: A Mini Review. *Adv. Colloid Interface Sci.* **2006,** *128–130*, 249–256.

26. Han, J. H.; Floros, J. D. Casting Antimicrobial Packaging Films and Measuring Their Physical Properties and Antimicrobial Activity. *J. Plast. Film Sheeting* **1997,** *13*, 287–298.
27. Gontard, N.; Guilbert, S. Biopackaging: Technology and Properties of Edible and/or Biodegradable Material of Agricultural Origin. In *Food Packaging and Preservation;* Mathlouthi, M., Ed.; Blackie Academic & Professional: New York, NY, 1994; pp 159–181.
28. Bolin, H. R.; Huxsoll, C. C. Control of Minimally Processed Carrot (*Dascus carota*) Surface Discoloration Caused by Abrasion Peeling. *J. Food Sci.* **1991,** *56*, 416–418.
29. Mei, Y.; Zhao, Y. Barrier and Mechanical Properties of Milk Protein-based Edible Films Containing Nutraceuticals. *J. Agric. Food Chem.* **2003,** *51*, 1914–1918.
30. Romero-Bastida, C. A.; Bello-Pérez, L. A.; García, M. A.; Martino, M. N.; Solorza-Feria, J.; Zaritzky, N. E. Physicochemical and Microstructural Characterization of Films Prepared by Termal and Cold Gelatinization from Non-conventional Sources of Starches. *Carbohydr. Polym.* **2005,** *60*, 235–244.
31. ASTM. Standard Methods of Test for Water Vapor Transmission of Materials in Sheet form (E 96-00). In *Annual Book of ASTM Standards*, American Society for Testing and Material: Philadelphia, PA, 2001; pp 1048–1053.
32. Labuza, T. P.; Kaanane, A.; Chen, J. Y. Effect of Temperature on the Moisture Sorption Isotherms and Water Activity Shift of Two Dehydrated Foods. *J. Food Sci.* **1985,** *5*, 385–391.
33. ASTM D 82–91. Standard Test Methods for Tensile Properties of Thin Plastic Sheeting. In *Annual Book of ASTM Standards;* American Society for Testing & Materials: Philadelphia, PA, 1991.
34. Pereda, M.; Ponce, A. G.; Marcovich, N. E.; Ruseckaite, R. A.; Martucci, J. F. Chitosan-gelatin Composites and Bilayer Films with Potential Antimicrobial Activity. *Food Hydrocoll.* **2011,** *25*, 1372–1381.
35. Matta Fakhoury, F.; Martelli, S. M.; Canhadas Bertan, L.; Yamashita, F.; Innocentini Mei, L. H.; Collares Queiroz, F. P. Edible Films Made from Blends of Manioc Starch and Gelatin-influence of Different Types of Plasticizer and Different Levels of Macromolecules on Their Properties. *LWT Food Sci. Technol.* **2012,** *49*, 149–154.
36. Cerqueira, M. A.; Souza, B. W. S.; Teixeira, J. A.; Vicente, A. A. Effect of Glycerol and Corn Oil on Physicochemical Properties of Polysaccharide Films: A Comparative Study. *Food Hydrocoll.* **2012,** *27*, 175–184.
37. Mu, C.; Guo, J.; Li, Z.; Lin, W.; Li, D. Preparation and Properties of Dialdehyde Carboxymethyl Cellulose Crosslinked Gelatin Edible Films. *Food Hydrocoll.* **2012,** *27*, 22–29.
38. Zhang, Y.; Han, J. Plasticization of Pea Starch Films with Monosaccharides and Polyols. *J. Food Sci.* **2006,** *71*, E253–E261.
39. Ekrami, M.; Emam-Djomeh, Z. Water Vapor Permeability, Optical and Mechanical Properties of Salep-based Edible Film. *J. Food Process. Pres.* **2013,** *38*, 1812–1820.
40. Kurt, A.; Kahyaoglu, T. Characterization of a New Biodegradable Edible Film Made from Salep Glucomannan. *Carbohydr. Polym.* **2014,** *104*, 50–58.
41. Jouki, M.; Khazaei, N.; Ghasemlou, M.; HadiNezhad, M. Effect of Glycerol Concentration on Edible Production from Cress Seed Carbohydrate Gum. *Carbohydr. Polym.* **2013,** *96*, 39–46.
42. Ramos, O. L.; Reinas, I.; Silva, S. I.; Fernandes, J. C.; Cerqueira, M. A.; Pereira, R. N.; Vicente, A. A.; Fatima Pocas, M.; Pintado, M. E.; Xavier Malcata, F. Effect of Whey

Protein Purity and Glycerol Content upon Physical Properties of Edible Films Manufactured Therefrom. *Food Hydrocoll.* **2013,** *30,* 110–122.
43. Mali, S.; Grossmann, M. V. E.; Garcia, M. A.; Martino, M. N.; Zaritzky, N. E. Barrier, Mechanical and Optical Properties of Plasticized Yam Starch Films. *Carbohydr. Polym.* **2004,** *56,* 129–135.
44. Aguirre, A.; Borneo, R.; León, A. E. Properties of Triticale Protein Films and Their Relation to Plasticizing-antiplasticinzing Effects of Glycerol and Sorbitol. *Ind. Crops Prod.* **2013,** *50,* 297–303.
45. Viña, S. Z.; Mugridge, A.; García, M. A.; Ferreyra, R. M.; Martino, M. N.; Chaves, A. R.; Zaritzky, N. E. Effects of Polyvinylchloride Films and Edible Starch Coatings on Quality Aspects of Refrigerated Brussels Sprouts. *Food Chem.* **2007,** *103,* 701–709.
46. Reddy, N.; Yang, Y. Completely Biodegradable Soyprotein-jute Biocomposites Developed Using Water without any Chemicals as Plasticizer. *Ind. Crops Prod.* **2011,** *33,* 35–41.
47. Osés, J.; Fabregat-Vázquez, M.; Pedroza-Islas, R.; Tomás, S. A.; Cruz-Orea, A.; Maté, J. I. Development and Characterization of Composite Edible Films Based on Whey Protein Isolate and Mesquite Gum. *J. Food Eng.* **2009,** *92,* 56–62.
48. Cubero, N.; Monferrer, A.; Villalta, J. Food Technology Collection. In *Food Additives;* Mundi-Prensa: Madrid, Spain, 2002; pp 133–134.
49. Al-Hassan, A. A.; Norziah, M. H. Starch-Gelatin Edible Films: Water Vapor Permeability and Mechanical Properties as Affected by Plasticizers. *Food Hydrocoll.* **2012,** *26,* 108–117.
50. Mali, S.; Grossmann, M. V. E.; Garcia, M. A.; Martino, M. N.; Zaritzky, N. E. Microstructural Characterization of Yam Starch Films. *Carbohydr. Polym.* **2002,** *50,* 379–386.
51. Banegas, R. S.; Zornio, C. F.; Borges, A. M. G.; Porto, L. C.; Soldi, V. Preparation, Characterization and Properties of Films Obtained from Cross-linked Guar Gum. *Polimeros* **2013,** *23,* 182–188.
52. Leceta, I.; Guerrero, P.; Ibarburu, I.; Dueñas, M. T.; de la Caba, K. Characterization and Microbial Analysis of Chitosan-based Films. *J. Food Eng.* **2013,** *116,* 889–899.
53. Lazaridou, A.; Biliaderis, C. G. Thermophysical Properties of Chitosan, Chitosan-starch and Chitosan-pullulan Films near the Glass Transition. *Carbohydr. Polym.* **2002,** *48,* 179–190.
54. Kester, J. J.; Fennema, O. R. Edible Films and Coatings: A Review. In *Food Technol.* **1986,** *40,* 47–59.
55. Srichuwong, S.; Snuarti, T. C.; Mishima Isonoa, T. N.; Hisamatsu, M. Starches from Different Botanical Sources II: Contribution of Starch Structures to Swelling and Pasting Properties. *Carbohydr. Polym.* **2005,** *62,* 25–34.
56. Srinivasa, P. C.; Ramesh, M. N.; Tharanathan, R. N. Food Hydrocolloid. Properties and Sorption Study on Chitosan-polyvinyl Alcohol Blends Films. *Carbohydr. Polym.* **2003,** *52,* 431–438.
57. Perez-Gallardo, A.; Bello-Perez, L. A.; Garcia-Almendarez, B.; Montejano-Gaitan, G.; Barbosa-Canovas, G.; Regalado, C. Effect of Structural Characteristics of Modified Waxy Corn Starches on Rheological Properties, Film-forming Solutions, and on Water Vapor Permeability, Solubility and Opacity of Films. *Starch Starke* **2012,** *64,* 27–36.

CHAPTER 3

FOODBORNE PARASITES: ONE HEALTH PERSPECTIVE

K. PORTEEN[1*], S. WILFRED RUBAN[2], and NITHYA QUINTOIL[3]

[1]*Department of Veterinary Public Health and Epidemiology, Madras Veterinary College, Chennai 600007, India*

[2]*Department of Livestock Products Technology, Veterinary College, Hebbal, Bangalore 560024, India*

[3]*Department of Veterinary Public Health, Rajiv Gandhi Institute of Veterinary Education and Research, Puducherry, India*

[*]*Corresponding author. E-mail: rajavet2002@gmail.com*

CONTENTS

ABSTRACT

Parasitic foodborne diseases are generally under recognized; however, they are becoming more common. Globalization, international travel, increase in the population of highly susceptible persons, and change in culinary habits are various factors responsible for the emergence of foodborne parasitic diseases. With the increase in concern of these agents in foods and constraints related to their diagnosis and clinical management, the control and prevention of these diseases is often difficult, because it requires the disruption of a complex transmission chain, involving vertebrate hosts and invertebrate hosts, which interact in a constantly changing environment. However, in recent years with the advent of improved molecular biology-based diagnostic tools, appropriate communication/reporting will be the key factor associated with the increased diagnosis of foodborne parasitic diseases worldwide. This chapter discusses in detail various foodborne parasitic diseases, namely, protozoan, cestode, trematode, and nematode that are transmitted to humans and emphasizes the importance of a One Health approach, calling physicians and veterinarians to unify their efforts in the management of these diseases, several of which are zoonoses.

3.1 INTRODUCTION

Parasites are organisms that obtain their food from other living creatures. A "good" or well-adapted parasite does not kill its host because it depends on the host for a steady supply of food over a long period of time. Foodborne parasitic diseases show high degree of inter-connectedness of wildlife, livestock human health, and ecosystem thereby warrants a need for multidisciplinary approach, most commonly termed One Health to curtail the incidence and for effective control strategies. "One Health" proposes the unification of medical and veterinary sciences with the establishment of collaborative ventures in clinical care, surveillance and control of cross-species disease, education, and research into disease pathogenesis, diagnosis, therapy, and vaccination. Foodborne parasitic diseases affect both humans and animals which involves a complex interplay between parasites, arthropod vectors, environmental influence on vector distribution, companion animal and farm animal reservoir of infection, and susceptible human populations. The effective control of foodborne parasites will essentially involve interdisciplinary teams of microbiologists, parasitologists, entomologists, ecologists, epidemiologists, immunologists, veterinarians, public health officers, and human physicians. More importantly, the One Health approach fits perfectly the requirements for surveillance and control of these foodborne parasitic infections.

3.2 FACTORS CONTRIBUTING TO EMERGENCE OF FOODBORNE PARASITES

In the past, the risk of human infection with parasites was considered to be limited to distinct geographic regions because of parasites' adaptations to specific definitive hosts, selective intermediate hosts, and particular environmental conditions. These barriers are slowly being breeched—first by international travel developing into a major industry, and then by rapid, refrigerated food transport which became available to an unprecedented needs of the consumers.[1]

Other factors that may explain the emergence of some zoonotic parasitic diseases are the increase of the population of highly susceptible persons because of ageing, malnutrition, HIV infection and other underlying medical conditions, and changes in lifestyle, such as the increase in the number of people eating meals prepared in restaurants, canteens and fast food outlets as well as from street food vendors who do not always respect food safety, and the increase of eating raw or undercooked meats.

There are about 107 known species of parasites that are foodborne. While not all species are reported to infest domestic food sources or infect consumers all around the world, the likelihood of this possibility has significantly increased in recent years with the emergence of a truly global market place.

3.3 ROUTES OF ENTRY

Parasites enter the food production process via three main routes:

- Through contamination of food ingredients or raw materials on the farm;
- Through contaminated water included in the final product for product processing or washing, or used for cleaning processing equipment;
- Through transfer or spread via infected food handlers or food preparers in production, food service, or domestic settings.

Parasites are becoming more of a concern for the following reasons:

1. Increasing imports of fruits, vegetables, and ethnic foods, some of which originate in countries without modern sanitary facilities and inspection systems, may introduce parasites.
2. Immigrants from underdeveloped countries may be infected with parasites that could be transmissible to others, particularly during food preparation.
3. The popularity of raw or lightly cooked foods, such as sushi and raw pork sausages, may increase exposure to parasites.
4. As our population ages and more people have deficient immune systems, parasitic infections may have more severe consequences.
5. An interesting issue being considered by some parasitologists is the potential spread of parasitic diseases as global warming proceeds. Some diseases are presently confined to tropical and subtropical areas because cysts or intermediate hosts are not cold hardy. But if lake temperatures warm up and winter temperatures moderate, some diseases may encroach on temperate areas (Table 3.1).

TABLE 3.1 Parasites Found in Different Foods.

Foods	Protozoa	Nematodes	Tapeworms	Flukes
Beef	–	–	*Taenia saginata*	–
Pork	*Toxoplasma*	*Trichinella*	*Taenia solium*	–
Other meat	*Toxoplasma* and *Cryptosporidium* boar (lamb and mutton)	*Trichinella* (cougar, walrus, bear, horse, and wild boar)	*Gnathostoma* (frogs and snakes)	*Paragonimus* (wild boar and guinea pigs)
Milk	*Cryptosporidium*	–	–	–
Fish	–	*Anisakis* and *Gnathostoma*	*Diphyllobothrium*	*Clonorchis*
Clams, mussels, and oysters	*Cryptosporidium* and *Toxoplasma*	–	–	–
Water	*Cyclospora*, *Cryptosporidium*, *Giardia*, and *Toxoplasma*	*Ascaris* and *Gnathostoma*	*Echinococcus*	*Fasciola* and *Fasciolopsis*

Parasites of concern to food safety professionals include several worms, ranging from a few centimeters to several meters in length, and protozoa, single-celled organisms. Many parasitic infections are asymptomatic, others cause acute short-lived effects, and still others may persist in the body causing chronic effects.

3.3.1 PROTOZOAN PARASITES

Over recent decades, parasitic protozoa have been recognized as having great potential to cause waterborne and foodborne disease. Protozoan parasites in humans typically cause mild to moderate diarrhea although malnourished children, the elderly, and the immunocompromised may suffer prolonged and intense gastrointestinal symptoms that can be life-threatening. The organisms of greatest concern in food production worldwide are *Cryptosporidium*, *Cyclospora*, *Giardia*, and *Toxoplasma*.

3.3.1.1 CRYPTOSPORIDIUM PARVUM

C. parvum is an obligate, intracellular protozoan parasite first recognized as a human pathogen in 1976. The organism is transmitted via oocysts (i.e., the infectious stage in the organism's life cycle and resting stage equivalent to a bacterial spore) and shed in feces. The oocyst survives on stainless steel for several hours if kept wet, survives heating at 60°C for 1 min and resistant to chlorine (5% bleach solution). *Cryptosporidium* can cause mild or severe symptoms depending on the dose of oocysts ingested, the virulence of the strain of *C. parvum* and the immunocompetence of the affected individuals. In healthy adult volunteers, the median infective dose was determined to be 10–120 oocysts.[2]

A number of other *Cryptosporidium* species (*Cryptosporidium canis*, *Cryptosporidium felis*, *Cryptosporidium meleagridis*, and *Cryptosporidium muris*) can infect humans; however, such infections are rare and usually are detected and/or isolated from immunocompromised persons or children. Outside of humans, *C. parvum* has also been isolated from cattle, sheep, and goats. This broader host range translates into more opportunities for pathogen spread and occurrence in the environment. *Cryptosporidium hominis* resembles *C. parvum* in appearance and life cycle characteristics but infects only humans.

Foods at Risk: Foodborne outbreaks of *Cryptosporidium* have been associated with raw produce apparently contaminated by infected food handlers, cider, and unpasteurized milk that may have come into contact with cattle feces. The following foods are at great risk:

- raw milk,
- raw sausages (non-fermented),
- any food touched by a contaminated food handler, and
- salad vegetables fertilized with manure.

Disease: The disease caused by *Cryptosporidium is called cryptosporidiosis.* Onset of illness follows an incubation period of 7–10 days.

Illness/complications: Intestinal cryptosporidiosis is self-limiting in most otherwise healthy people. Some infected people are asymptomatic; in others, symptoms may range from mild to profuse diarrhea, with passage of 3–6 L of watery stool per day accompanied by abdominal pain, nausea, and vomiting. Dehydration is a major concern, particularly for pregnant women young children, and immunocompromised people in whom the infection becomes chronic.

Transmission: Person-to-person transmission, by ingestion of contaminated food and waterborne transmission.

Mechanism: Cryptosporidiosis is acquired through ingestion of the oocyst, the organism's infective stage. The oocyst is 4–6 μm in diameter, about half the size of a red blood cell. After being ingested, *C. parvum* oocysts attach themselves to gastrointestinal epithelial cells, where reproduction takes place. The zygotes become one of two types of sporulated oocysts: one with a thin wall, which exists in the gastrointestinal tract and can cause continued infection of the host, and the other with a thick wall, which sheds in the feces and infects other hosts. The mechanism by which the organism causes illness—for example, whether or not a toxin is present—is not fully understood. The mechanisms underlying extraintestinal cryptosporidiosis also are unclear; however, it is believed that the intestines are the originating site.

Prevention and Control:

- pasteurization,
- temperature above 73°C render the oocyst non-infectious,
- inactivated by freezing at −15°C,
- acid conditions (pH 4) results in loss of oocyst viability,
- sensitive to drying,
- sensitive to ultraviolet light,
- 0.35% peracetic acid inactivates oocysts,
- avoid cross contamination from raw to ready-to-eat foods, and
- thoroughly wash fruits and vegetables.

Diagnosis: Enzyme immunoassay (EIA) and immuno fluorescent antibody (IFA) (both direct and indirect methods) can detect *Cryptosporidium*. Molecular-based tests (i.e., polymerase chain reaction (PCR)) have also been developed and successfully implemented in some laboratories.

3.3.1.2 CYCLOSPORA CAYETANENSIS

It is a protozoan parasite that causes the gastrointestinal illness cyclosporiasis. *C. cayetanensis* is a protozoan parasite which belongs to the phylum *Apicomplexa*, subclass *Coccidiaina* and family *Eimeriidae*. Many species of *Cyclospora* have been identified in animals. However, *C. cayetanensis* is the only species identified in humans and appears to be restricted to this host.[6] Oocysts are resistant to pesticides commonly used on farms and to sanitizers used by the food industry.

Infective Dose: The minimum infective dose of oocysts, the oocyst sporulation rate, and their survival under different environmental conditions are unknown. However, it is presumed to be low (possibly as low as 10 oocysts) on the basis of data from outbreak investigations.

Source: It is likely that the most significant transmission occurs where sewage, or water contaminated by human feces, has been applied to horticultural crops. This could occur via the use of contaminated water for the application of pesticides, sprinkling contaminated water on horticultural produce to maintain freshness or washing the produce in contaminated water drawn from ponds, lakes, or rivers. Contaminated hands of food handlers, baskets, and containers in markets could also lead to contaminated produce.[7]

Onset: The onset of illness from infection with *C. cayetanensis* is usually 7–10 days from the time of ingestion.

Symptoms of Disease: *Cyclospora* infection has a range of outcomes from no clinical symptoms of disease (asymptomatic infection) to severe diarrhea resulting in dehydration and weight loss. Other symptoms can include anorexia, nausea, vomiting, abdominal bloating, cramping, fatigue, body aches, and low-grade fever.[4] The onset of illness is 2–14 days (average of 7 days). In untreated individuals, a cycle of remitting and relapsing symptoms can occur that lasts for weeks to months. Shedding of oocysts occurs during the illness and can continue for several weeks after symptoms have abated.[5] Infection with *C. cayetanensis* can lead to long-term sequelae including malabsorption, biliary disease, Reiter's syndrome (reactive arthritis), and Guillain–Barré syndrome (a peripheral nervous system disorder that causes paralysis).[6]

Mode of Transmission: *C. cayetanensis* is transmitted via the fecal-oral route by consumption of contaminated food or water. Direct person-to-person transmission is unlikely as the oocysts shed from individuals are not infectious and require extended periods of time outside the host to sporulate.[5]

Diagnosis: Identification of this parasite is made through symptoms and through microscopic examination of stool specimens. Shape and size characteristics of immature (unsporulated) oocysts present in the stool help to confirm a *cyclospora* infection. *C. cayetanensis* oocysts are perfectly rounded and 8–12 μm in diameter. In addition, when viewed by ultraviolet fluorescence microscopy, *cyclospora* oocysts have the appearance of a pale-blue halo.

Mechanism: The complete life cycle of *C. cayetanensis* is not known, although it is established that the parasite multiplies in the cells lining the small intestine of the host.

3.3.1.3 TOXOPLASMA GONDII

T. gondii is obligate intracellular parasites that belong to the family *Sarcocystidae*. *T. gondii* is ubiquitous and is found around the world. According to the FAO/WHO global ranking of foodborne parasites, this parasite was ranked 4th from an international food safety perspective. *T. gondii* can virtually infect all warm-blooded vertebrates, including humans, whereas cats and the other felids are the only known definitive hosts.

T. gondii has a complex life cycle that includes a separate asexual and sexual cycle and involves a definitive and an intermediate host. The asexual cycle occurs in a wide range of intermediate hosts that include warm-blooded vertebrates, such as birds, carnivores, rodents, pigs, primates, and humans. The sexual cycle occurs exclusively in the definitive host, the wild and domestic feline. Domesticated cats are the principal definitive hosts and shed infectious oocysts in their feces. Oocysts are resistant to the effects of many environmental factors and can persist for several years, under certain conditions.

Routes of Transmission:

People can get infected through

- Ingestion of infectious oocysts shed by cat feces, for example, via consumption of contaminated vegetables, fruits, or water,

- Ingestion of tissue cysts contained in undercooked meat, parasite-encysted (with bradyzoites) meats (from sheep, goats, pigs, wildlife, and horses; cattle),
- Transplacental transmission, and
- Organ transplantation and blood transfusion.

Food at Risk:

- **Meat:** Toxoplasmosis outbreaks associated with raw meat consumption. Epidemiological studies in humans suggest that 30–63% of cases are attributable to meat.[3]
- Water—nfection via oocysts from cat feces—important source in developing countries.
- Vegetables—oocyst contamination.
- Shellfish—filtration of water contaminated with oocysts.
- Milk and milk products (role unknown).

Onset: Five to 23 days after ingestion from contaminated food, water, and fingers.

Illness: Women who become infected during pregnancy typically are asymptomatic, although the parasite (tachyzoites) can cross the placenta. In infected pregnant women, the fate of the fetus falls into three possibilities: miscarriage or stillbirth; head deformities; or brain or eye damage. Some of the clinical manifestations of *T. gondii* infections in the unborn include hydrocephalus or microcephaly, intracranial calcification, and chorioretinitis.

A newborn may become infected at birth with *T. gondii*, but not show any apparent symptoms of infection. However, latent infections in these individuals may cause loss of vision, when they are adults (ocular toxoplasmosis); seizures; or mental disabilities. In addition, immunocompromised individuals may develop pneumonitis, retinochoroiditis, brain lesions, and central nervous system diseases.

Symptoms: In acute toxoplasmosis, sore lymph nodes and muscle pains develop in 10–20% of patients and can last for several weeks, after which symptoms no longer are exhibited. Symptoms of ocular toxoplasmosis are blurred or reduced vision, tearing of or redness in the eye, pain, and sensitivity to light. In healthy individuals, toxoplasmosis usually is asymptomatic. In some cases, flu-like symptoms may appear, such as swollen lymph glands, fever, headache, and muscle aches. Death is rare in acute cases.

Mechanism: Very complex life cycle, in which cats (or felids) are the primary or "definitive" host; and they become infected with *T. gondii* by eating sporulated (mature) oocysts from contaminated environmental sources or *T. gondii* encysted tissue sources (such as rodents). Intermediate hosts, that is, humans, birds, pigs, sheep, goats, and cattle, become infected after eating food or drinking from contaminated water sources. Infected cats will begin shedding within 3–10 days, and this will last for 10–14 days. In the intermediate host, parasites will penetrate the intestinal epithelium and migrate throughout the host as tachyzoites. In acute infections in humans, intestinal epithelial cells are the primary site of invasion, with potential subsequent spread to other sites, such as the brain, heart, and skeletal muscle.

Diagnosis:

Method	**Sensitivity**
Histology, immunohistochemistry	+
In vitro culture	+
Bioassay (golden standard)	
mouse inoculation; 3–4 weeks; serology; examine brains	++
cat feeding; oocyst secretion in feces	++++
PCR	+++

Prevention and Control:

Controlling *Toxoplasma* presence in meat

- Preharvest measures
  a. farm management (cat and rodent control, feed),
  b. vaccination (sheep), and
  c. *Toxoplasma* monitoring at the slaughterhouse.

- Postharvest measures
  a. alternative processing of meat from infected carcasses (freezing),
  b. retailer (cooking instructions on package), and
  c. consumer education.

Inactivation of *Toxoplasma* tissue cysts in meat

- Heat treatment,
- Freezing, 2 days at < −12°C,
- Gamma irradiation (0.5 kGy),
- High pressure (300 MPa),
- Acidity and enhancing solutions.

3.3.1.4 GIARDIA LAMBLIA

G. lamblia (also referred to as *Giardia intestinalis* or *Giardia duodenalis*) is a single-celled, enteric protozoan parasite that moves with the aid of five flagella, which also assist with attachment to intestinal epithelium. *Giardia* is infective in the cyst stage, when it is also extremely resistant to environmental stressors, including cold temperatures and chemicals.

Infective Dose: Ingestion of one or more cysts may cause disease.

Onset: Usually 1–2 weeks after ingestion of a cyst(s).

Illness/complications: Giardiasis is self-limiting in most people. Severe dehydration due to loss of fluids is a major concern, especially in young children. Malabsorption of vitamins, protein, and iron all are possible with chronic infections, and it has been suggested that, in children, this can result in stunted growth and development. Chronicity of infection is correlated with an absence of secretory IgA in the intestinal lumen.

Symptoms: Infections sometimes are asymptomatic. When symptoms are present, they generally consist of especially malodorous diarrhea, malaise, abdominal cramps, flatulence, and weight loss.

Duration: Generally 2–6 weeks, unless the illness becomes chronic, in which case it may last for months or years and may become difficult to treat.

Route of Entry: oral

Sources: Infection typically results after ingestion of soil, water, or food contaminated with feces of infected humans or animals. Giardiasis is most frequently associated with consumption of contaminated water. *Giardia* cysts are not killed by chlorine levels typically used to rinse produce post-harvest and are especially difficult to wash off of complex food surfaces.

Diagnosis: *G. lamblia* is frequently diagnosed by visualizing the organism, either the trophozoite (active reproducing form) or the cyst (the resting stage that is resistant to adverse environmental conditions) in stained preparations or unstained wet mounts of liquid stool, with the aid of a microscope. *Giardia* cysts are 10–20 μm in length and are easily distinguished from much smaller *Cryptosporidium* oocysts. Commercial direct fluorescence antibody kits are available to stain the organism, with reported sensitivities and specificities reaching 100%. Organisms may be concentrated by sedimentation or flotation; however, these procedures reduce the number of recognizable organisms in the sample. Therefore, a single stool specimen is usually insufficient for diagnosis. Enzyme-linked immunosorbent assays (ELISAs) that detect excretory–secretory products of the organism, as well as cyst wall proteins, are also available. In addition, non-enzymatic immunoassays exist. When compared with microscopy, such tests have sensitivities and specificities ranging from 85% to 100%.

3.3.2 PARASITIC WORMS: NEMATODES

3.3.2.1 ANISAKIS

Fish is the most commonly identified host for *Anisakis* include mackerel, squid, sardines, salmon, and raw or pickled herring. Dolphins and whales are the normal definitive hosts for *Anisakis*. Adult worms in these marine mammals produce eggs that pass out with the feces, hatch, and the larvae are consumed by shrimp. When fish or squid eat the shrimp, the larvae are released, bore through the stomach wall, and may remain in the abdominal cavity or penetrate nearby muscles. The life cycle is completed when infected fish or squid are eaten by marine mammals. Humans are an accidental host and these larvae cannot mature in the human gut. Instead, the worms burrow into the intestinal or stomach wall and may wander to the liver, lungs, or other tissues, causing gastric disturbances and allergic reactions. In fact, some people who appear to be allergic to "fish" are actually allergic to *Anisakis* in the fish muscle. It should be noted that even if larvae are killed by cooking, this will not completely eliminate the potential for causing allergic reactions in sensitive people.

3.3.2.2 ASCARIS

Ascaris lumbricoides is a common intestinal roundworm parasite infecting an estimated one-quarter of the world's population. Babies may become infected within months after birth due to inadequate hygiene and the subsequent growth of the worms stunts growth and contributes to diarrheal infections and early childhood mortality. Many infected adults do not exhibit symptoms although these worms are known to irritate the intestinal lining and interfere with the absorption of fats and protein. In some cases, *Ascaris* causes more severe infections in the liver or lungs. Humans are the only known host for this roundworm. Eggs passed out with feces may be ingested by the same or another person who drinks contaminated water, eats with dirty hands, or eats uncooked vegetables that have been fertilized with contaminated human wastes. Upon ingestion, the eggs hatch in the intestine and the worms may migrate to the lungs or liver before returning to the intestine and maturing.

3.3.2.3 GNATHOSTOMA

Nematodes of the genus *Gnathostoma* cause a foodborne human illness characterized by creeping skin eruptions that are sometimes accompanied by erythema and local edema. Occasionally the worms migrate to the eye or other internal organs and may cause meningitis and other serious illness. Gnathostomes usually utilize two intermediate hosts. Adult worms living in wild or domestic cats and dogs release eggs with the host feces. Larvae hatch in fresh water and are ingested by small crustaceans that are then eaten by fish, frogs, or other animals. In these animals, the larvae migrate out of the gut and encyst in muscles. Consumption of raw or lightly cooked, infected fish, shrimp, frogs, etc., introduces the parasite into humans or other animals. Gnathostomes are active little parasites and migrate out of the gut to various tissues. In cats and dogs, they eventually migrate back to the stomach and complete their life cycle while in humans they often migrate to the skin, causing a creeping eruption.

3.3.2.4 TRICHINELLA SPIRALIS (TRICHINOSIS)

Nematodes of the genus *Trichinella* are one of the most widespread zoonotic pathogens in the world. Infection by *Trichinella* spp. has been detected in domestic and wild animals of all continents, with the exception of Antarctica.[10] Trichinellosis in humans occurs with the ingestion of *Trichinella* larvae that are encysted in muscle tissue of domestic or wild animal meat. The most important source of human infection worldwide is the domestic pig. However, meats of wild boars and horses have played a significant role during outbreaks within the past decades.[9] The occurrence of trichinellosis in humans is strictly related to cultural food practices, including the consumption of raw or undercooked meat.

Trichinella completes one round of its life cycle in one host animal. After infective larvae in meat are ingested, they encyst in the intestine, mature, and a new generation of infective larvae is produced. They burrow out of the intestine and travel throughout the body. Highest concentrations of larvae in pigs are found in the diaphragm and tongue but they are also present in various skeletal muscles. The encapsulated larvae may remain alive for many years, but often they become calcified and die within 6–12 months. In order for the larvae to mature, the infected muscle must be eaten by another carnivore.

3.3.3 TREMATODES

3.3.3.1 FASCIOLA HEPATICA (LIVER FLUKE)

F. hepatica is a well-known parasite of domesticated ruminants causing significant economic losses in the cattle and sheep industries of some countries. Human cases usually occur in countries where sanitary facilities are inadequate and clean water is not available. Disease symptoms include fever, abdominal pain, weight loss, and enlarged liver. Adult *Fasciola* inhabits the liver, producing eggs that pass out with the feces. The eggs hatch in water; larvae penetrate snails and undergo further development before leaving the snail and encysting in water or on vegetation in or near the water. Humans often become infected from eating variety of other plants growing near contaminated fields and from drinking contaminated water.

3.3.3.2 FASCIOLOPSIS BUSKI (FASCIOLOPSIASIS AND INTESTINAL FLUKE)

F. buski is the largest trematode infecting humans, with adult worms typically 8–10 cm long. Anemia, headache, and gastric distress characterize mild infections, while heavier infections cause severe abdominal pain, malnutrition, edema, and sometimes intestinal blockage. This parasite requires a single intermediate host. Eggs are deposited in feces, hatch in water, and the larvae penetrate snails and undergo development. After 4–6 weeks, the parasites emerge from the snails and encyst in water or on aquatic plants. Consumption of contaminated water or of raw aquatic vegetables allows for completion of the life cycle.

3.3.3.3 PARAGONIMUS (LUNG FLUKE)

Paragonimus is a foodborne parasite which causes subacute to chronic inflammatory disease of lungs in human. Ten species of the trematode *Paragonimus* are reported to infect humans and the most common being *Paragonimus westermani*. Symptoms of diarrhea, abdominal pain, and fever may occur early after infection and progress to coughing and thoracic pain as the worms settle in the lungs. The life cycle of this parasite includes two or more intermediate hosts: a freshwater snail, then a crab or crayfish, and sometimes an animal which eats the crabs and then it is consumed by humans. Most commonly, people become infected by eating raw or pickled crabs. Neither pickling nor salting destroys *Paragonimus*, but adequate cooking will make crabs, crayfish, and meat safe to eat.

3.3.3.4 CLONORCHIS/OPISTHORCHIS

In Asia, several related parasitic worms of the genera *Clonorchis* and *Opisthorchis* lodge in the liver of infected humans and other animals causing blockage and hyperplasia of the bile passages. Liver flukes have a complex life cycle involving two intermediate hosts, snails, and fish. Humans and other fish-eating animals complete the life cycle by eating raw, infected fish and digesting out the cysts. Then the larvae migrate to the liver, mature, and produce eggs.

3.3.4 CESTODES

3.3.4.1 TAENIASIS

Tapeworms of the genus *Taenia* may be the most familiar of foodborne parasitic worms. The terms cysticercosis and taeniasis refer to foodborne zoonotic infections with larval and adult tapeworms, respectively. Two species are of primary concern to humans and all have intermediate stages in important domestic animals: *Taenia saginata* in cattle and *Taenia solium* in hogs. Consumption of raw or inadequately cooked, infected beef or pork introduces the larvae into the human intestinal tract where they mature into adult worms. Infections may be asymptomatic or may generate non-specific complaints such as altered appetite, abdominal pain, diarrhea, or constipation. Humans may also serve as the intermediate host for *T. solium* (but not for *T. saginata*). The most serious consequences occur when the larvae reach the brain, causing neurocysticercosis that often triggers headaches, seizures, and other neurological symptoms.

3.3.4.2 DIPHYLLOBOTHRIUM (FISH TAPEWORM)

Diphyllobothriosis, a parasitosis caused by flatworms of the genus *Diphyllobothrium*, is contracted by consuming raw or undercooked fish. Humans are one of the primary definitive hosts and the adult tapeworm may grow to a length of ten meters in the intestine.

3.3.4.3 ECHINOCOCCUS

Echinococcosis in humans occurs as a result of infection by the larval stages of taeniid cestodes of the genus *Echinococcus*. The infection is acquired through accidental ingestion of parasite eggs shed by a carnivore final host. These eggs may contaminate fruits and vegetables resulting in a food-borne infection.[11] Six species have been recognized, of which four are of public health concern: *Echinococcus granulosus* (cause of cystic echinococcosis), *Echinococcus multilocularis* (cause of alveolar echinococcosis) *Echinococcus oligarthus*, and *Echinococcus vogeli* (causes of polycystic echinococcosis).

The adult stage of this tapeworm lives in dogs, foxes, and other canids and intermediate stages normally infect sheep, goats, pigs, horses, and cattle. Humans can also serve as an intermediate host if they ingest tapeworm eggs in contaminated water or on raw, contaminated vegetables. The larval tapeworms form fluid-filled cysts (called hydatid cysts) in the liver, lungs, and other organs of intermediate hosts.

3.4 CONCLUSIONS

The rise in general public concern over security of the food chain and food safety has helped to focus more attention on zoonotic parasites. However, for many of the zoonotic parasites, the systems for routine diagnosis, monitoring or reporting are inadequate, or even non-existing. As a consequence, the incidence of human disease and parasite occurrence in food is underestimated.

The risk of foodborne parasites is increasing worldwide, and this situation seems to be driven by several interacting factors. Wildlife populations can naturally migrate, bringing foodborne pathogens from one area to another. Human travelers may also play a role in the translocation of wildlife species and in the introduction of exotic parasitic species into previously free areas. Foodborne parasitic infections are on the rise and the unification of health professional efforts is vital to rectify this situation and to reduce their current burden in terms of morbidity and mortality. Attitude change and increased communication between physicians and veterinarians could play a pivotal role in this process, and it is evident that there is a need for a One Health approach toward a better management. The diagnosis of parasitic diseases is challenging and information exchange from physicians to veterinarians, and vice versa is beneficial for both sides, but mainly for patients. National and local conferences on One Health, where veterinarians and physicians could sit around the same table, should be promoted worldwide.[8] Role of pet and domiciled animals in the transmission of zoonotic disease is very well documented therefore veterinarians may play a vital role by notifying public health authorities for developing effective control strategies at national level.

KEYWORDS

- **parasites**
- **protozoan**
- **cestode**
- **trematode**
- **nematode**

REFERENCES

1. Orlandi, P. A.; Chu, D-M. T.; Bier, J. W.; Jackson, G. J. Parasites and the Food Supply. *Food Technol.* **2002,** *56*, 72–81.
2. DuPont, H. L.; Chappell, C. L.; Sterling, C. R.; Okhuysen, P. C.; Rose, J. B.; Jakubowski, W. The Infectivity of *Cryptosporidium parvum* in Healthy Volunteers. *New Engl. J. Med.* **1995,** *332*, 855–859.
3. Cook, A. J.; Gilbert, R. E.; Buffolano, W.; Zufferey, J.; Petersen, E.; Jenum, P. A.; Foulon, W.; Semprini, A. E.; Dunn, D. T. Sources of Toxoplasma Infection in Pregnant Women: European Multicentre Case–Control Study. European Research Network on Congenital Toxoplasmosis. *BMJ* **2000,** *321*, 142–147.
4. FDA. *Bad Bug Book: Foodborne Pathogenic Microorganisms and Natural Toxins Handbook,* 2nd ed.; US Food and Drug Administration: Silver Spring, 2012, pp 127–129. http://www.fda.gov/Food/FoodborneIllnessContaminants/CausesOfIllness-BadBugBook/ucm2006773.htm.
5. Hall, R. L.; Jones, J. L.; Hurd, S.; Smith, G.; Mahon, B. E.; Herwaldt, B. L. Population-based Active Surveillance for *Cyclospora* Infection: United States, Foodborne Diseases Active Surveillance Network (FoodNet), 1997–2009. *Clin. Infect. Dis.* **2012,** *54* (5), S411–S417.
6. Ortega, Y. R.; Sanchez, R. Update on *Cyclospora cayetanensis*, a Food-borne and Waterborne Parasite. *Clin. Microbiol. Rev.* **2010,** *23* (1), 218234.
7. Tram, N. T.; Hoang, L. M. N.; Cam, P. D.; Chung, P. T.; Fyfe, M. W.; Issac-Renton, J. L.; Ong, C. S. L. *Cyclospora* spp. in Herbs and Water Samples Collected from Markets and Farms in Hanoi, Vietnam. *Trop. Med. Int. Health.* **2010,** *13* (11), 1415–1420.

8. Zinsstag, J.; Schelling, E.; Waltner-Toews, D.; Tanner, M. From "One Medicine" to "One Health" and Systemic Approaches to Health and Well-Being. *Prev. Vet. Med.* **2011,** *101*, 148–156.
9. Gottstein, B.; Pozio, E.; Nöckler, K. Epidemiology, Diagnosis, Treatment, and Control of Trichinellosis. *Clin. Microbiol. Rev.* **2009,** *22*, 127–145.
10. Pozio, E.; Murrell, K. D. Systematics and Epidemiology of Trichinella. *Adv. Parasitol.* **2006,** *63*, 367–439.
11. Moro, P.; Schantz, P. M. Echinococcosis: A Review. *Int. J. Infect. Dis.* **2009,** *13*, 125–133.

CHAPTER 4

EFFECT OF CALCIUM CONTENT ON GELATION OF LOW METHOXY CHICKPEA PECTIN

V. URIAS-ORONA, A. RASCÓN-CHU*, J. MÁRQUEZ-ESCALANTE, K. G. MARTÍNEZ-ROBINSON, and A. C. CAMPA-MADA

Research Center for Food and Development, CIAD, AC., Hermosillo 83000, Sonora, Mexico

Corresponding author. E-mail: arascon@ciad.mx

CONTENTS

ABSTRACT

The aim of this study was to investigate the effect of calcium content on the gelation of low methoxy chickpea pectin. Pectin was acid extracted from chickpea husk and presented an esterification degree of 41%, an intrinsic viscosity (η) of 374 mL/g and a molecular weight of 670 kDa. Rheological changes occurring during 1% (w/v) pectin gelation at different calcium/pectin molar ratios (0.75, 0.86, 0.97, and 1.08) were investigated. Gelation time decreased from 9 to 2 min and elastic modulus (G') of the gel increased from 12 to 30 Pa when the calcium/pectin molar ratios augmented from 0.75 to 1.08. The results attained suggest that chickpea husk can be a potential source of calcium depending gelling pectin for food applications.

4.1 INTRODUCTION

The term pectin groups a family of complex heteropolysaccharides consisting of homogalacturonan α-(1→4) or α-(1→2)-linked-D-GalA*p*-2-*(1→2)* units and rhamnogalacturonan, GalA*p*--(1→2)-Rha*p*-α-(1→4)-GalA*p*-α-(1→2)-Rha*p* regions. In the latter, neutral sugar side chains containing mainly L-arabinose, D-galactose, and D-xylose are covalently attached to the rhamnosyl residues of the backbone.[1] Pectins have the capability to form gels, but the gelation mechanism varies depending on the degree of esterification (DE). Pectin containing more than 50% of their carboxyl groups substituted with methoxy groups is considered high methoxy pectin (HM). This type of pectins form gels in presence of relatively high concentration of soluble solids, usually sucrose, and low pH values. Pectin containing less than 50% of their carboxyl groups substituted with methoxy groups are classified as low methoxy (LM). In this regard, LM pectin forms gels by interaction with calcium ions by the "egg box" mechanism.[2] Both HM and LM pectin are widely used as gelling and stabilizers in food products.[3]

A large amount of by-products is produced during chickpea processing in regions where this is a major food legume. However, to our knowledge, detailed information about the effect of calcium content on the gelling capability of this polysaccharide from chickpea husk has not been yet reported elsewhere. The aim of this work was to investigate the effect of calcium content on the gelation of LM chickpea pectin.

4.1.1 PECTIN EXTRACTION

After enzymatic treatment to eliminate starch and protein, pectin was acid extracted by using 100 g sample and 700 mL of HCl 0.05 N at 80°C for 1 h at 80 rpm. The extract was centrifuged at 10,000 rpm for 10 min and supernatant was precipitated in 65% (v/v) ethanol treated for 4 h at 4°C, pectin was then collected and freeze-dried.[4]

4.1.2 CARBOHYDRATES

Monosaccharide composition was determined after pectin hydrolysis with CF_3COOH, 4 N at 120°C for 4 h. The reaction was stopped on ice, the extracts were evaporated under air at 40°C. The extract was rinsed twice with 200 μL of water and resuspended in 500 μL of water. All samples were filtered through 0.45 μm (Whatman) and analyzed by high-performance liquid chromatography (HPLC) using a Supelcogel Pb column (300 × 7.8 mm^2; Supelco Inc., Bellefonte, PA) eluted with H_2SO_4, 5 mM (filtered 0.2 μm, Whatman) at 0.6 mL/min and 50°C. A refractive index detector Star 9040 (Varian, St. Helens, Australia) and a Star Chromatography Workstation system control version 5.50 were used. The internal standard was inositol.[5]

4.1.3 DEGREE OF ESTERIFICATION

A calibration curve was developed for DE determination by using pectin commercial standards with a known DE (60% and 85%). Subsequent pectin standards with a known DE 65%, 70%, 75%, and 80% were prepared by mixing appropriate amounts of the two commercial standards. DE value of chickpea pectin was inferred using linear fit equation of calibration curve obtained by correlation of the ratio: area of esterified carboxyl groups/(area of esterified carboxyl groups + area of non-esterified carboxyl groups) of pectin commercial standard with their corresponding known DE value.[6] The analysis was performed by FT-IR spectroscopy (Nicolet Instrument Corp., Madison, WI). All standards and samples were dried and stored in desiccators prior to FT-IR analysis. Samples were incorporated into KBr and pressed into a 1 mm pellet.

4.1.4 INTRINSIC VISCOSITY

Relative viscosity was measured by registering pectin solutions flow time in an Ubbelohde capillary viscometer (OB size, CANNON Instrument Company, State College, PA) at 20 ± 0.1°C, immersed in a temperature controlled Koehler bath. The intrinsic viscosity (η) and viscosimetric molecular weight were determined by the Mead, Kraemer, and Fouss method.[7,8]

4.1.5 PECTIN GELATION

The gelling pectin–calcium mixtures were prepared as previously described.[4] After preparation of the sample, the mixture was transferred onto the rheometer. Rheological test was performed by small amplitude oscillatory shear by using a strain-controlled rheometer (AR-1500ex, TA Instruments, USA) in oscillatory mode. A cone and plate geometry (5.0 cm in diameter, 0.04 rad in cone angle) was used and exposed edges of the sample were covered with mineral oil fluid to prevent evaporation during measurements. Pectin gelation kinetic was monitored at 25°C for 30 min by following the storage (G') and loss (G") modulus. All measurements were carried out at 0.25 Hz and 2.5% strain.

4.2 RESULTS AND DISCUSSION

The pectin yield found in this study was 41% (w pectin/w chickpea husk) dry basis, which is close to the values reported in others husk tissues like sunflower (7–11%).[9] Chickpea husk pectin contained 67% (w/w) of galacturonic acid. Residues of galactose, arabinose, rhamnose, glucose, mannose, and xylose were also detected (Table 4.1). Pectin presented an intrinsic viscosity (η) of 374 mL/g and a viscosimetric molecular weight (Mw) of 670 kDa. Chemical composition (η) and Mw values found in this study are in the range reported for other pectins.[9–11]

Chickpea pectin FTIR spectrum is presented in Figure 4.1. The peaks shown between 950 and 1200 cm^{-1} are considered as the "finger print" region for carbohydrates, which allows the identification of major chemical groups in polysaccharides: bands position and intensity are specific for every polysaccharide.[12] The bands between 3600 and 2500 cm^{-1} are related to the presence of O—H stretching whereas absorbance at 1730–1760 and 1600–1630 cm^{-1} is attributed to the ester carbonyl (C=O) groups and carboxyl ion-stretching band (COO–), respectively.[13] FTIR spectrum of chickpea husk pectin presented lower absorbance at 1750 cm^{-1} than at 1650 cm^{-1}, strongly suggesting LM pectin.

TABLE 4.1 Composition of Chickpea Husk Pectin.

Galacturonic acid	67.0 + 0.4
Arabinose	7.7 + 0.3
Galactose	12.3 + 0.5
Glucose	1.6 + 0.2
Xylose	0.4 + 0.1
Mannose	0.6 + 0.1
Rhamnose	10.4 + 0.7

Results are expressed in g/100 g pectin.

All results are obtained from triplicates.

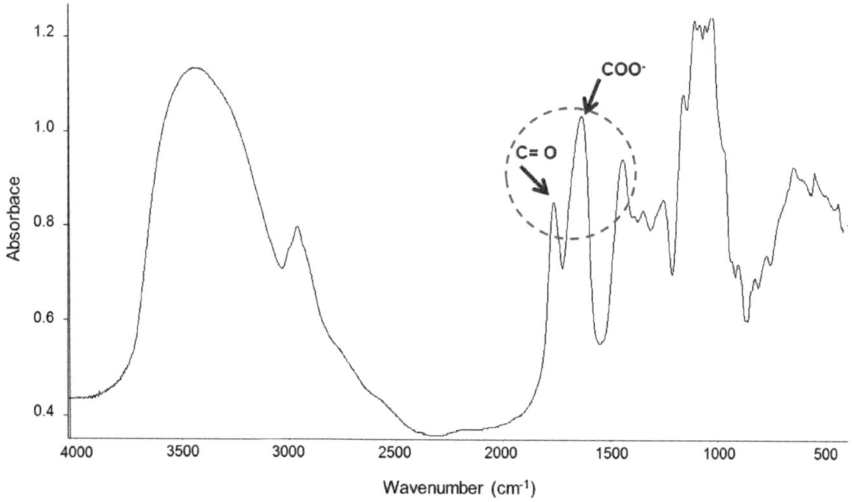

FIGURE 4.1 FTIR spectra of chickpea husk pectin.

In order to determinate the DE in husk chickpea pectin, FTIR spectra of pectin standards were recorded and the ratio of $Area_{1750}/(Area_{1750} + Area_{1650})$ calculated. These data were used to obtain a calibration curve (Fig. 4.2) which was used to calculate the DE of husk chickpea pectin, obtaining a value of 41%.

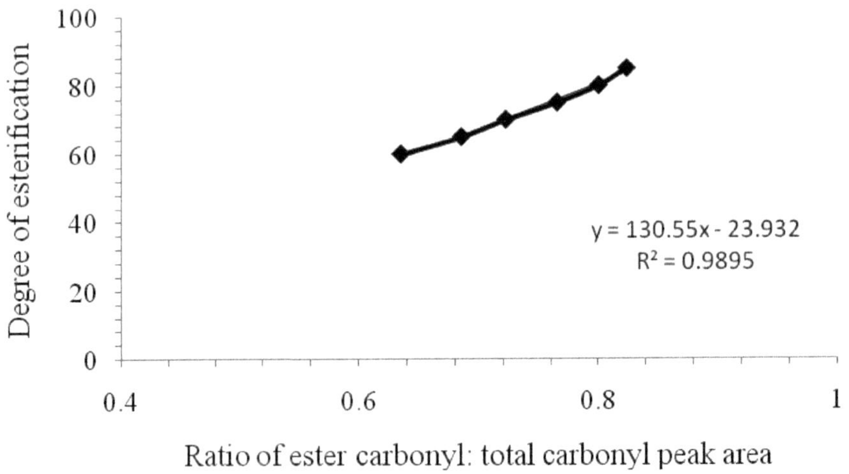

FIGURE 4.2 Degree of esterification of pectin standards versus the ratio of $Area_{1750}/(Area_{1750} + Area_{1650})$.

Husk chickpea pectin formed gels under calcium exposure. Rheological changes occurring during 1% (w/v) pectin gelation at different calcium/pectin molar ratios (0.75, 0.86, 0.97, and 1.08) were investigated. It was found that gelation time (tg) decreased from 9 to 2 min and elastic modulus (G') increased from 12 to 30 Pa when the calcium/pectin molar ratios augmented from 0.67 to 1.0 (Figs. 4.3 and 4.4, respectively). The gel set time was determined as the time at which the elastic (G') and viscous (G") moduli intersected at the study frequency (0.25 Hz). The gel time deduced from rheological measurements corresponded roughly to the moment when a test sample, placed in the same experimental conditions, stopped flowing. Thus, the measurement method (in particular the frequency of measurement) did not significantly affect the gel time.

In spite of the presence of the same pectin concentration, at lower calcium content the number of cross-links established is lower, resulting in a weaker gel structure. Nevertheless, the G' values of cured gels were higher than those reported for other pectin gels.[14–16] The higher G' values found in this tests, compared to literature data, could be related to longer de-esterified galacturonic acid blocks within the extracted chickpea pectin chains. Some other intermolecular interactions like hydrogen bonds could be formed, but they are much weaker as compared to the ionic cross-links formed by carboxyl groups. In LM pectin, the number of sequences of non-methoxylated galacturonic acid residues is long enough for the formation of the so-called "egg boxes" as cross-linking regions.

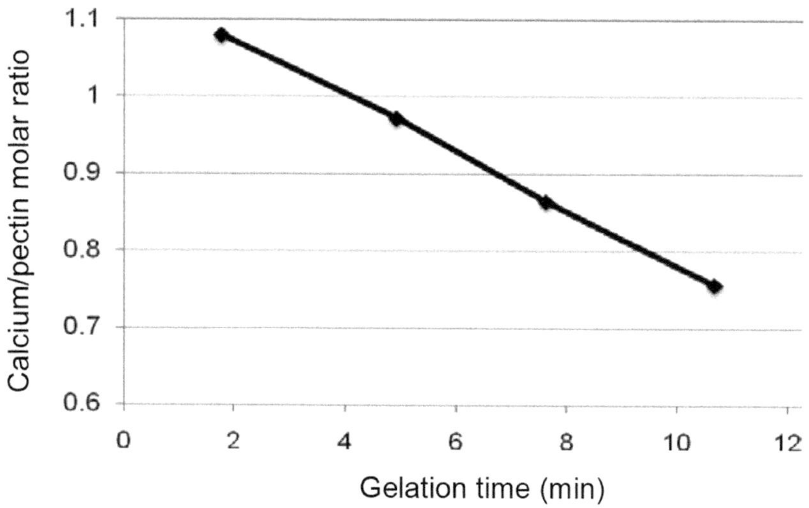

FIGURE 4.3 Effect of calcium/pectin molar ratios on the gelation time (tg) of 1% chickpea pectin solutions.

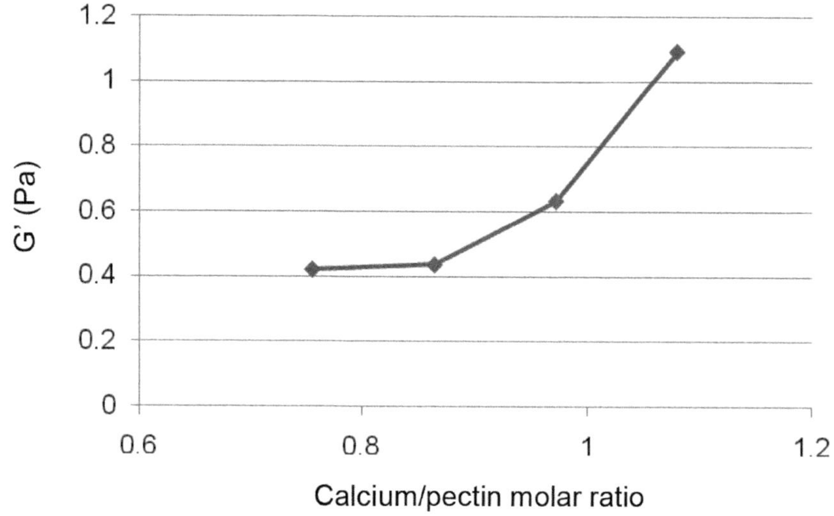

FIGURE 4.4 Effect of calcium/pectin molar ratios on the elastic modulus (G') of 1% chickpea pectin gels.

The "egg box" structure is a junction zone resulting of the ionic interaction of the non-methoxylated galacturonic acid blocks that form well-adapted cavities where the calcium ions fit in allowing the linking of two polysaccharide segments.[15] Therefore, the formation of higher amounts of egg box structures results in stronger gels.

4.3 CONCLUSIONS

The pectin extracted from chickpea husk is LM pectin capable to form elastic gels at different calcium/pectin molar ratios. Shorter gelation time and higher elastic modulus values are obtained by increasing the calcium content in the system. The results attained suggest that chickpea husk can be a potential source of calcium depending gelling pectin for food applications, where low sugar is desired. Further research is undergoing in order to explore the microstructural characteristics of the gels formed.

ACKNOWLEDGMENTS

The authors are pleased to acknowledge Alma C. Campa-Mada (CIAD) for their technical assistance.

KEYWORDS

- **pectin gels**
- **rheology**
- **food**
- **chickpea pectin**
- **galactose**
- **calcium content**
- **rheological changes**

REFERENCES

1. Voragen, A. G. J.; Pilnik, W.; Thibault, J. F.; Axelos, M. A. V.; Renard, C. M. G. C. Pectins. In *Food polysaccharides;* Stephen, A. M., Ed.; Marcel Dekker: New York, NY, 1995; pp 287–339.
2. Willats, W. G. T.; Paul Knox, J.; Mikkelsen, J. D. Pectin: New Insights into an Old Polymer are Starting to Gel. *Trends Food Sci. Technol.* **2006,** *17*, 97–104.

3. Ström, A.; Ribelles, P.; Lundin, L.; Norton, I.; Morris, E. R.; Williams, A. K. Influence of Pectin Fine Structure on the Mechanical Properties of Calcium-pectin and Acid-pectin Gels. *Biomacromolecules* **2007,** *8*, 2668–2674.
4. Urias-Orona, V.; Rascón-Chu, A.; Lizardi-Mendoza, J.; Carvajal-Millán, E.; Gardea, A. A.; Ramírez-Wong, B. A Novel Pectin Material: Extraction, Characterization, and Gelling Properties. *Int. J. Mol. Sci.* **2010,** *11* (10), 3686–3695.
5. Carvajal-Millan, E.; Rascón-Chu, A.; Márquez-Escalante, J.; Ponce de León, N.; Micard, V.; Gardea, A. Maize Bran Gum: Extraction, Characterization and Functional Properties. *Carbohydr. Polym.* **2007,** *69*, 280–285.
6. Chatijigakis, A. K.; Pappas, C.; Proxenia, N.; Kalantzi, O.; Rodis, P.; Polissiou, M. FT-IR Spectroscopic Determination of the Degree of Esterification of Cell Wall Pectin from Stored Peaches and Correlation to Textural Changes. *Carbohydr. Polym.* **1998,** *37*, 395–408.
7. Mead, D. J.; Fouss, R. M. Viscosities of Solutions of Polyvinyl Chloride. *J. Am. Chem. Soc.* **1942,** *64*, 277–282.
8. Kraemer, E. O. Molecular Weight of Celluloses and Cellulose Derivatives. *Ind. Eng. Chem. Res.* **1938,** *30*, 1200–1203.
9. Monsoor, M. A.; Proctor, A. Preparation and Functional Properties of Soy Hull Pectin. *J. Am. Oil Chem. Soc.* **2001,** *78*, 709–713.
10. Yapo, B. M.; Robert, C.; Etienne, I.; Wathelet, B.; Paquot, M. Effect of Extraction Conditions on the Yield, Purity and Surface Properties of Sugar Beet Pulp Pectin Extracts. *Food Chem.* **2007,** *100*, 1356–1364.
11. Rascón-Chu, A.; Martínez-López, A. L.; Carvajal-Millán, E.; Ponce de León-Renova, N.; Márquez-Escalante, J.; Romo-Chacón, A. Pectin from Low Quality 'Golden Delicious' Apples: Composition and Gelling Capability. *Food Chem.* **2009,** *116*, 101–103.
12. Coimbra, M. A.; Barroso, A.; Barrosa, A.; Rutledgeb, D. N.; Delgadilloa, I. FTIR Spectroscopy as a Tool for the Analysis of Olive Pulp Cell-wall Polysaccharide Extracts. *Carbohydr. Res.* **1999,** *317*, 145–154.
13. Kačuráková, M.; Capeka, P.; Sasinková, V.; Wellnerb, N.; Ebringerová, A. FT-IR Study of Plant Cell Wall Model Compounds: Pectic Polysaccharides and Hemicelluloses. *Carbohydr. Polym.* **2000,** *43*, 195–203.
14. Cardoso, S. M.; Coimbra, M. A.; Lopes da Silva, J. A. Calcium-mediated Gelation of an Olive Pomace Pectin Extract. *Carbohydr. Polym.* **2003,** *52*, 125–133.
15. Durand, D.; Bertrand, C.; Clark, A.H.; Lips, A. Calcium-induced Gelation of Low Methoxy Pectin Solutions-thermodynamic and Rheological Considerations. *Int. J. Biol. Macromol.* **1990,** *12*, 14–18.
16. Braccini, I.; Pérez, S. Molecular Basis of Ca-induced Gelation in Alginates and Pectins: The Egg-box Model Revisited. *Biomacromolecules* **2001,** *2*, 1089–1096.

CHAPTER 5

ANTIOXIDANT ACTIVITY AND GELLING CAPABILITY OF β-GLUCAN FROM A DROUGHT-HARVESTED OAT

NANCY RAMÍREZ-CHAVEZ[1], JUAN SALMERÓN-ZAMORA[1], ELIZABETH CARVAJAL-MILLAN[2*], KARLA MARTÍNEZ-ROBINSON[2], RAMONA PÉREZ-LEAL[1], and AGUSTIN RASCÓN-CHU[3]

[1]*Faculty of Agro-Technological Sciences, Autonomous University of Chihuahua, Chihuahua 31125, Mexico*

[2]*Biopolymers, CTAOA, Centro de Investigación en Alimentación y Desarrollo, AC., Carretera a la Victoria Km 0.6, Hermosillo, Sonora, México*

[3]*Biotechnology, CTAOV, Centro de Investigación en Alimentación y Desarrollo, AC., Carretera a la Victoria Km 0.6, Hermosillo, Sonora, México*

[*]*Corresponding author. E-mail: ecarvajal@ciad.mx*

CONTENTS

ABSTRACT

β-glucan from a drought-harvested oat variety was investigated for the first time as a potential antioxidant and gelling agent. β-glucan presented an intrinsic viscosity (η) of 315 mL/g and a viscosimetric molecular weight of 369 kDa. The identity of β-glucan was confirmed by Fourier-transform infrared spectroscopy (FTIR). β-glucan exhibited a dose-dependent free radical scavenging activity, as shown by its 2,2-diphenyl-1-picrylhydrazyl (DPPH) radical inhibition. At 1.0 mg/mL β-glucan exhibited a scavenging rate of 25% on DPPH radicals. Gelling capability of β-glucan at 10% (w/v) was investigated by rheological measurements. After an induction period, elasticity and viscosity attained plateau values of 50 and 3 Pa, respectively. The results attained suggest that β-glucan from drought-harvested oat presents good potential as antioxidant and gelling agent.

5.1 INTRODUCTION

Free radicals can be induced during inflammatory processes, cardiovascular problems, allergic reactions, and among other diseases as a mechanism of biological membrane destruction and the endogenous antioxidant system cannot totally prevent the development of oxidative stress. Therefore, research on exogenous substances with antioxidant properties that might be used for prophylactics or therapy of diseases where free radicals are activated has been intensified in the last years.[1] Polysaccharides have been used in the food industry and in medicine for a long time due to their biological activities. It has been reported that, polysaccharides in general have antioxidant activities, and can be explored as novel potential antioxidants.[2] Due to their antioxidant activity, polysaccharides extracted from fungal, bacterial, and plant sources have been proposed as therapeutic agents.[3,4] Antioxidant activity of polysaccharides has been related to the reduction power and free radical scavenging activity of these molecules.[5,6]

Research on oat grain properties (*Avena sativa*) has received increasing attention as it has been reported to reduce serum cholesterol levels and attenuate postprandial blood glucose and insulin responses, which has been related to the presence of β-glucan. β-glucan is a linear polysaccharide made up entirely of sequences of (1-4)-linked D-glucopyranosyl units separated by single (1-3)-β-linked units.[7] On the other hand, β-glucan presents a high potential application as texturizing, fat-mimetic, and gelling agent.[8] It has been reported that barley glucan present antioxidant activity,[9] a property that would increase the utility of β-glucan as a supplement or additive. However, the antioxidant activity of β-glucan from oat has not been extensively investigated.

Oat is planted as a forage crop in Northern Mexico, where rainfall has an erratic distribution and, therefore, seeds yields and quality are low.[10] These drought-harvested oat seeds have been studied on the basis of forage yield and nutritional value.[11] However, to our knowledge, detailed information on the free radical scavenging activity and the viscoelastic properties of β-glucan has not been yet reported elsewhere.

The aim of this study was to investigate the antioxidant activity and gelling capability of β-glucan from a drought-harvested oat variety in order to assess potential new functional properties of this polysaccharide.

5.2 RESULTS AND DISCUSSION

5.2.1 EXTRACTION AND CHARACTERIZATION

β-glucan yield was 2.4% (w β-glucan/w oat seeds, dry basis db), which means that around 60% of the β-glucan initially present in the oat seed (4.0% w β-glucan/w oat seed db) was recovered. This yield value was in agreement with that previously reported for β-glucan extracted from oat seeds under similar conditions to those used in the present research.[12] Chemical composition of β-glucan is presented in Table 5.1. The extraction protocol adopted in this study provided a β-glucan of high purity (β-glucan content of 92% db). Arabinose, xylose, protein, ash, and ferulic acid were also detected in this sample.

TABLE 5.1 Composition (%) of Oat β-glucan.

β-glucan	92.00 ± 1.50
Arabinose	0.75 ± 0.01
Xylose	1.18 ± 0.01
Protein	4.10 ± 0.30
Ash	2.30 ± 1.20
Ferulic acid	0.03 ± 0.001

All results were obtained from triplicate experiments.

The intrinsic viscosity and the viscosimetric molecular weight values of β-glucan were 315 mL/g and 369 kDa, respectively. These values were similar to those reported for oat β-glucan extracted under similar conditions.[13] Intrinsic viscosity provides a convenient measure of the space-occupancy of individual polymer coils. The molecular size plays an important role on the solubility, conformation, and rheological properties of the molecule in solution.[8]

5.2.2 FOURIER-TRANSFORM INFRARED SPECTROSCOPY

In order to confirm the identity of oat β-glucan, the sample was analyzed by FTIR (Fig. 5.1). It was found that β-glucan spectrum exhibited similarities in its absorption pattern to that reported by other authors.[14] FTIR spectrum presents the main band centered at 1035 cm^{-1} which could be assigned to —OH bending, with shoulders at 1158, 995, and 897 cm^{-1} that were related to the antisymmetric C—O—C stretching mode of the glycosidic link and β(1-4) linkages. Absorbance was observed at 1720 cm^{-1} indicating a low degree of esterification with aromatic esters such as hydroxycinnamic acids,[15] which could be related to the low amount of ferulic acid detected in this oat β-glucan (Table 5.1). Bands at 1648 and 1541 cm^{-1} are related to protein content.[14] The protein signal detected could be related to protein residues.

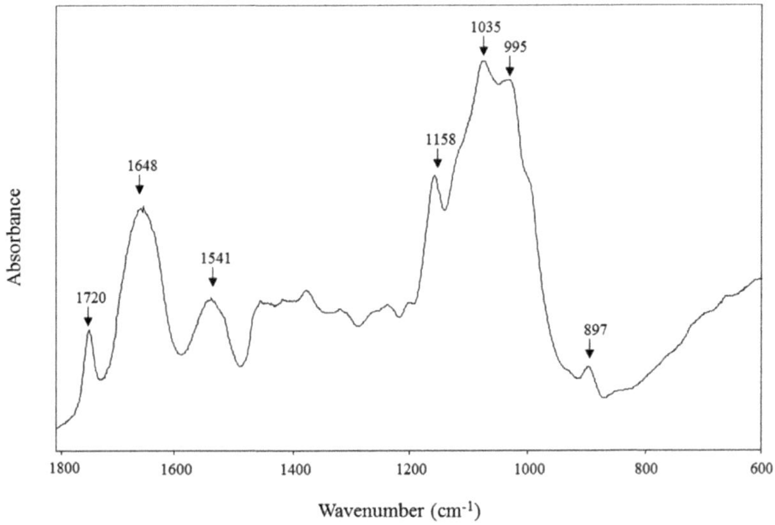

FIGURE 5.1 Fourier-transform infrared spectroscopy spectrum of oat β-glucan.

5.2.3 ANTIOXIDANT ACTIVITY

DPPH radicals have been widely used as a model system to study the scavenging activity of different natural compounds. The antiradical performance of β-glucan with respect to DPPH radicals was measured. The color of the system changes from purple to yellow when the absorbance at 517 nm decreases as a result of the formation of DPPH-H through donation of hydrogen by antioxidants.[16] Figure 5.2 shows that the scavenging activity of β-glucan on DPPH radicals is related to the polysaccharide concentration. At the concentration of 1.0 mg/mL, the antioxidant activity was 25%. A recent study reported that β-glucan extracted from barley and oat scavenged 60% and 10% of the hydroxyl radicals, respectively.[9] Nevertheless, in that investigation, the authors used a higher polysaccharide concentration (10 mg/mL) and a different antioxidant activity assay. In addition, in that report, the composition of the investigated oat β-glucan was reported as unknown. On the other hand, the scavenging activity of oat β-glucan against DPPH found in this study was higher than that reported for other polysaccharides such as commercial pectin (10%) under the same concentration (1.0 mg/mL).5 It has been previously suggested that a relatively low molecular weight in polysaccharides appeared to increase the antioxidant activity.[17] It has also been reported that β-glucan antioxidant activity may be associated with the source and extraction method of obtaining β-glucan.[9] However, the mechanisms by which β-glucan scavenges hydroxyl radicals are not yet clarified.

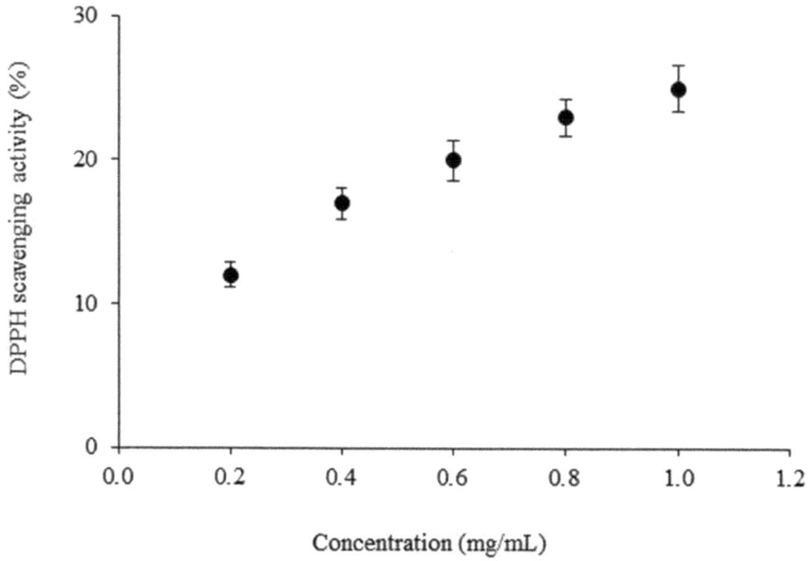

FIGURE 5.2 Scavenging activity of oat glucan against DPPH radicals. Each value is presented as mean ± standard error ($n = 3$).

5.2.4 GELLING

The gelling capability of aqueous β-glucan dispersions was monitored as a function of time at 25°C. The time-dependent evolution of the storage (G') and loss (G") modulus was monitored periodically at 0.25 Hz frequency and strain level of 1% (Fig. 5.3). During gelation G' and G" increased with time and attained a plateau value with the sample adopting gel properties. The initial rise of G', which is sharper than the rise of G", involves the establishment of a three-dimensional network structure. The subsequent slower rise in G' could be attributed to further cross-linking and rearrangement of cross-links. The transition from the stage of rapidly increasing modulus to the plateau stage occurs when most of the sol fraction has been converted into the gel phase.[18] β-glucan gels belong to the category of physically cross-linked gels whose three-dimensional structure is stabilized mainly by multiple inter and intra chain hydrogen bonds in the junction zones of the polymeric network.[8] At the end of the gel curing experiment the mechanical spectra became typical of elastic gel networks (G'>>G" and G' nearly independent on frequency). The condition of linear viscoelasticity was fulfilled during the gel cure experiments as demonstrated by strain sweeps at 0.25 Hz performed before and after the time dependence experiments (Fig. 5.4).

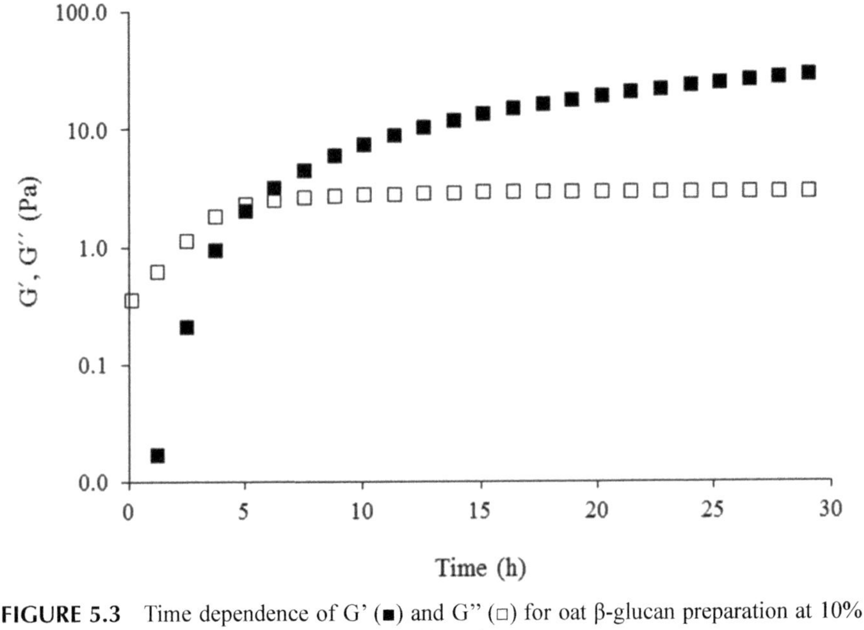

FIGURE 5.3 Time dependence of G' (■) and G" (□) for oat β-glucan preparation at 10% (w/v). Data obtained at 25°C, 0.25 Hz, and 1% strain.

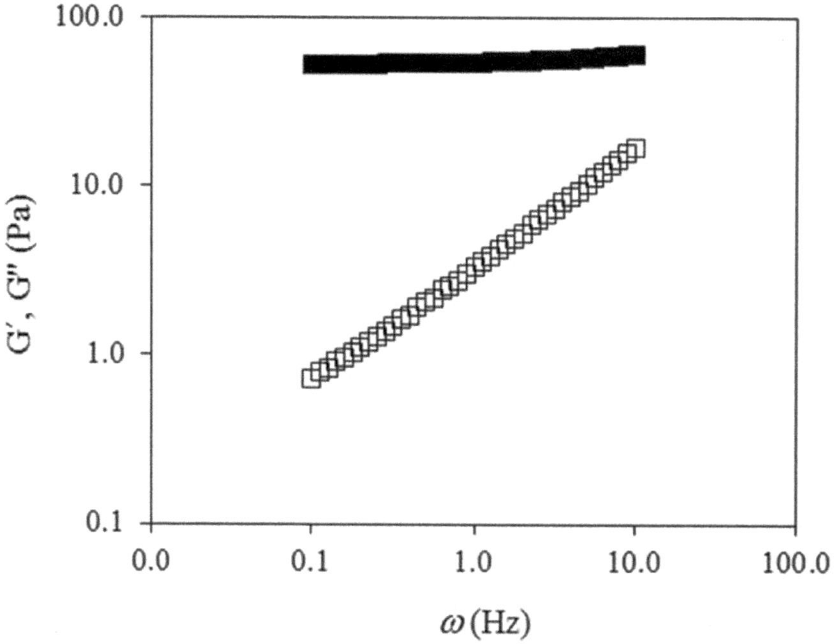

FIGURE 5.4 Mechanical spectrum of oat β-glucan gels at 10% (w/v) G' (■) and G'' (□) after gel network development (30 h). Data obtained at 25°C and 1% strain.

5.3 EXPERIMENTAL

5.3.1 MATERIALS

Whole oat seeds from *A. sativa* cultivar Karma harvested under drought conditions were provided by the National Institute for Investigation in Forestry, Agriculture and Animal Production in Mexico (INIFAP). The seed hull was removed by hand and seeds were milled down to 0.84 mm particle size using a M20 Universal Mill (IKA®, Werke Staufen, Germany). All chemical products were purchased from Sigma Chemical Co. (St. Louis, MO, USA).

5.3.2 METHODS

β-glucan was water extracted from milled oat seeds (1 Kg/3 L) for 15 min at 25°C. The water extract was then centrifuged (12096 × g, 20°C, 15 min) and supernatant recovered. Starch was then enzymatically degraded

(Termamyl®120, 100°C, 30 min, 2800 U/g of sample and, amyloglucosidase, 3 h, 50°C, pH 5, 24 U/g of sample). The destarched extract was then deproteinized with pronase (pH 7.5, 20°C, 16 h followed for 100°C, 10 min, 0.4 U/ g of sample). Supernatant was precipitated in 65% ethanol treated for 4 h at 4°C. Precipitate was recovered and dried by solvent exchanged (80%, v/v ethanol, absolute ethanol, and acetone) to give oat β-glucan.

5.3.3 CHEMICAL ANALYSIS

The β-glucan content was determined by the method of McCleary and Gennie-Holmes[19] using Megazyme® mixed linkage β-glucan assay kit. Sugar composition was determined after β-glucan hydrolysis with 2 N trifluoroacetic acid at 120ºC for 2 h. The reaction was stopped on ice, the extract was evaporated under air at 40ºC and rinsed twice with water (200 L) and resuspended in water (500 μL). All samples were filtered through 0.45 m (Whatman) and analyzed by high-performance liquid chromatography (HPLC) using a Supelcogel Pb column (300 × 7.8 mm^2; Supelco Inc., Bellefonte, PA) eluted with 5 mM H_2SO_4 (filtered 0.2 m, Whatman) at 0.6 mL/min and 50ºC.[20] A Varian 9012 HPLC with Varian 9040 refractive index detector (Varian, St. Helens, Australia) and a Star Chromatography Workstation system control version 5.50 were used. A series of sugar calibration standards were prepared in HPLC grade water at concentrations appropriate for creating a calibration curve for each sugar of interest in the range of 0.2–12.0 mg/mL. The internal standard was inositol. Protein content was determined according to Bradford.[21] Ash content was according to the AACC approved method.[22]

5.3.4 INTRINSIC VISCOSITY AND VISCOSIMETRIC MOLECULAR WEIGHT

Specific viscosity (η_{sp}) was measured by registering β-glucan solutions flow time in an Ubbelohde capillary viscometer at 25 ± 0.1°C, immersed in a temperature-controlled bath. β-glucan solutions were prepared at different concentrations, dissolving dried β-glucan in water for 18 h with stirring at room temperature. β-glucan solutions and solvent were filtered using 0.45 μm membrane filters before viscosity measurements. The intrinsic viscosity (η) and the viscosimetric molecular weight were estimated from viscosity measurement of β-glucan solutions by extrapolation of Kraemer and Mead and Fouss curves to "zero" concentration.[23]

5.3.5 FTIR SPECTROSCOPY

Infrared spectra were collected using a Nicolet FTIR spectrometer (Nicolet Instrument Corp., Madison, USA). The samples were pressed into KBr pellets. A blank KBr disk was used as background. In order to obtain more exact band positions Fourier self-deconvolution was applied. Spectra were recorded between 600 and 1800 cm^{-1}.

5.3.6 DPPH SCAVENGING ACTIVITY

The antioxidant activity was assessed as described before.[24] A 0.1 mM solution of DPPH (2,2-diphenyl-1-picrylhydrazyl) in ethanol was prepared and this solution (1 mL) was added to sample (10–50 μg/mL, 2 mL) in water. After 30 min, absorbance was measured at 517 nm. Vitamin C was used as a positive control. The inhibition of DPPH radicals by the pectin samples was calculated according to the following equation:

$$\text{DPPH-scavenging activity (\%)} = [1 - (\text{Asample517 nm} - \text{Ablank517 nm})/\text{Acontrol517 nm}] \times 100$$

The measurements were performed using a spectrophotometer Lambda 25 UV/VIS (PerkinElmer, USA).

5.3.7 GELLING

A β-glucan solution at 10% (w/v) was prepared by gentle stirring samples in double distilled water at 75°C until complete solubilization of the material. After preparation, the sample was transferred onto the rheometer. Rheological tests were performed by small amplitude oscillatory shear by using a strain-controlled rheometer (AR-1500ex, TA Instruments, USA) in oscillatory mode. Plate geometry (5.0 cm in diameter) was used and exposed edges of the sample were covered with mineral oil fluid to prevent evaporation during measurements. β-glucan gelation kinetic was monitored at 25°C for 30 h by following the storage (G') and loss (G") modulus. All measurements were carried out at a frequency of 0.25 Hz and 1% strain (in linear domain). The mechanical spectrum of β-glucan gels was obtained by frequency sweep from 0.1 to 100 Hz at 1% strain. All measurements were at 25°C.

5.3.8 STATISTICAL ANALYSIS

All determinations were made in triplicates and the coefficients of variation were lower than 8%. All results are expressed as mean values.

5.4 CONCLUSIONS

The ability to scavenge free radicals is a precious property for the prevention of various diseases and aging. In this study, it was shown that β-glucan extracted from a drought-harvested oat exerts antioxidant activity and present gelling capability. The present finding provides a basis for further structural analysis and evaluation of the bioactivities of the drought-harvested oat glucan for its application in food and medicinal fields.

ACKNOWLEDGMENTS

The authors are pleased to acknowledge Alma C. Campa-Mada from CIAD for their technical assistance.

KEYWORDS

- **polysaccharide**
- **antioxidant activity**
- **free radicals**
- **gels**
- **xylose**
- **rheological measurements**

REFERENCES

1. Proctor, P. H.; Reynolds, E. S. Free Radicals and Disease in Man. *Physiol. Chem. Phys. Med. NMR* **1984,** *16*, 175–195.
2. Tomida, H.; Yasufuku, T.; Fujii, T.; Kondo, Y.; Kai, T.; Anraku, M. Polysaccharides as Potential Antioxidative Compounds for Extended-release Matrix Tablets. *Carbohydr. Res.* **2010,** *345*, 82–86.

3. Li, S. P.; Zhang, G. H.; Zeng, Q.; Huang, Z. G.; Wang, Y. T.; Dong, T. T.; Tsim, K. W. Hypoglycemic Activity of Polysaccharide, with Antioxidation, Isolated from Cultured Cordyceps Mycelia. *Phytomedicine* **2006,** *13*, 428–433.
4. Urias-Orona, V.; Huerta-Oros, J.; Carvajal-Millan, E.; Lizardi-Mendoza, J.; Rascón-Chu, A.; Gardea, A. A. Component Analysis and Free Radicals Scavenging Activity of *Cicer arietinum* L. Husk Pectin. *Molecules* **2010,** *15*, 6948–6955.
5. Mateos-Aparicio, I.; Mateos-Peinado, C.; Jiménez-Escrig, A.; Rupérez, P. Multifunctional Antioxidant Activity of Polysaccharide Fractions from the Soybean Byproduct Okara. *Carbohydr. Polym.* **2010,** *82*, 245–250.
6. Yang, S. S.; Cheng, K. T.; Lin, Y. S.; Liu, Y. W.; Hou, W. C. Pectin Hydroxamic Acid Exhibit Antioxidant Activities In Vitro. *J. Agric. Food Chem.* **2004,** *52*, 4270–4273.
7. Izydorczyk, M. S.; Biliaderis, C. G. *Functional Food Carbohydrates,* 1st ed.; CRC Press: Boca Raton, FL, 2007; pp 1–72.
8. Cui, S. W. *Polysaccharide Gums from Agricultural Products. Processing, Structures and Functionality;* Technomic Publishing: Lancaster, PA, 2001; pp 252–258.
9. Kofuji, K.; Aoki, A.; Tsubaki, K.; Konishi, M.; Isobe, T.; Murata, Y. Antioxidant Activity of β-Glucan. *ISRN Pharm.* **2012,** *2012*, 125864.
10. Ramos-Chavira, N.; Carvajal-Millan, E.; Marquez-Escalante, J.; Santana-Rodriguez, V.; Rascon-Chu, A.; Salmerón-Zamora, J. Characterization and Functional Properties of an Oat Gum Extracted from a Drought Harvested *A. sativa. Food Sci. Biotech.* **2009,** *18*, 900–903.
11. Ramos-Chavira, N. C. Physicochemical and Functional Characterization of β-glucans from Oat Varieties Developed in Chihuahua State. Dissertation, University of Chihuahua/Center for Food and Development, CIAD, AC, 2008.
12. Beer, M. U.; Arrigoni, E.; Amado, R. Extraction of Oat Gum from Oat Bran: Effects of Process on Yield, Molecular Weight Distribution, Viscosity and (1-3),(1-4)-β-D-glucan Content of the Gum. *Cereal Chem.* **1996,** *73*, 58–62.
13. Doublier, J. L.; Wood, P. J. Rheological Properties of Aqueous Solutions (1-3),(1-4)-β-D-glucan from Oats (*Avena Sativa L.*). *Cereal Chem.* **1995,** *72*, 335–340.
14. Barron, C.; Rouau, X. FTIR and Raman Signatures of Wheat Grain Peripheral Tissues. *Cereal Chem.* **2008,** *85*, 619–625.
15. Séné, C. F. B.; McCann, M. C.; Wilson, R. H.; Grinter, R. Fourier-transform Raman and Fourier-transform Infrared Spectroscopy. An Investigation of Five Higher Plant Cell Walls and Their Components. *Plant Physiol.* **1994,** *106*, 1623–1631.
16. Zoete, V.; Vezin, H.; Bailly, F.; Vergoten, G.; Catteau, J. P.; Bernier, J. L. 4-Mercapto-imidazoles Derived from Naturally Occurring Antioxidant Ovothiols 2. Computational and Experimental Approach of the Radical Scavenging Mechanism. *Free Radical Res.* **2000,** *32*, 525–533.
17. Chen, H. X.; Zhang, M.; Xie, B. J. Quantification of Uronic Acids in Tea Polysaccharides Conjugates and Its Antioxidant Properties. *J. Agric. Food Chem.* **2004,** *52*, 3333–3336.
18. Ross-Murphy, S. B. *Biophysical Methods in Food Research;* Blackwell: Oxford, UK, 1984; pp 138–199.
19. McCleary, B. V.; Gibson, T. S.; Mugford, D. C. Measurement of Total Starch in Cereal Products by Amyloglucosidase-a-amylase Method: Collaborative Study. *J. Assoc. Official Anal. Chem. Int.* **2007,** *80*, 571–579.
20. Carvajal-Millan, E.; Rascón-Chu, A.; Márquez-Escalante, J.; Ponce de León, N.; Micard, V.; Gardea, A. Maize Bran Gum: Extraction, Characterization and Functional Properties. *Carbohydr. Polym.* **2007,** *6*, 280–285.

21. Bradford, M. A Rapid and Sensitive Method for the Quantification of Microgram Quantities of Protein Utilizing the Principle of Protein-dye Binding. *Analyt. Biochem.* **1976,** *72,* 248–254.
22. AACC. *Approved Methods of the American Association of Cereal Chemists;* The American Association of Cereal Chemists: St. Paul, MN, 1998.
23. Carvajal-Millan, E.; Guigliarelli, B.; Belle, V.; Rouau, X.; Micard, V. Storage Stability of Arabinoxylan Gels. *Carbohydr. Polym.* **2005,** *59,* 181–188.
24. Hou, W.; Hsu, F. L.; Lee, M. H. Yam (*Dioscorea batatas Deene*) Tuber Mucilage Exhibited Antioxidant Activities In Vitro. *Planta Med.* **2002,** *68,* 1072–1076.

PART II
Food Science and Technology Research

CHAPTER 6

MICROSTRUCTURE AND SWELLING OF WHEAT WATER EXTRACTABLE ARABINOXYLAN AEROGELS

JORGE A. MARQUEZ-ESCALANTE*

Center for Food and Development, CIAD, AC., Hermosillo 83000, Sonora, Mexico

*E-mail: marquez1jorge@gmail.com

CONTENTS

ABSTRACT

Wheat water extractable arabinoxylan (WEAX) aerogels were prepared by supercritical drying (SC-CO_2). Scanning electron microscopy analysis of these WEAX aerogels showed a homogenous microporous structure different to that reported for lyophilized WEAX hydrogels reported in the literature which exhibit a honeycomb-like structure. WEAX aerogels were swollen and their structural parameters were calculated. Rehydrated WEAX aerogels presented a molecular weight between two cross-links (Mc), a cross-linking density (ρc), and a mesh size (ξ) of 122×10^3 g/mol, 11.8×10^{-6} mol/cm^3, and 480 nm, respectively, which are in the range reported for WEAX hydrogels in the literature. WEAX aerogels presenting these microstructural and swelling characteristics will open new applications in food, cosmetics, biomedical, and catalysis, among others.

6.1 INTRODUCTION

An aerogel is a highly porous solid material with exceptional surface area, suitable for loading active compounds. Aerogels are produced by drying hydrogels, usually employing supercritical drying with CO_2 (SC-CO_2), which is able to avoid the structural collapse of material and maintains the porous texture of wet material.[1] All materials that can be obtained as hydrogels from solutions are potential candidates to form aerogels.[2] Aerogels can be produced from inorganic or organic materials. Among organic materials, natural polysaccharides have interesting properties such as stability, availability, renewability, low toxicity, and biodegradability.[3,4] Arabinoxylans are polysaccharides capable of form aerogels.[5] These polysaccharides have been classified as water extractable (WEAX) or water unextractable (WUAX). One of the most important properties of WEAX is the ability to form hydrogels by covalent cross-linking involving FA oxidation by either chemical (ferric chloride and ammonium persulfate) or enzymatic (peroxidase/H_2O_2 and laccase/O_2) free radical-generating agents.[6] This oxidation allows the coupling of WEAX chains through the formation of dimers and trimers of FA (di-FA and tri-FA), generating an aqueous three-dimensional network. Furthermore, there are physical interactions between AX chains that contribute to the stability of the network.[6] WEAX hydrogels are little affected by changes in temperature, ionic strength, and pH.[7] In addition, WEAX hydrogels are neutral and have no odor or color.[8] Because of these characteristics and the macroporous structure of WEAX hydrogels, they have been evaluated as matrices for controlled release of therapeutic proteins to be administered orally and further absorbed in the colon.[7,9] However, WEAX hydrogels have some disadvantages such as low stability in storage due to the reversible action of the cross-linking enzyme and the low loading capacity.[10] An option that could eliminate such disadvantage is the formation of aerogels as they would present a greater surface area for absorption of active compounds. The objective of this study was to investigate the microstructure and swelling of WEAX aerogels.

6.2 EXPERIMENTAL

6.2.1 MATERIALS

WEAX were isolated from wheat flour and characterized as previously reported.[11] WEAX presented a pure of 67% (xylose + arabinose), a FA and di-FA content of 0.45 and 0.05 μm/mg of WEAX, respectively, and an A/X ratio of 0.67. Laccase (benzenediol: oxygen oxidoreductase, EC 1.10.3.2) from *Trametes versicolor* and other chemical products were purchased from Sigma Chemical Co. (St. Louis, MO, USA).

6.2.2 METHODS

6.2.2.1 AEROGELS PREPARATION

WEAX aerogels were prepared as previously described.[11] A WEAX solution at 2% (w/v) was prepared in 0.05 M citrate phosphate buffer at pH 5. Laccase was used as cross-linking agent. After WEAX hydrogels were formed, a WEAX alcogel was prepared by solvent exchange. WEAX alcogels were dried by SC-CO_2. Drying process was prepared in Dense Gas Management System (Marc Sims, Berkeley, CA, USA).

6.2.2.2 SCANNING ELECTRON MICROSCOPY (SEM)

WEAX aerogels for SEM were disposed on aluminum stand employed conductive self-adhesive carbon label. Samples were examined at low voltage (1.8 kV) in a JSM-7401F field emission SEM (JEOL, Tokyo, Japan). SEM images were obtained in secondary electrons modes.[12]

6.2.2.3 SWELLING

WEAX gels were allowed to swell in 20 mL of 0.02% (w/v) sodium azide solution to avoid microbial contamination. During 10 h the samples were blotted and weighted. After weighted, a new aliquot of sodium azide solution was added to the gels. Gels were maintained at 25°C during the test. The analysis was carried out by triplicate.[7]

6.3 RESULTS AND DISCUSSION

6.3.1 WEAX AEROGEL

Figure 6.1 shows the WEAX aerogel prepared by SC-CO_2, which can be observed as a porous material in pill form, the textural properties of this material were previously reported.[13] Table 6.1 presents the dimensions of WEAX aerogel and the mass of polysaccharide in this material. WEAX aerogel dimensions were 5.57 ± 0.07 and 2.58 ± 0.15 mm for diameter (D) and height (h), respectively. A reduction in gel dimensions occurred during solvent exchange, D was reduced by 45.7% and h was reduced by 39.7%. This behavior can be explained as consequence of high polarity of polysaccharides and high affinity toward water. Therefore, in nonpolar solvents as ethanol, polysaccharide present higher affinity toward polysaccharide chains than toward solvent molecules, which promote the aggregation and the formation of a dense structure.[14] On the other hand, the reduction of dimensions during supercritical drying was lower (39.7% and 27.6% for D and h, respectively). The supercritical conditions of the mixture ethanol-CO_2 avoid the presence of liquid between polysaccharide chains achieving the absence of superficial tension that collapses the gel structure.[15] Others aerogels from polysaccharides as chitin nanowhiskers,[16] wheat starch,[17] potato starch,[15] cellulose,[18] alginate,[19,20] pectin,[21] and β-glucan[22] have reported that the shrinkage occurs during the solvent exchange and/or SC-CO_2. The size and shape of hydrogel, the chemical nature and concentration of polysaccharide, the concentration and type of solvent, the step number in solvent exchange, and supercritical conditions are factors that determine the shrinkage grade of the obtained aerogels.

The mass of WEAX in the aerogel was 15.17 ± 0.21 mg, which corresponds to 54.17% of original mass original in the hydrogel. This mass recovery could be attributed to the polysaccharide moisture removal during the drying process[22] and/or the possible transfer of non-cross-linked polysaccharide chains toward the solvent during solvent exchange.

FIGURE 6.1 WEAX aerogel in pill form.

TABLE 6.1 Characteristics of WEAX Gels.

Gel type	Dimensions (mm)		Mass of WEAX (mg)[a]
	Diameter	Height	
Hydrogel	14 ± 0.00[b]	9.33 ± 0.00[b]	28
Alcogel	6.40 ± 0.09	2.78 ± 0.19	n.d.[c]
Aerogel	5.57 ± 0.07	2.58 ± 019	15.17 ± 0.27

[a]Original mass of WEAX in solution.

[b]Nominal dimension given by the mold.

[c]Not determined.

Measurements carried out on 10 samples.

6.3.2 SCANNING ELECTRON MICROSCOPY

The secondary electron SEM micrographs in Figure 6.2a–c show the external and internal structure of WEAX aerogels. SEM micrographs indicate an external microstructure (Fig. 6.2a) densely packed and quite different to a microporous internal microstructure (Fig. 6.2b). This effect may in part be consequence of the sample preparation. Figure 6.2c shows detailed

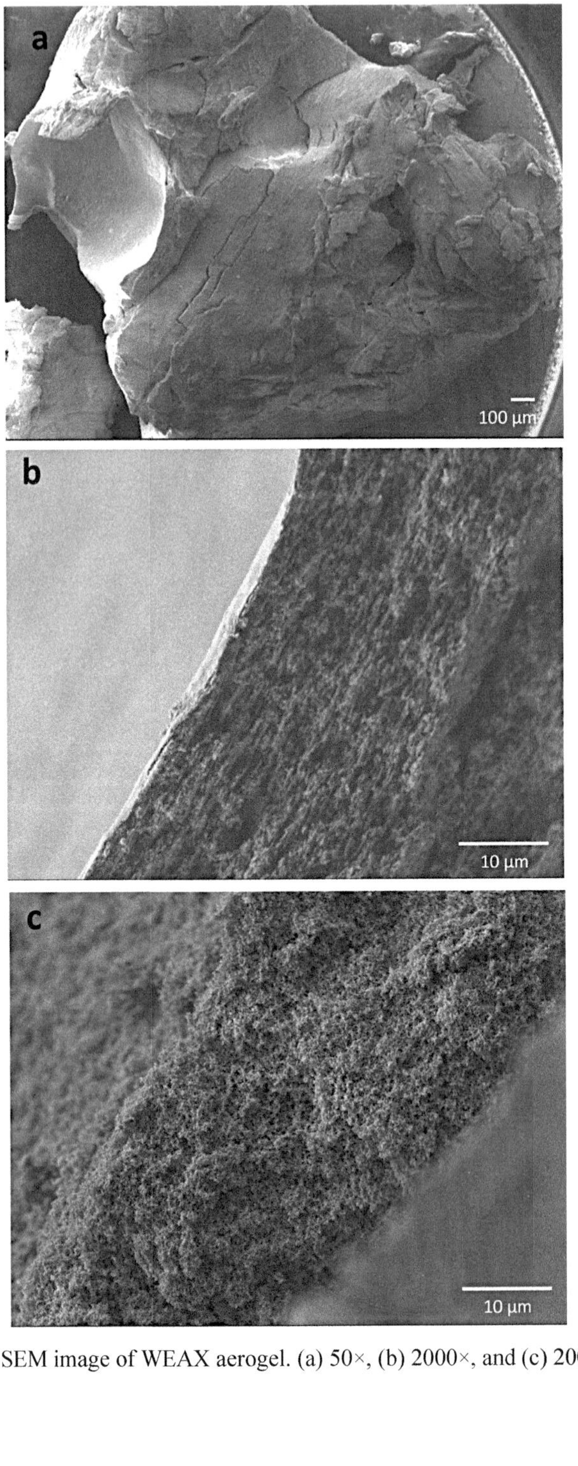

FIGURE 6.2 SEM image of WEAX aerogel. (a) 50×, (b) 2000×, and (c) 2000×.

internal microstructure of WEAX aerogel, which is composed of interconnected particles at nano level, constituting chains that produce a microporous network.[13] This microporous network is quite different to that reported for lyophilized WEAX hydrogels which exhibit a honeycomb-like structure with porous at micro level.[12,23,24] In WEAX like in any polysaccharide, the fine structure, concentration, freezing rate, and drying type are factors that influence the internal structure of the gel. Moreover, it is known that the SC-CO_2 preserves better the fine structure of the gels than freeze drying (lyophilization); because the evaporation of solvent with low surface tension avoids the disruption structural arrangement of gel. Instead, during freeze-drying the disruption of the structural arrangement is given by the growth of ice crystal inside the sample.

6.3.3 SWELLING

The swelling ratio value (*q*) in rehydrated WEAX aerogel was 29.9 g water/g WEAX (Fig. 6.3), which is lower than the *q* value previously reported for wheat WEAX hydrogels (69 g water/g WEAX).[7] This lower water uptake could be explained in terms of the existence of short sections of

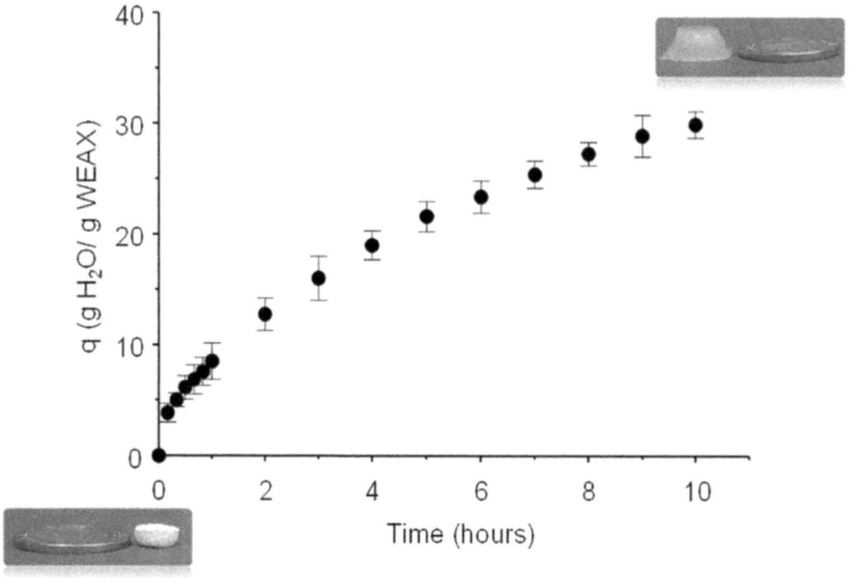

FIGURE 6.3 Swelling of WEAX aerogel in sodium azide at 0.2% (w/v).

non-cross-linked WEAX chains in the network. Non-cross-linked polymer chains sections in the gel can expand easily, leading to a higher water uptake.[8] The above was observed in the swelling of maize WEAX hydrogel containing caffeine and induced by peroxidase, nevertheless, the swelling ratio was not calculated in that report.[12]

The structural parameters of rehydrated WEAX aerogel (Table 6.2) were different than those reported in the literature for wheat WEAX hydrogels induced by laccase (119×10^3 g/mol, 14×10^{-6} mol/cm^3 and 201 nm for *Mc*, *ρc*, and *ξ*, respectively).[11] Differences in chemical structure (FA initial content, arabinose to xylose ratio, and molecular weight) between these WEAX could explain these results. However, in general, the *q*, *Mc*, *ρc*, and *ξ* values in rehydrated WEAX aerogels were in the range of those reported for other WEAX and WUAX hydrogels (18–69 for *q*, 31–119 × 10^3 g/mol for *Mc*, 14–49 mol/cm^3 for *ρc*, and 80–301 nm for *ξ*).[7] Clearly, complementary studies of WEAX aerogels are needed in order to investigate the loading and controlled release capability of this new material.

TABLE 6.2 Structural Parameters of Rehydrated WEAX Aerogel.

$M_c^{a} \times 10^{-3}$ (g/mol)	122.0 ± 2.20
$\rho_c^{b} \times 10^{-6}$ (mol/cm^3)	11.8 ± 0.01
ξ^{c} (nm)	480 ± 15.7

[a]Molecular weight between two cross-links. [b]Cross-linked density. [c]Mesh size.
All values are means of three repetitions.

6.4 CONCLUSIONS

WEAX aerogels present a homogenous microporous structure different to that reported for lyophilized WEAX hydrogels reported in the literature which exhibit a honeycomb-like structure. Swollen WEAX aerogels presented structural parameters (molecular weight between two cross-links, cross-linking density, and mesh size) in the range reported for WEAX and WUAX hydrogels. Using WEAX for making high-added-value aerogels with specific characteristics will open new applications of these materials in food, cosmetics, bio-medical applications, and catalysis, among others.

KEYWORDS

- **laccase**
- **WEAX aerogel**
- **polysaccharide**
- **swelling**
- **water extractable**
- **microporous structure**

REFERENCES

1. Akimov, Y. K. Fields of Application of Aerogels (Review). *Instrum. Exp. Tech.* **2003,** *46*, 287–299.
2. Pierre, A. C.; Pajonk, G. M. Chemistry of Aerogels and Their Applications. *Chem. Rev.* **2002,** *102*, 4243–4266.
3. Mnasri, N.; Moussaoui, Y.; Elaloui, E.; Salem, R. B.; Lagerge, S.; Douillard, J. M.; de Menorval, L. C. Study of Interaction between Chitosan and Active Carbon in View of Optimising Composite Gels Devoted to Heal Injuries. *EPJ Web Conf.* **2012,** *29*, 28.
4. Hüsing, N.; Schubert, U. Aerogels—Airy Materials: Chemistry, Structure, and Properties. *Angew. Chem. Int. Ed.* **1998,** *37*, 22–45.
5. Carvajal-Millan, E.; Guilbert, S.; Doublier, J. L.; Micard, V. Arabinoxylan/protein Gels: Structural, Rheological and Controlled Release Properties. *Food Hydrocoll.* **2006,** *20*, 53–61.
6. Vansteenkiste, E.; Babot, C.; Rouau, X.; Micard, V. Oxidative Gelation of Feruloylated Arabinoxylan as Affected by Protein. Influence on Protein Enzymatic Hydrolysis. *Food Hydrocoll.* **2004,** *18*, 557–564.
7. Carvajal-Millan, E.; Guilbert, S.; Morel, M. H.; Micard, V. Impact of the Structure of Arabinoxylan Gels on Their Rheological and Protein Transport Properties. *Carbohydr. Polym.* **2005,** *60*, 431–438.
8. Meyvis, T. K. L.; De Smedt, S. C.; Demeester, J.; Hennink, W. E. Influence of the Degradation Mechanism of Hydrogels on Their Elastic and Swelling Properties During Degradation. *Macromolecules* **2000,** *33*, 4717–4725.
9. Berlanga-Reyes, C. M.; Carvajal-Millán, E.; Lizardi-Mendoza, J.; Rascón-Chu, A.; Marquez-Escalante, J. A.; Martínez-López, A. L. Maize Arabinoxylan Gels as Protein Delivery Matrices. *Molecules* **2009,** *14*, 1475–1482.
10. Carvajal-Millan, E.; Guigliarelli, B.; Belle, V.; Rouau, X.; Micard, V. Storage Stability of Laccase Induced Arabinoxylan Gels. *Carbohydr. Polym.* **2005,** *59*, 181–188.
11. Carvajal-Millan, E.; Landillon, V.; Morel, M. H.; Rouau, X.; Doublier, J. L.; Micard, V. Arabinoxylan Gels: Impact of the Feruloylation Degree on Their Structure and Properties. *Biomacromolecules* **2005,** *6*, 309–317.

12. Iravani, S.; Fitchett, C. S.; Georget, D. M. R. Physical Characterization of Arabinoxylan Powder and Its Hydrogel Containing a Methyl Xanthine. *Carbohydr. Polym.* **2011,** *85*, 201–207.
13. Marquez-Escalante, J.; Carvajal-Millan, E.; Miki-Yoshida, M.; Alvarez-Contreras, L.; Toledo-Guillén, A.; Lizardi-Mendoza, J.; Rascón-Chu, A. Water Extractable Arabinoxylan Aerogels Prepared by Supercritical CO2 Drying. *Molecules* **2013,** *18*, 5531–5542.
14. Ghafar, A.; Gurikov, P.; Raman, S.; Parikka, K.; Tenkanen, M.; Smirnova, I.; Mikkonen, K. S. Mesoporous Guar Galactomannan Based Biocomposite Aerogels Through Enzymatic Crosslinking. *Compos. Part A Appl. Sci. Manuf.* **2016,** *94*, 93–103.
15. García-González, C. A.; Alnaief, M.; Smirnova, I. Polysaccharide-based Aerogels—Promising Biodegradable Carriers for Drug Delivery Systems. *Carbohydr. Polym.* **2011,** *86*, 1425–1438.
16. Heath, L.; Zhu, L.; Thielemans, W. Chitin Nanowhisker Aerogels. *ChemSusChem.* **2013,** *6*, 537–544.
17. Ubeyitogullari, A.; Ciftci, O. N. Formation of Nanoporous Aerogels from Wheat Starch. *Carbohydr. Polym.* **2016,** *147*, 125–132.
18. Wang, X.; Zhang, Y.; Jiang, H.; Song, Y.; Zhou, Z.; Zhao, H. Fabrication and Characterization of Nano-cellulose Aerogels via Supercritical CO_2 Drying Technology. *Mater. Lett.* **2016,** *183*, 179–182.
19. Mallepally, R. R.; Bernard, I.; Marin, M. A.; Ward, K. R.; McHugh, M. A. Superabsorbent Alginate Aerogels. *J. Supercrit. Fluids* **2013,** *79*, 202–208.
20. Martins, M.; Barros, A. A.; Quraishi, S.; Gurikov, P.; Raman, S. P.; Smirnova, I.; Duarte, A. R. C.; Reis, R. L. Preparation of Macroporous Alginate-based Aerogels for Biomedical Applications. *J. Supercrit. Fluids* **2015,** *106*, 152–159.
21. Veronovski, A.; Tkalec, G.; Knez, Ž.; Novak, Z. Characterisation of Biodegradable Pectin Aerogels and Their Potential Use as Drug Carriers. *Carbohydr. Polym.* **2014,** *113*, 272–278.
22. Comin, L. M.; Temelli, F.; Saldaña, M. D. A. Flax Mucilage and Barley Beta-glucan Aerogels Obtained Using Supercritical Carbon Dioxide: Application as Flax Lignan Carriers. *Innov. Food Sci. Emerg. Technol.* **2015,** *28*, 40–46.
23. Morales-Ortega, A.; Carvajal-Millan, E.; López-Franco, Y.; Rascón-Chu, A.; Lizardi-Mendoza, J.; Torres-Chavez, P.; Campa-Mada, A. Characterization of Water Extractable Arabinoxylans from a Spring Wheat Flour: Rheological Properties and Microstructure. *Molecules* **2013,** *18*, 8417–8428.
24. Martinez-Lopez, A. L.; Carvajal-Millan, E.; Lopez-Franco, Y.; Lizardi-Mendoza, J.; Rascon-Chu, A. Antioxidant Activity of Maize Bran Arabinoxylan Microspheres. In *Food Composition and Analysis Methods and Strategies;* Haghi, A. K., Carvajal-Millan, E., Eds.; Apple Academic Press: Hoboken, NJ, 2014.

CHAPTER 7

ASSESSMENT OF QUERCETIN ISOLATED FROM *ENICOSTEMMA LITTORALE* AGAINST A FEW CANCER TARGETS: AN IN SILICO APPROACH

R. SATHISHKUMAR*

Department of Botany, PSG College of Arts and Science, Coimbatore, India

**E-mail: sathishbioinf@gmail.com*

CONTENTS

ABSTRACT

Cancer, being a malignant disease, the cure and best treatment method has not been identified yet. Although, the drugs are reported to have narrow therapeutic index and are often palliative as well as unpredictable, in the last few decades, cancer chemotherapy has been developed as one of the major medical advancement in treating cancer types. To infer the existing difficulties and since the present era thrust in finding the alternatives from medicinal plants, quercetin, a compound isolated from *Enicostemma littorale* was studied using molecular docking with its ability to predict the binding site and efficiency, could be used in designing a drug. A few proteins involve in the cause of cancer are selected in this study and the docking analysis was compared between the interactions of native ligands and quercetin. Among 10 proteins, the quercetin was best interacting with Bcl2 protein of G-score value −5.74 Kcal/mol and 9 number of interaction. Molecular dynamics analysis also indicated the stability of Bcl2-quercetin complex at 74th sample among 100 samples generated during the simulation period of 1 ns. The reduction in potential energy was observed from −14,658.9 to −14,707.3 ê of 0–100th sample.

7.1 IMPORTANCE OF CANCER SPECIFIC PROTEINS

7.1.1 ACUTE LYMPHOBLASTIC LEUKEMIAS

Acute lymphoblastic leukemias (ALLs) are a form of leukemia or cancer of the white blood cells characterized by excess lymphoblasts. Malignant, immature white blood cells continuously multiply and overproduce in the bone marrow causing damage and death by crowding out normal cells in the bone marrow, and by spreading (infiltrating) to other organs. ALL is most common in childhood with a peak incidence at 2–5 years of age, and another peak in old age. The overall cure rate in children is about 80%, and about 45–60% of adults have long-term disease-free survival.[69]

In general, cancer is caused by damage to DNA that leads to uncontrolled cellular growth and spread throughout the body, either by increasing chemical signals that cause growth, or interrupting chemical signals that control growth.[73] Damage can be caused through the formation of fusion genes, as well as the deregulation of a proto-oncogene via juxtaposition of the promoter of another gene, for example, the T-cell receptor gene. This damage may be caused by environmental factors, such as chemicals, drugs, or radiation.

7.1.1.1 SYMPTOMS AND TREATMENT

The signs and symptoms of ALL are variable but follow from bone marrow replacement and/or organ infiltration. It causes generalized weakness and fatigue, anemia, frequent or unexplained fever and infection, weight loss and/or loss of appetite, excessive and unexplained bruising, bone pain and joint pain (caused by the spread of "blast" cells to the surface of the bone or into the joint from the marrow cavity), breathlessness, enlarged lymph nodes, liver and/or spleen, pitting edema (swelling) in the lower limbs and/or abdomen, and petechia, which are tiny red spots or lines in the skin due to low platelet levels.

Prednisolone, an active metabolite of prednisone which is a corticosteroid drug with predominant glucocorticoid receptor (GR or GCR), and low mineralocorticoid activity, making it useful for the treatment of a wide range of inflammatory and auto-immune conditions, such as asthma ulcerative colitis, temporal arteritis and Crohn's disease, Bell's palsy, multiple sclerosis, cluster headaches, vasculitis, ALL and autoimmune hepatitis, systemic lupus erythematosus, and dermatomyositis.[16] With high affinity prednisolone irreversibly binds GR α and β. Virtually, α-GR and β-GR are found in all tissues varies in numbers between 3000 and 10,000 per cell depending on the tissue involved.[39] Prednisolone can activate and influence biochemical behavior of most cells. The steroid/receptor complexes dimerize and interact with cellular DNA in the nucleus, binding to steroid-response elements, and modifying gene transcription. They both induce synthesis of some genes and, therefore, some proteins and inhibit synthesis of others.[60]

Resistance or sensitivity to glucocorticoids is considered to be of crucial importance for disease prognosis in childhood ALL. Prednisolone exerted a delayed biphasic effect on the resistant CCRF-CEM leukemic cell line, necrotic at low doses, and apoptotic at higher doses. At low doses, prednisolone exerted a pre-dominant mitogenic effect despite its induction on total cell death, while at higher doses, prednisolones mitogenic and cell death effects were counter balanced. Early gene microarray analysis revealed notable differences in 40 genes. The mitogenic/biphasic effects of prednisolone are of clinical importance in the case of resistant leukemic cells. This approach might lead to the identification of gene candidates for future molecular drug targets in combination therapy with glucocorticoids, along with early markers for glucocorticoid resistance.[39]

7.1.1.2 GLUCOCORTICOID RECEPTORS

The GR also known as NR3C1 (nuclear receptor subfamily 3, group C, and member 1) is a receptor that cortisol and other glucocorticoids bind to. The GR is expressed in almost every cell in the body which regulates genes controlling the development, metabolism, and immune response.[45] Because the receptor gene is expressed in several forms, it has many different (pleiotropic) effects in different parts of the body.[62] Like the other steroid receptors, the GR is modular in structure and contains the following domains (labeled A–F):

- A/B—N-terminal regulatory domain
- C—DNA-binding domain (DBD)
- D—hinge region
- E—ligand-binding domain (LBD)
- F—C-terminal domain

7.1.2 BREAST CANCER

Breast cancer also known as malignant breast neoplasm cancer originates from breast tissue, most commonly from the inner lining of milk ducts or the lobules that supply the ducts with milk. Cancers originating from ducts are known as ductal carcinomas, whereas those originating from lobules are known as lobular carcinomas.[65] The size, stage, rate of growth, and other characteristics of the tumor determine the kinds of treatment. Treatment may include surgery, drugs (hormonal therapy and chemotherapy), radiation, and/or immunotherapy. Surgical removal of the tumor provides the single largest benefit, with surgery alone being capable of producing a cure in many cases. To somewhat increase the likelihood of long-term disease-free survival, several chemotherapy regimens are commonly given in addition to surgery.[21] Most forms of chemotherapy kill cells that are dividing rapidly anywhere in the body and as a result cause temporary hair loss and digestive disturbances. Radiation may be added to kill any cancer cells in the breast that were missed by the surgery which usually extends survival somewhat although radiation exposure to the heart may cause heart failure in the future. Some breast cancers are sensitive to hormones such as estrogen and/or progesterone, which make it possible to treat them by blocking the effects of these hormones.[6]

Prognosis and survival rate vary and these greatly depend on the cancer type and staging. With best treatment and dependent on staging, 5-year relative survival varies from 98% to 23% with an overall survival rate of 85%. Worldwide, breast cancer comprises 22.9% of all non-melanoma skin cancers in women. In 2008, breast cancer caused 458,503 deaths worldwide, 13.7% of cancer deaths in women alone which was 100 times more common than men, although males tend to have poorer outcomes due to delays in diagnosis. The first noticeable symptom of breast cancer is typically a lump that feels different from the rest of the breast tissue. More than 80% of breast cancer cases are discovered when the woman feels a lump.[38] The earliest breast cancers are detected by a mammogram and lumps found in lymph nodes located in the armpits can also indicate breast cancer. Indications of breast cancer other than a lump may include changes in breast size or shape, skin dimpling, nipple inversion, or spontaneous single-nipple discharge.

Occasionally, breast cancer occurs as metastatic disease, that is, cancer that has spread beyond the original organ which causes symptoms depending on the location of metastasis where common sites include bone, liver, lung, and brain. Unexplained weight loss can occasionally herald an occult breast cancer with symptoms of fevers or chills and sometimes with bone or joint, jaundice, or neurological symptoms.[25] These symptoms are "non-specific" as they can also be manifested in many other illnesses. Most symptoms of breast disorder do not turn out to represent underlying breast cancer. Benign breast diseases such as mastitis and fibroadenoma of the breast are more common causes of breast disorder symptoms.[82] The appearance of a new symptom should be taken seriously by both patients and their doctors, because of the possibility of an underlying breast cancer at almost any age.

7.1.2.1 HORMONE BLOCKING THERAPY

Some breast cancers require estrogen to continue growing. They can be identified by the presence of estrogen receptors (ER) and progesterone receptors (PR+) on their surface (sometimes referred together as hormone receptors).[14] These ER cancers can be treated with drugs that either block the receptors, for example, tamoxifen, or alternatively block the production of estrogen.

7.1.2.2 ESTROGEN RECEPTOR

ER refers to a group of receptors that are activated by the hormone 17β-estradiol (estrogen). Two types of estrogen receptor exist as ER, which is a member of the nuclear hormone family of intracellular receptors and as estrogen G protein-coupled receptor GPR30 (GPER) which is a G protein-coupled receptor. ERs are widely expressed in different tissue types, however, exhibits some notable differences in their expression patterns in which ERα is found in breast cancer cells, respectively.[41] Therefore, tamoxifen is an antagonist in breast and used in breast cancer treatment.

7.1.2.3 ACTION OF TAMOXIFEN IN ESTROGEN RECEPTOR

Not all breast cancers respond to tamoxifen and many develop resistance despite initial benefit. Massarweh et al.[49] used an in vivo model of ER-positive breast cancer (MCF-7 xenografts) to investigate mechanisms of this resistance and developed strategies to circumvent it. Epidermal growth factor receptor (EGFR) and human epidermal growth factor receptor 2 (HER2), which were barely detected in controlling estrogen-treated tumors, increased slightly with tamoxifen and were markedly increased when tumors became resistant. Gefitinib, which inhibits EGFR/HER2, improved the antitumor effect of tamoxifen and delayed acquired resistance, but had no effect on estrogen-stimulated growth. Phosphorylated levels of *p42/44* and *p38* mitogen-activated protein kinases (both downstream of EGFR/HER2) were increased in the tamoxifen-resistant tumors and were suppressed by gefitinib. There was no apparent increase in phosphorylated AKT (also downstream of EGFR/HER2) in resistant tumors, but it was nonetheless suppressed by gefitinib. Phosphorylated insulin-like growth factor-IR (IGF-IR), which can interact with both EGFR and membrane ER, was elevated in the tamoxifen-resistant tumors compared with the sensitive group. However, ER-regulated gene products, including total IGF-IR itself and PR+ remained suppressed even at the time of acquired resistance. Tamoxifen's antagonism of classic ER genomic function was retained in these resistant tumors and even in tumors that over expressed HER2 (MCF-7 HER2/18). In conclusion, EGFR/HER2 may mediate tamoxifen resistance in ER-positive breast cancer despite continued suppression of ER genomic function by tamoxifen. IGF-IR expression remains dependent on ER, but it is activated in the tamoxifen-resistant tumors. Their study provides a rationale to combine HER inhibitors with tamoxifen in clinical studies, even in tumors that do not initially over express EGFR/HER2.[49]

7.1.3 CNS LYMPHOMA

A primary central nervous system lymphoma (PCNSL) also known as microglioma and primary brain lymphoma, is a primary intracranial tumor appearing mostly in patients with severe immunosuppression (typically patients with AIDS).[20] PCNSLs represent around 20% of all cases of lymphomas in HIV infections (other types are Burkitt's lymphomas and immunoblastic lymphomas). PCNSL is highly associated with Epstein–Barr virus (EBV) infection (>90%) in immunodeficient patients (such as those with AIDS and those iatrogenically immunosuppressed) and does not have a predilection for any particular age group. In the immunocompetent population, PCNSLs typically appear in older patients in their 50s and 60s.[18] Importantly, the incidence of PCNSL in the immunocompetent population has been reported to have increased more than 10-fold from 2.5 to 30 cases per 10 million populations.[13]

Surgical resection is usually ineffective because of the depth of the tumor. Treatment with irradiation and corticosteroids often only produces a partial response and tumor recurs in more than 90% of patients. Median survival is 10–18 months in immunocompetent patients, and less in those with AIDS. The addition of IV methotrexate (MTX) and folinic acid (leucovorin) may extend survival to a median of 3.5 years. If radiation is added to MTX, median survival time may increase beyond 4 years.[30] However, radiation is not recommended in conjunction with MTX because of an increased risk of leukoencephalopathy and dementia in patients older than 60. Dihydrofolate reductase (DHFR) is the intracellular protein involves in homocysteine remethylation in CNS through the synthesis of 5,10-methyltetra-hydrofolate reductase which imparts both nucleic acid synthesis and homocysteine remethylation via 5,10-methyltetrahydrofolate reductase and 5-methyltetrahydrofolate-homocysteine S-methyltransferase (MTR).[50]

7.1.3.1 DIHYDROFOLATE REDUCTASE

DHFR is an enzyme that reduces dihydrofolic acid to tetrahydrofolic acid using NADPH as electron donor which can be converted to the kinds of tetrahydrofolate cofactors used in 1-carbon transfer chemistry. In humans, the DHFR enzyme is encoded by the *DHFR* gene.[10] Found in all organisms, DHFR has a critical role in regulating the amount of tetrahydrofolate in the cell. Tetrahydrofolate and its derivatives are essential for purine and thymidylate synthesis (TS) which are important for cell proliferation and cell growth.[9] DHFR plays a central role in the synthesis of nucleic acid precursors and it has been shown that mutant cells that completely lack DHFR require glycine, a purine, and thymidine to grow.[23]

7.1.3.2 METHOTREXATE

MTX formerly known as amethopterin is an antimetabolite and antifolate drug. It is used in the treatment of cancer, autoimmune diseases, ectopic pregnancy, and for the induction of medical abortions by inhibiting the metabolism of folic acid.[54] It competitively inhibits DHFR, an enzyme that participates in the tetrahydrofolate synthesis. The affinity of MTX for DHFR is about one thousand-fold that of folate.[67] DHFR catalyzes the conversion of dihydrofolate to the active tetrahydrofolate. Folic acid is needed for the de novo synthesis of the nucleoside thymidine, required for DNA synthesis. Also folate is needed for purine base synthesis; therefore, purine synthesis will be inhibited. Therefore, MTX inhibits the synthesis of DNA, RNA, thymidylates, and proteins.[58]

7.1.4 LUNG CANCER

Lung cancer is a disease that consists of uncontrolled cell growth in tissues of the lung. This growth may lead to metastasis which is the invasion of adjacent tissue and infiltration beyond the lungs.[52] The vast majority of primary lung cancers is carcinomas derived from epithelial cells and it is the most common cause of cancer-related death in men and women which is roughly estimated to cause about 1.3 million deaths worldwide.

Lung cancer may be seen on chest radiograph and computed tomography (CT scan). The diagnosis is confirmed through biopsy where treatment and prognosis depend on the histological type of cancer, the stage (degree of spread), and the patient's performance status.[77] Possible treatments include surgery, chemotherapy, and radiotherapy, however, survival depends on stage, overall health, and other factors, but overall only 14% of people diagnosed with lung cancer survive five years after the diagnosis. Symptoms include dyspnea (shortness of breath), hemoptysis (coughing up blood), chronic coughing or change in regular coughing pattern, wheezing, chest pain or pain in the abdomen, cachexia (weight loss), fatigue and loss of appetite, dysphonia (hoarse voice), clubbing of the fingernails (uncommon), and dysphagia (difficulty swallowing).[26]

Many of the symptoms of lung cancer (bone pain, fever, and weight loss) are nonspecific; in the elderly, these may be attributed to comorbid illness. In many patients, the cancer has already spread beyond the original site by the time they have symptoms and seek medical attention. Common sites of metastasis include the brain, bone, adrenal glands, contralateral (opposite) lung, liver, pericardium, and kidneys.[8] The *Bcl-2* gene has been implicated in lung carcinomas as well as schizophrenia and autoimmunity. It is also thought to be involved in resistance to conventional cancer treatment. This supports a role for decreased apoptosis in the pathogenesis of cancer.

7.1.4.1 BCL-2

Bcl-2 (B-cell lymphoma 2) is the founding member of Bcl-2 family of apoptosis regulator proteins encoded by the *Bcl-2* gene. Bcl-2 derives its name from B-cell lymphoma 2 as it is the second member of a range of proteins initially described in chromosomal translocations involving chromosomes 14 and 18 in follicular lymphomas. Bcl-2 protein located on outer membrane of mitochondria, consisting 239 amino acids which suppress apoptosis, regulates cell death by controlling the mitochondrial membrane permeability. It also inhibits caspase activity either by preventing the release of cytochrome c from the mitochondria and/or by binding to the apoptosis activating factor (APAF-1). Hypermethylation at the 5th carbon of the cytosine residues would lead to the origin of cancer.[66]

7.1.4.2 DOCETAXEL

Docetaxel (as generic or with trade name Taxotere) is a clinically well-established anti-mitotic chemotherapy medication (it interferes with cell division). It is used mainly for the treatment of breast, ovarian, and non-small cell lung cancer.[11] Docetaxel has an FDA approved claim for treatment of patients who have locally advanced, or metastatic breast or non-small-cell lung cancer that have undergone anthracycline-based chemotherapy and failed to stop cancer progression or relapsed (US-FDA, 2006). Docetaxel is of the chemotherapy drug class; taxane, and is a semi-synthetic analogue of paclitaxel (Taxol), an extract from the bark of the rare Pacific yew tree *Taxus brevifolia.*[11] Due to scarcity of paclitaxel, extensive research was carried out leading to the synthesis of docetaxel—an esterified product of 10-deacetyl baccatin III, which is extracted from the renewable and readily available European yew tree.

7.1.4.3 ACTION OF DOCETAXEL IN LUNG CANCER

Docetaxel affects microtubule polymerization and surprisingly differences in tumor sensitivity to the taxanes have also been observed. Docetaxel was superior to paclitaxel in inhibiting in vivo growth of human lung and prostate but not breast cancer models. They compared drug cytotoxicity effects on tubulin isoforms, markers of apoptosis and proteomic profiles in human prostate (LNCaP), lung (SK-MES, MV-522), and breast (MCF-7, MDA-231) cancer cell lines in vitro. Cytotoxicity was found in the order of SK-MES<MV-522<LNCaP<MCF-7<MDA-MB-231 in which docetaxel was more effective. Cytotoxicity was directly correlated with Bcl-2 expression in vitro, whereas inversely correlated with docetaxel sensitivity in vivo. Proteomic profiling identified a protein expressed in lung and prostate cells, which was differentially regulated by docetaxel and paclitaxel in SK-MES. The superior activity of docetaxel in tumors with low Bcl-2 warrants further studies on biomarkers for drug sensitivity and investigation of docetaxel in combination with drugs that reduce *Bcl-2* gene expression.[32]

7.1.5 ORAL CANCER

Oral cancer is a subtype of head and neck cancer said to be arising as a primary lesion originating in any of the oral tissues, by metastasis from a distant site of origin or by extension from a neighboring anatomic structure, such as the nasal cavity. It may originate in the tissues of the mouth and may be of varied histologic types, such as teratoma, adenocarcinoma derived from a major or minor salivary gland, lymphoma from tonsillar or other lymphoid tissue, or melanoma from the pigment producing cells of the oral mucosa.[80] There are several types of oral cancers, but around 90% are squamous cell carcinoma (SCC) originating in the tissues that line the mouth and lips. Oral or mouth cancer most commonly involves the tongue and it may also occur on the floor of the mouth, cheek lining, gingiva (gums), lips, or palate (roof of the mouth). Most oral cancers look very similar under the microscope and are called SCC.[64]

Skin lesion, lump, or ulcer on the tongue, lip, or other mouth areas is usually small and most often look pale colored, dark, or discolored where early sign may be a white patch (leukoplakia) or a red patch (erythroplakia) on the soft tissues of the mouth which are initially painless may develop with burning sensation or pain when the tumor is advanced. Additional symptoms that may be associated with this disease are tongue problems, swallowing difficulty, and mouth sores. EGFR is overexpressed in the cells of certain types of human carcinomas—for example, in oral and breast cancers, whereas high expression of EGFR is frequently observed in many solid tumor types including oral squamous cell carcinomas (OSCC). This study investigated whether treatment with gefitinib would inhibit the metastatic spread in OSCC cells. This was evaluated using orthotopic xenografts of highly metastatic OSCC. There were observed Metastasis in six of 13 gefitinib treated animals (46.2%), compared with all of 12 control animals (100%). After exposure to gefitinib, OSCC cells showed a marked reduction in cell adhesion ability to fibronectin and in the expression of integrin α3, α, β1, β 4, β5, and β6.[68]

7.1.5.1 EPIDERMAL GROWTH FACTOR

EGF is a growth factor that plays an important role in the regulation of cell growth, proliferation, and differentiation by binding to its receptor EGFR. Human EGF is a 6045 Da protein with 53 amino acid residues and 3 intramolecular disulfide bonds.[5]

EGF is a low molecular weight polypeptide first purified from the mouse submandibular gland, but since then found in many human tissues including submandibular gland and parotid gland.[12] Salivary EGF is regulated by dietary inorganic iodine which plays an important physiological role in the maintenance of oro-esophageal and gastric tissue integrity.[29] The biological effects of salivary EGF includes healing of oral and gastroesophageal ulcers, inhibition of gastric acid secretion, stimulation of DNA synthesis as well as mucosal protection from intraluminal injurious factors, such as gastric acid, bile acids, pepsin, and trypsin and to physical, chemical, and bacterial agents.[78]

7.1.5.2 GEFITINIB

Gefitinib is a drug used in the treatment of certain types of cancer and like erlotinib; it is an EGFR inhibitor, which interrupts signaling through the EGFR in target cells. Gefitinib is the first selective inhibitor of EGFR's tyrosine kinase domain.[75] The target protein (EGFR) is also sometimes referred to as Her1 or ErbB-1 depending on the literature source. This leads to inappropriate activation of the anti-apoptotic Ras signaling cascade, eventually leading to uncontrolled cell proliferation.[56] Research on gefitinib-sensitive non-small cell lung cancers has shown that a mutation in the EGFR tyrosine kinase domain is responsible for activating anti-apoptotic pathways. These mutations tend to confer increased sensitivity to tyrosine kinase inhibitors such as gefitinib and erlotinib.[70] Gefitinib inhibits EGFR tyrosine kinase by binding to the adenosine triphosphate (ATP)-binding site of the enzyme.[53] Thus, the function of the EGFR tyrosine kinase in activating the anti-apoptotic Ras signal transduction cascade is inhibited, and malignant cells are inhibited.

7.1.6 PROSTATE CANCER

Prostate cancer is a form of cancer that develops in the prostate gland of the male reproductive system. Most prostate cancers are slow growing; however, there are cases of aggressive prostate cancers. The cancer cells may metastasize (spread) from the prostate to other parts of the body, particularly the bones and lymph nodes. Prostate cancer may cause pain, difficulty in urinating, problems during sexual intercourse, or erectile dysfunction. Other symptoms can potentially develop during later stages of the disease.

Rates of detection of prostate cancers vary widely across the world, with South and East Asia detecting less frequently than in Europe and especially the United States. Prostate cancer tends to develop in men over the age of fifty and although it is one of the most prevalent types of cancer in men, many never have symptoms, undergo no therapy, and eventually die of other causes.[76] This is because cancer of the prostate is, in most cases, slow-growing, symptom-free, and since men with the condition are older they often die of causes unrelated to the prostate cancer, such as heart/circulatory disease, pneumonia, other unconnected cancers, or old age. About 2/3 of cases are slow growing, the other third more aggressive and fast developing. Many factors, including genetics and diet, have been implicated in the development of prostate cancer. The presence of prostate cancer may be indicated by symptoms, physical examination, prostate-specific antigen (PSA), or biopsy.[17]

Early prostate cancer usually causes no symptoms. Often it is diagnosed during the workup for an elevated PSA noticed during a routine checkup. Its changes within the gland, therefore, directly affect urinary function. Because the "vas deferens" deposits seminal fluid into the prostatic urethra, and secretions from the prostate gland itself are included in semen content, prostate cancer may also cause problems with sexual function and performance, such as difficulty in achieving erection or painful ejaculation.[51] Advanced prostate cancer can spread to other parts of the body, possibly causing additional symptoms. The most common symptom is bone pain, often in the vertebrae (bones of the spine), pelvis, or ribs. Spread of cancer into other bones such as the femur is usually to the proximal part of the bone. Prostate cancer in the spine can also compress the spinal cord, causing leg weakness and urinary and fecal incontinence.

The androgen receptor (AR) helps prostate cancer cells to survive and is a target for many anti-cancer research studies; so far, inhibiting the AR has only proven to be effective in mouse studies. Prostate-specific membrane antigen (PSMA) stimulates the development of prostate cancer by increasing folate levels for the cancer cells to survive and grow; PSMA increases available folates for use by hydrolyzing glutamated folates.[83] Andarine is an orally active partial agonist for ARs. It has been shown to inhibit the development, progression, and metastasis as well in autochthonous transgenic adenocarcinoma of the mouse prostate (TRAMP) model, which spontaneously develops prostate cancer.[27]

7.1.6.1 ANDROGEN RECEPTOR

The AR also known as NR3C4 (nuclear receptor subfamily 3, group C, member 4) is a type of nuclear receptor that is activated by binding either of the androgenic hormones testosterone or dihydrotestosterone in the cytoplasm and then translocating into the nucleus. The AR is most closely related to the PR+ and the progestin in higher dosages has the ability to block AR.[61] Androgens cause slow epiphysis or maturation of the bones, but more of the potent epiphysis effect comes from the estrogen produced by aromatization of androgens. Steroid users of teen age may find that their growth had been stunted by androgen and/or estrogen excess.[22]

7.1.6.2 ANDARINE

Andarine (GTx-007, S-4) is an investigational selective AR modulator (SARM) developed by GTX, Inc. for treatment of conditions, such as muscle wasting, osteoporosis, and benign prostatic hypertrophy, using the non-steroidal androgen antagonist bicalutamide as a lead compound. Andarine is an orally active partial agonist for ARs. It is less potent in both anabolic and androgenic effects than other SARMs. Perhaps, in an animal model of benign prostatic hypertrophy, andarine was shown to reduce prostate weight with similar efficacy to finasteride, but without producing any reduction in muscle mass or anti-androgenic side effects. This suggests that it is able to competitively block binding of dihydrotestosterone to its receptor targets in the prostate gland, but its partial agonist effects at ARs prevent the side effects associated with the anti-androgenic drugs traditionally used for treatment of benign prostatic hyperplasis (BPH).

7.1.7 RECTAL CANCER

Colorectal cancer, less formally known as bowel cancer, is a cancer characterized by neoplasia in the colon, rectum, or vermiform appendix. Colorectal cancer is clinically distinct from anal cancer, which affects the anus. Colorectal cancers start in the lining of the bowel. If left untreated, it can grow into the muscle layers underneath, and then through the bowel wall. Most begin as a small growth on the bowel wall: a colorectal polyp or adenoma. These mushroom-shaped growths are usually benign, but some develop into cancer over time. Localized bowel cancer is usually diagnosed through colonoscopy. Invasive cancers that are confined within the wall of the colon [tetranitromethane (TNM) stages I and II] are often curable with surgery. Colorectal cancer is the third most commonly diagnosed cancer in the world, but it is more common in developed countries. More than half of the people who die of colorectal cancer live in a developed region of the world (http://globocan.iarc.fr/). GLOBOSCAN estimated that, in 2008 reported 1.23 million cases with colorectal cancer of which more than 600,000 people died.

7.1.7.1 SYMPTOMS

Symptoms include blood in the stool, narrower stools, a change in bowel habits and general stomach discomfort. As an anti-rectal cancer, chemotherapy targets TS such as fluorinated pyrimidine fluorouracil or certain folate analogues to inhibit rectal cancer.

7.1.7.2 THYMIDYLATE SYNTHASE

TS is an enzyme used to generate thymidine monophosphate (dTMP), which is subsequently phosphorylated to thymidine triphosphate for use in DNA synthesis and repair. By means of reductive methylation, deoxyuridine monophosphate (dUMP), and N5, N10-methylene tetrahydrofolate together used formed dTMP, yielding dihydrofolate as a secondary product. This provided the sole de novo pathway for production of dTMP and, therefore, is the only enzyme in folate metabolism in which the 5, 10-methylenetetrahydrofolate is oxidized during one-carbon transfer.[28]

The enzyme is essential for regulating the balanced supply of the four DNA precursors in normal DNA replication. However, defects in the enzyme activity, affects the regulation process causing various biological and genetic abnormalities such as thymineless death[35] is an important target for certain chemotherapeutic drugs. Therefore, it is about 30–35 KDa in most species except in protozoan and plants where it exists as a bifunctional enzyme that includes a DHFR domain. A cysteine residue is involved in the catalytic mechanism (it covalently binds the 5, 6-dihydro-dUMP intermediate). The sequence around the active site of this enzyme is conserved from phages to vertebrates.

7.1.7.3 FOLINIC ACID

Folinic acid or leucovorin generally administered as calcium or sodium folinate (or leucovorin calcium/sodium) is an adjuvant used in cancer chemotherapy involving the drug MTX which works in synergy combination with the chemotherapy, 5-fluorouracil. Folinic acid is a 5-formyl derivative of tetrahydrofolic acid and is readily converted to other reduced folic acid derivatives (e.g., tetrahydrofolate) and, therefore, vitamin activity is equivalent to folic acid. Since it does not require the action of TS for its conversion, its function as a vitamin is unaffected by inhibition of this enzyme by drugs such as MTX (Therapeutic Information Resources Australia, 2004). Therefore, folinic acid allows purine/pyrimidine synthesis to occur in the presence of DHFR inhibition, so that some normal DNA replication and RNA transcription processes can proceed without interruption. Folate metabolism supports the synthesis of nucleotides as well as the transfer of methyl groups. Polymorphisms in folate-metabolizing enzymes have been shown to affect risk of colorectal neoplasia and other malignancies.

7.1.8 RENAL CANCER

Renal cell carcinoma (RCC, also known as hypernephroma) is a kidney cancer that originates in the lining of the proximal convoluted tubule, the very small tubes in the kidney that filter the blood and remove waste products. RCC is the most common type of kidney cancer in adults, responsible for approximately 80% of cases.[55] It is also known to be the most lethal of all the genitourinary tumors. Initial treatment is most commonly a radical or partial nephrectomy and remains the mainstay of curative treatment.[63] Where the tumor is confined to the renal parenchyma, 5-year survival rate is 60–70%, but this is lowered considerably where metastases have spread. It is resistant to radiation therapy and chemotherapy, although some cases respond to immunotherapy. Targeted cancer therapies, such as sunitinib, temsirolimus, bevacizumab, interferon-alpha, and possibly sorafenib have improved the outlook for RCC (progression-free survival), although they have not yet demonstrated improved survival. It is reported that in a population of 58,240 (35,370 men and 22,870 women) around 13,040 men and women died of kidney and renal cancer.[33] A wide range of symptoms has been encountered with renal carcinoma depending on the areas of body it affects. The classic triad is hematuria (blood in the urine), flank pain, and an abdominal mass. This triad only occurs in 10–15% of cases, and is generally indicative of more advanced disease. Today, the majorities of renal tumors are asymptomatic and are detected incidentally on imaging, usually for an unrelated cause. Signs may include abnormal urine color (dark, rusty, or brown) due to blood in the urine (found in 60% of cases), join pain (found in 40% of cases), abdominal mass (25% of cases), malaise, weight loss or anorexia (30% of cases), polycythemia (5% of cases), and anemia resulting from depression of erythropoietin (30% of cases).[37] Also, there may be erythrocytosis (increased production of red blood cells) due to increased erythropoietin secretion.

The presenting symptom may be due to metastatic disease, such as a pathologic fracture of the hip due to a metastasis to the bone varicocele, the enlargement of one testicle, usually on the left (2% of cases). This is due to blockage of the left testicular vein by tumor invasion of the left renal vein; this typically does not occur on the right as the right gonadal vein drains directly into the inferior vena cava. Vision abnormalities, pallor or plethora, hirsutism–excessive hair growth (females), constipation and hypertension (high blood pressure) resulting from secretion of renin by the tumor (30% of cases), elevated calcium levels (hypercalcemia), Stauffer syndrome-paraneoplastic, non-metastatic liver disease, night sweats, and severe weight loss.

7.1.8.1 MTOR KINASE

The mammalian target of rapamycin (mTOR) also known as mechanistic target of rapamycin or FK506 binding protein 12-rapamycin associated protein 1 (FRAP1) is a protein encoded by *FRAP1* gene. mTOR is a serine/threonine protein kinase that regulates cell growth, cell proliferation, cell motility, cell survival, protein synthesis, and transcription. mTOR belongs to the phosphatidylinositol 3-kinase-related kinase protein family. mTOR integrates the input from upstream pathways, including insulin, growth factors (such as IGF-1 and IGF-2), and amino acids. mTOR also senses cellular nutrient and energy levels and redox status. The mTOR pathway is dysregulated in human diseases, especially certain cancers. Rapamycin is a bacterial product that can inhibit mTOR by associating with its intracellular receptor FKBP12. The FKBP12-rapamycin complex binds directly to the FKBP12-rapamycin binding (FRB) domain of mTOR. mTOR is the catalytic subunit of two molecular complexes.[4]

7.1.8.2 TORISEL

Temsirolimus (CCI-779) is an intravenous drug for the treatment of RCC, developed by Wyeth Pharmaceuticals and approved by the US Food and Drug Administration (FDA),[19] and was also approved by the European Medicines Agency (EMEA) during November 2007. It is a derivative of sirolimus and sold as Torisel. Temsirolimus is a specific inhibitor of mTOR and interferes with the synthesis of proteins that regulate proliferation, growth, and survival of tumor cells. Treatment with temsirolimus leads to cell cycle arrest in the G1 phase, and also inhibits tumor angiogenesis by reducing synthesis of vascular endothelial growth factor (VEGF).[79]

The potent inhibitor of the mTOR is temsirolimus which is comprised for cell cycle, angiogenesis, and proliferation and has been proven beneficial in the treatment of advanced RCC. Temsirolimus is officially approved for first line therapy in high-risk previously untreated mRCC patients.[24]

7.1.9 SQUAMOUS CELL CARCINOMA

SCC or SqCC, occasionally rendered as "squamous-cell carcinoma," is a histological form of cancer that arises from the uncontrolled multiplication of transformed malignant cells showing squamous differentiation and tissue architecture. SCC is one of the most common histological forms of cancer, and frequently forms in a large number of body tissues and organs, including skin, lips, mouth, esophagus, urinary bladder, prostate, lung, vagina, and cervix, among others. Despite the common name, SCCs that present in different sites often show large differences in their presentation[47] It is reported that 42,610 men and 31,400 women around 11,790 men and women died of cancer of the skin in US alone.[7]

Symptoms are highly variable depending on the involved organs. SCC of the skin begins as a small nodule and as it enlarges the center becomes necrotic and sloughs and the nodule turns into an ulcer. The lesion caused by SCC is often asymptomatic, ulcer or reddish skin plaque that is slow growing, intermittent bleeding from the tumor, especially on the lip. The clinical appearance is highly variable; usually the tumor presents an ulcerated lesion with hard, raised edges. The tumor may be in the form of a hard plaque or a papule, often with an opalescent quality, with tiny blood vessels. The tumor can lie below the level of the surrounding skin, and eventually ulcerates and invades the underlying tissue. The tumor grows relatively slowly, unlike basal cell carcinoma (BCC), SCC has a substantial risk of metastasis. Imiquinoid has been used with success for SCC in situ of the skin and the penis, but the morbidity and discomfort of the treatment is severe. An advantage is the cosmetic result: after treatment, the skin resembles normal skin without the usual scarring and morbidity associated with standard excision. Imiquinoid is not FDA-approved for any SCC.

7.1.9.1 TOLL-LIKE RECEPTOR 7

Toll-like receptor (TLR) 7 also known as TLR7 is an immune gene possessed by humans and other mammals. The protein encoded by this gene is a member of the TLR family which plays a fundamental role in pathogen recognition and activation of innate immunity. TLRs are highly conserved from drosophila to humans and share structural and functional similarities. They recognize pathogen-associated molecular patterns (PAMPs) that are expressed on SCC, and mediate the production of cytokines necessary for the development of effective immunity. The various TLRs exhibit different patterns of expression. This gene is predominantly expressed in lung, placenta and spleen, and lies in close proximity to another family member.[43]

7.1.9.2 IMIQUIMOD

Imiquimod marketed by Meda AB, Graceway Pharmaceuticals, and iNova Pharmaceuticals under the trade names Aldara, Zyclara, and Mochida as Beselna activates immune cells through the TLR7, commonly involved in pathogen recognition.[36]

TLR7 activation by imiquimod has pleiotropic effects on innate immune cells, but its effects on T cells remain largely uncharacterized. Because tumor destruction and formation of immunological memory are ultimately T-cell-mediated effects, SCC treated with imiquimod before excision contained dense T-cell infiltrates associated with tumor cell apoptosis and histologically evidenced tumor regression. Effector T cells from treated SCC produced more IFN-γ, granzyme, and perforin and less IL-10, and transforming growth factor-β (TGF-β) than T cells from untreated tumors. Treatment of normal human skin with imiquimod induced activation of resident T cells and reduced IL-10 production but had no effect on IFN-γ, perforin, or granzyme, suggesting that these latter effects arise from the recruitment of distinct populations of T cells into tumors. Thus, imiquimod stimulates tumor destruction by recruiting cutaneous effector T cells from blood and by inhibiting tonic anti-inflammatory signals within the tumor.[74]

7.1.10 THYROID CANCER

Thyroid cancer is a thyroid neoplasm that is malignant. It can be treated with radioactive iodine or chemotherapy. Most often the first symptom of thyroid cancer is a nodule in the thyroid region of the neck. However, many adults have small nodules in their thyroids, but typically fewer than 5% of these nodules are found to be malignant. Sometimes the first sign is an enlarged lymph node, followed by pain in the anterior region of the neck and changes in voice.[31] Thyroid nodules are of particular concern when they are found in those under the age of 20. The presentation of benign nodules at this age is less likely, and thus the potential for malignancy is far greater. It is reported that in about 10,740 men and 33,930 women around 1690 men and women were died and diagnose of cancer during 2010. But as the cancer grows, difficulty in swallowing or breathing, a benign goiter, hyperthyroidism, or hypothyroidism will be developed.

7.1.10.1 RAF PROTEIN

Raf proto-oncogene serine/threonine is a specific protein kinase also known as proto-oncogene c-Raf or simply c-Raf, is an enzyme that in humans is encoded by the *Raf1* gene. Protein kinases are overactive in many of the molecular pathways that cause cells to become cancerous. These pathways include Raf kinase, PDGF (platelet-derived growth factor), VEGF receptor 2 and 3 kinases, and c Kit the receptor for Stem cell factor. The c-Raf protein functions in the MAPK/ERK signal transduction pathway as part of a protein kinase cascade[42] functions downstream of the Ras subfamily of membrane associated GTPases to which it binds directly. Once activated Raf-1 is phosphorylated and activate the dual specific protein kinases MEK1 and MEK2, which, in turn, phosphorylate to activate the serine/threonine-specific protein kinases ERK1 and ERK2. Activated ERKs are pleiotropic effectors of cell physiology and play an important role in the control of gene expression involved in the cell division cycle, apoptosis, cell differentiation, and cell migration.[71]

7.1.10.2 SORAFENIB

Sorafenib shows promise for thyroid cancer and are being used for some patients who do not qualify for clinical trials. Numerous agents are in phase II clinical trials and XL184 has started a phase III trial. Sorafenib (a bi-aryl urea) is a small molecular inhibitor of several tyrosine protein kinases (VEGFR and PDGFR) and Raf. Sorafenib is unique in targeting the Raf/Mek/Erk pathway (MAP kinase pathway) by inhibiting some intracellular serine/threonine kinases (e.g., Raf-1, wild-type B-Raf, and mutant B-Raf).[81]

However, the antiproliferative activity of sorafenib varies widely depending on the oncogenic signaling pathways driving proliferation. Sorafenib has also been shown recently to sequester Raf-1 and B-Raf in a stable inactive complex in treated tumor cell lines expressing wild-type B-Raf. This alteration of Raf-1 protein complexes by sorafenib may result in perturbation of other Raf-1 complexes with MST-1 and ASK-1, which are involved in thyroid cell survival signaling mechanisms.[81] In a netshield, over 10 different types of cancers (ALL, breast cancer, CNS lymphoma, lung, oral, prostate, renal, rectal cancer, SCC, and thyroid cancer) with their targets (GR, ER, DHFR, Bcl-2 protein, EFG, AR, mTOR, TS, TLR-7, and RAF protein), and their drugs which are in market have been identified. Since, quercetin a naturally extracted and purified small molecule have been achieved and projected over the previous chapters have to be evaluated for its efficiency as an anti-cancer agent, it was further examined with the above objective and compared against the synthetic molecules (drug in use) over in silico approaches.

Bioinformatics is the combination of biology and information technology. It records, annotates, stores, analyzes, and searches/retrieves the nucleic acid sequence, protein sequence and structures that provide new biological insights. Molecular docking plays an important role in analyzing the inhibitory action and interaction studies of small ligand molecules with protein structures[46] and it is a key tool in structural molecular biology and computer-assisted drug design. The goal of ligand-protein docking is to predict the predominant binding models of a ligand with a protein of known three-dimensional structure and to predict the affinity and activity of the small molecule in order to study whether it forms a stable complex. Therefore, docking plays an important role in the rational drug designing. The result obtained from the docking study would be useful in both understanding the inhibitory mode and in rapidly and accurately predicting the activities of newly designed inhibitors on the basis of docking scores.[72] Hence, the models also provide some beneficial clues in structural modification for designing new inhibitors for the treatment of diseases, such as cancer with much higher inhibitory activities against cancer inducing proteins. In our study, molecular interaction profile of the target proteins with the ligand obtained from the herb *E. littorale* was compared against the already available synthetic ligands. Infact, finally molecular dynamics (MD) of best G-scored complex was alone done for stabilization.

7.2 ADME-TOX PROPERTIES

ADME-Tox refers to absorption, distribution, metabolism, excretion, and toxicity properties which should be considered to develop a new drug, as they are the main cause of failures for candidate molecules in drug design. The early evaluation of these properties during drug design could save time and money because certain properties make a drug different from other compounds. An appropriate concentration of the drug must circulate in the body for a reasonable length of time to achieve a desired beneficial effect with minimum adverse effects. For this process, oral drugs have to dissolve or suspend in the gastrointestinal tract (GI) and be absorbed through the gut wall to reach the blood stream though the liver. From there, the drug will be distributed to various tissues and organs and finally binds to its molecular target and exert its desired action. The drug is then subjected to hepatic metabolism followed by its elimination as bile or *via* the kidneys. Several pharmacokinetics properties are involved in this mechanism.

Bioavailability depends on absorption and liver first-pass metabolism. The volume of its distribution, together with its clearance rate, determines the half-life of a drug, and, therefore, its dosage. Poor biopharmaceutical properties, such as poor aqueous solubility and slow dissolution rate can lead to poor oral absorption and hence low oral bioavailability. The conversion of active compounds into qualified clinical candidates has proved to be a challenge. At the molecular level, a coordinated system of transporters, channels, receptors, and enzymes act as gatekeepers to foreign molecules affecting the ADME-Tox properties of a given molecule in very different ways. The optimal approach for the ADME-Tox support of discovery will be one that uses both in vitro and in silico experiments in a complementary way ensuring that ADME-Tox is used and considered at almost every stage of the discovery process, from hit identification to lead optimization.[85]

In the hit identification stage, the primary goal of the in silico ADME-Tox models is to identify compounds or series of compounds with at least acceptable drug-like properties that are then disregarded. Another goal is to identify potential weaknesses and liabilities in the selected series highlighting the issues that will be focused in the improvement/optimization efforts. In the lead identification stage, the objective is to identify a small number of chemical series with the activity, selectivity, and drug-like properties required for a potential candidate. The application of in silico ADME-Tox should focus on predictions of chemical modifications of compounds that will improve ADME-Tox properties. In vitro assays are used to measure the ADME-Tox properties of the newly synthesized compounds. This information is valuable for the refinement of the in silico ADME-Tox models. Similar to lead identification, the lead optimization on ADME-Tox properties consist of an iterative workflow, starting from in silico prediction, to chemical synthesis, to experimental testing and confirmation, and to model refinement.[2] (Table 7.1).

Lipinski's rule of five is a rule of thumb to evaluate drug likeness, or determine if a chemical compound with a certain pharmacological or biological activity has properties that would make it a likely orally active drug in humans. The rule was formulated by Christopher A. Lipinski in 1997, based on the observation that most medication drugs are relatively small and lipophilic molecules.[44] Lipinski's rule says that, in general, an orally active drug has no more than one violation of the following criteria:

TABLE 7.1 ADME/Tox Properties Description by Qikprop.

Property or descriptor	Description	Range or recommended values
mol_MW	Molecular weight of the molecule	130.0–725.0
SASA	Total solvent accessible surface area (SASA) in square angstroms using a probe with a 1.4 Å radius	0.0–450.0
Volume	Total solvent-accessible volume in cubic angstroms using a probe with a 1.4 Å radius	500.0–2000.0
DonorHB	Estimated number of hydrogen bonds that would be donated by the solute to water molecules in an aqueous solution. Values are averages taken over a number of configurations, so they can be non-integer	0.0–6.0
accptHB	Estimated number of hydrogen bonds that would be accepted by the solute from water molecules in an aqueous solution. Values are averages taken over a number of configurations, so they can be non-integer	2.0–20.0
QP logPC 16	Predicted hexadecane/gas partition coefficient	4.0–18.0
QP logP oct	Predicted octanol/gas partition coefficient	8.0–35.0
QP logP w	Predicted water/gas partition coefficient	4.0–45.0
QP logP o/w	Predicted octanol/water partition coefficient	–2.0–6.5
QP logS	Predicted aqueous solubility, log S. S in mol dm–3 is the concentration of the solute in a saturated solution that is in equilibrium with the crystalline solid	–6.5–0.5
metab	Number of likely metabolic reactions	1–8
Human oral absorption	Predicted qualitative human oral absorption: 1, 2, or 3 for low, medium, or high. The text version is reported in the output. The assessment uses a knowledge-based set of rules, including checking for suitable values of percent human oral absorption, number of metabolites, number of rotatable bonds, log *P*, solubility, and cell permeability	>80% is high
Percent human oral absorption	Predicted human oral absorption on 0–100% scale. The prediction is based on a quantitative multiple linear regression model. This property usually correlates well with human oral absorption, as both measures the same property.	<25% is poor

TABLE 7.1 *(Continued)*

Property or descriptor	Description	Range or recommended values
QP PCaco	Predicted apparent Caco-2 cell permeability in nm/sec. Caco2 cells are a model for the gut-blood barrier. QikProp predictions are for non-active transport.	<25 poor, >500 great
QP log BB	Predicted brain/blood partition coefficient. Note: QikProp predictions are for orally delivered drugs so, for example, dopamine and serotonin are CNS negative because they are too polar to cross the blood–brain barrier	–3.0–1.2
QP P MDCK	Predicted apparent MDCK cell permeability in nm/sec. MDCK cells are considered to be a good mimic for the blood–brain barrier. QikProp predictions are for non-active transport.	<25 poor, >500 great
Rule of five	The rules are: mol_MW < 500, QPlogPo/w < 5, donorHB ≤ 5, accptHB ≤ 10. Compounds that satisfy these rules are considered drug-like. (The "five" refers to the limits, which are multiples of 5.)	Maximum is 3
Rule of three	The three rules are: QP logS >-5.7, QP PCaco >22 nm/s, #primary metabolites < 7. Compounds with fewer (and preferably no) violations of these rules are more likely to be orally available	Maximum is 4

- Not more than five hydrogen bond donors (nitrogen or oxygen atoms with one or more hydrogen atoms).
- Not more than 10 hydrogen bond acceptors (nitrogen or oxygen atoms).
- A molecular mass not greater than 500 Da.
- An octanol-water partition coefficient[40] log *P* not greater than 5.

Rule of 3 explained by Jorgensen are for the oral intake of the drug, compounds violating the rule were not considered to be administered orally. The rule says that

- QP logS >−5.7 represents the aqueous solubility of the drug molecule,
- QP PCaco >22 nm/s represents the permeability of drug molecule at the gut/brain barrier (Caco-2 cells),
- Primary metabolites <7.

7.2.1 MOLECULAR DYNAMICS

MD is a computer simulation of physical movements of atoms and molecules where they interacts each other for a period of time. Therefore, the movement of molecules and atoms are determined numerically by Newton's equations of motion, whereas the forces between the particles and potential energy are defined by molecular mechanics force fields (MMFFs). MD is also termed as "statistical mechanics by numbers" and "Laplace's vision of Newtonian mechanics" and the results are used to determine macroscopic thermodynamic properties. MD simulations have provided detailed information on the fluctuations and conformational changes of proteins and nucleic acids. These methods are now routinely used to investigate the structure dynamics and thermodynamics of biological molecules and their complexes. They are also used in the determination of structures from x-ray crystallography and from NMR experiments. Simply MD refers the understanding properties of assemblies of molecules in terms of their structure and the microscopic interactions between them. The two main families of simulation technique are MD and Monte Carlo (MC). Simulation act as a bridge between theory and experiment.[1] MD simulations are used in the field of kinetics and irreversible processes, equilibrium ensemble sampling, and modeling tools. Moreover, it is an important tool for understanding the physical basis of the structure and function of biological macromolecules. The early view of proteins as relatively rigid structures has been replaced by a dynamic model in which the internal motions and resulting conformational changes play an essential role in their function.[48]

7.2.1.1 MATERIALS AND METHODS

7.2.1.1.1 Databases

The databases Protein Data Bank (PDB) (http://www.pdb.org) was used to retrieve the three dimensional structures of the cancer proteins, such as GR, ER, DHFR, Bcl-2 protein, AR, TS, mTOR, TLR7, RAF protein, and EGFR for ALL, breast cancer, CNS lymphoma, lung cancer, prostate cancer, rectal cancer, renal cancer, SCC, thyroid cancer, and oral cancer, respectively, were retrieved with higher resolution >3.00 Å, respectively, and stored in pdb file formats (Table 7.2). The synthetic ligands noted form the literatures such as prednisolone (ALL), tamoxifen (breast cancer), MTX (CNS lymphoma), docetaxel (lung cancer), andarine (prostate cancer), folinic acid (renal cancer), Torisel (rectal cancer), imiquinoid (SCC), sorafenib (thyroid cancer), gefitinib (oral cancer), and isolated compound quercetin were retrieved from the PubChem (http://pubchem.ncbi.nlm.nih.gov/) (Table 7.3).

7.2.1.1.2 Protein and Ligand Processing

Ligand binding site for the selected proteins were identified using Q-Site finder. Toxicity prediction was carried out using QikProp module by Fast mode access for all 10 ligand to compare with quercetin compound in order to predict for central nervous system activity, metabolic activity, and logarithmic value of each atom predicted. Before docking process, the protein and ligand molecules are prepared. The preparation of a protein involves a number of steps, to include explicit hydrogens, refined and to hydrogenate the structures of ligand and the ligand-receptor complex to made it suitable for performing high accuracy docking in the Maestro graphical user interface.

Using LigPrep module the structures of prednisolone, tamoxifen, MTX, docetaxel, andarine, folinic acid, Torisel, imiquinoid, sorafenib, gefitinib and quercetin were prepared and refined by adding hydrogen atoms and minimized with OPLS_2005 force field. In both the case of protein and ligand, the resulting structure was energy minimized and used for docking studies. The grid was generated for which the retrieved structures might adopt more than one conformation on binding, and to ensure that possible actives are not missed. Docking was carried out using Glide module.

TABLE 7.2 List of Cancer with Specific Receptors Chosen as a Target for Investigation.

Sl. no.	Name of the cancer	Targets with PDB ID	Amino acid length	Ribbon representation of cancer-specific protein viewed using PyMol
1	Acute lymphoblastic leukemia	Glucocorticoid receptor (1R4O)	92	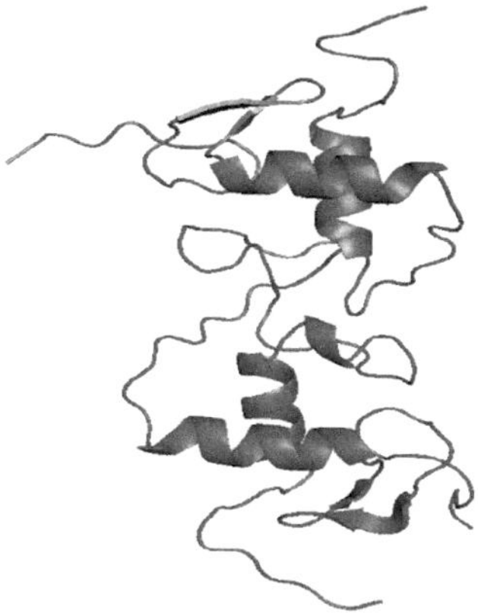
2	Breast cancer	Estrogen receptor (3ERD)	261	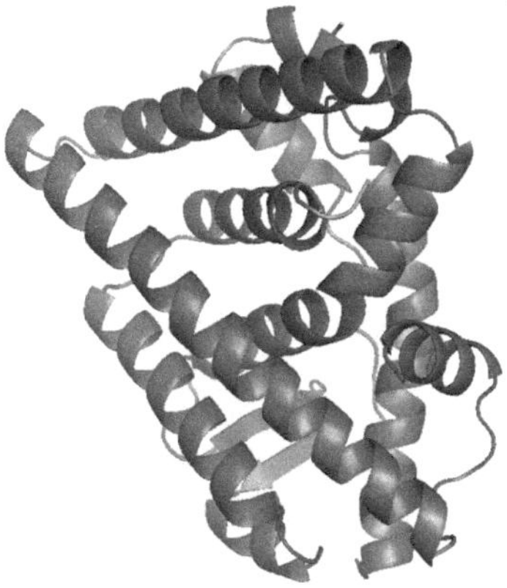
3	CNS lymphoma	Dihydrofolate receptor (1DRF)	186	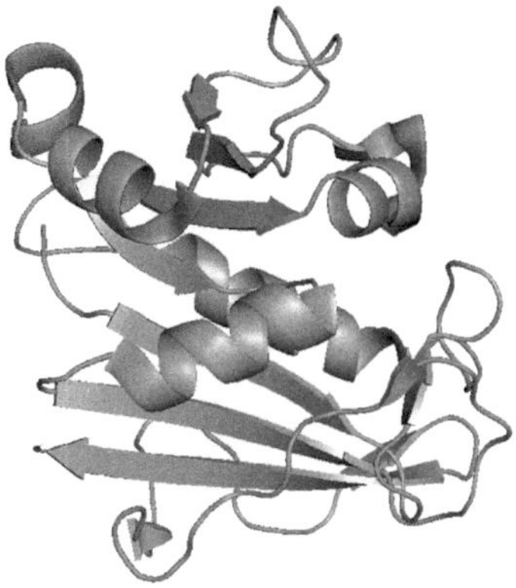

TABLE 7.2 *(Continued)*

Sl. no.	Name of the cancer	Targets with PDB ID	Amino acid length	Ribbon representation of cancer-specific protein viewed using PyMol
4	Lung cancer	Bcl-2 protein (1GJH)	166	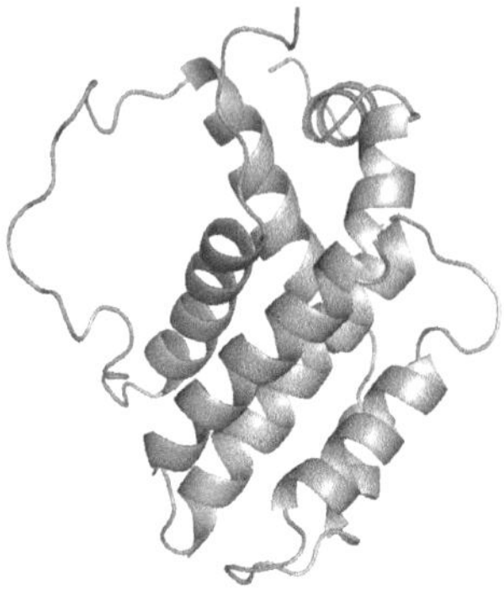
5	Oral cancer	Epidermal growth factor (1IVO)	622	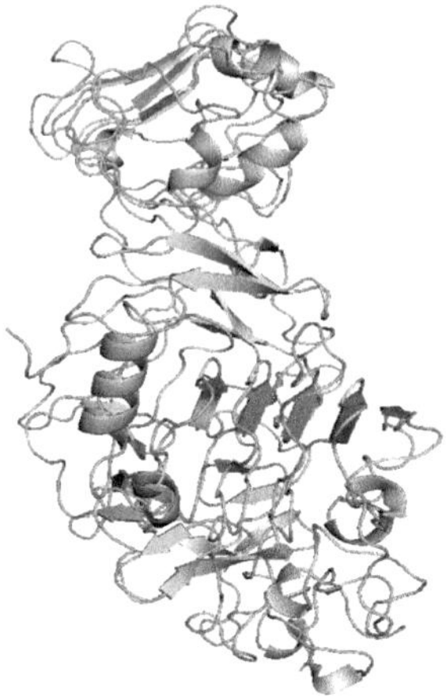
6	Prostate cancer	Androgen receptor (2AM9)	266	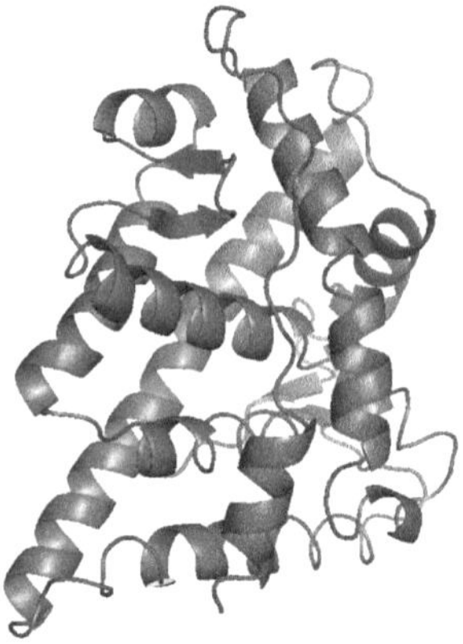

TABLE 7.2 *(Continued)*

Sl. no.	Name of the cancer	Targets with PDB ID	Amino acid length	Ribbon representation of cancer-specific protein viewed using PyMol
7	Renal cancer	mTOR (1AUE)	100	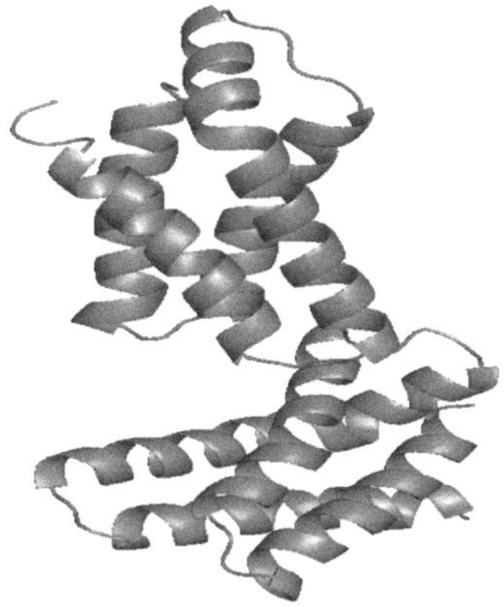
8	Rectal cancer	Thymidylate synthase (1HVY)	288	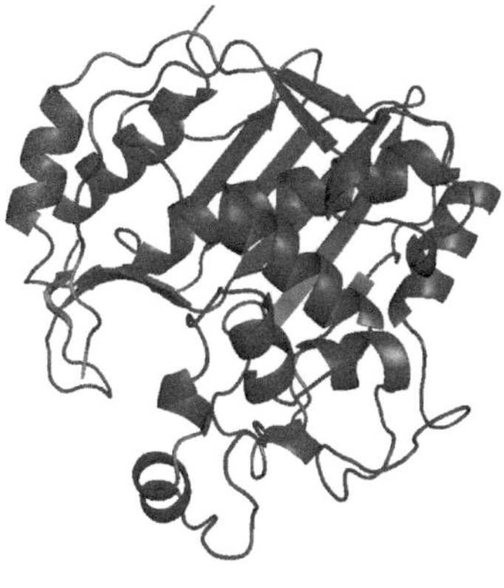
9	Squamous cell carcinoma	TLR7 (1ZIW)	680	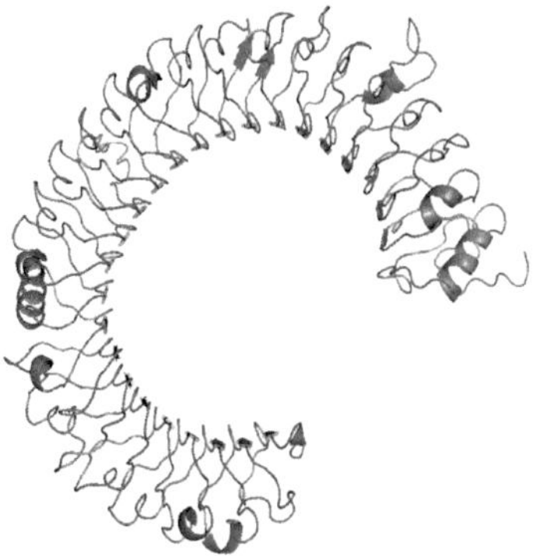

TABLE 7.2 *(Continued)*

Sl. no.	Name of the cancer	Targets with PDB ID	Amino acid length	Ribbon representation of cancer-specific protein viewed using PyMol
10	Thyroid cancer	RAF protein (1C1Y)	167	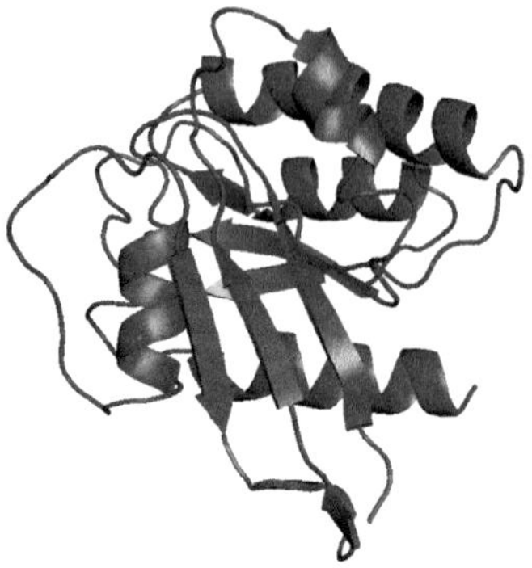

TABLE 7.3 List of Ligands in Use for Treatment.

Ligand name with PubChem ID	2D structure	Molecular formula	Molecular weight (g/mol)
Prednisolone (5755)	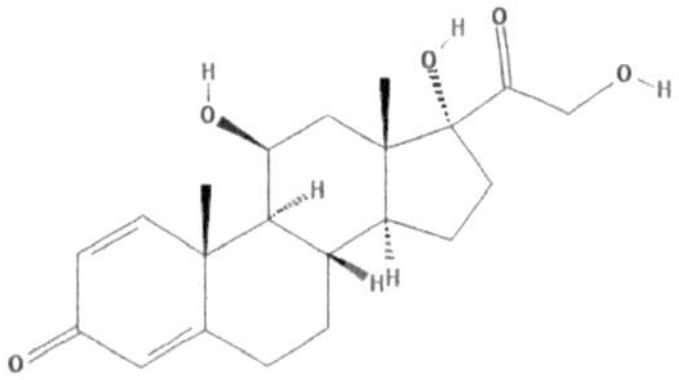	$C_{21}H_{28}O_5$	360.44402
Tamoxifen (2733526)	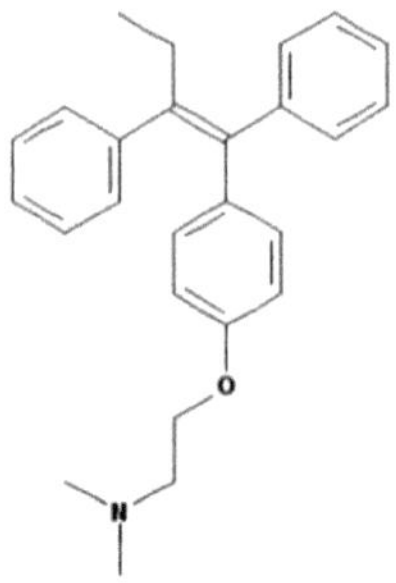	$C_{26}H_{29}NO$	371.51456
Methotrexate (126941)	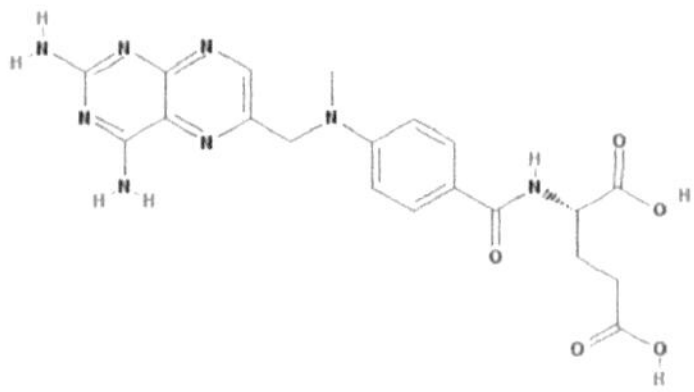	$C_{20}H_{22}N_8O_5$	454.43928

TABLE 7.3 *(Continued)*

Ligand name with PubChem ID	2D structure	Molecular formula	Molecular weight (g/mol)
Docetaxel (148124)	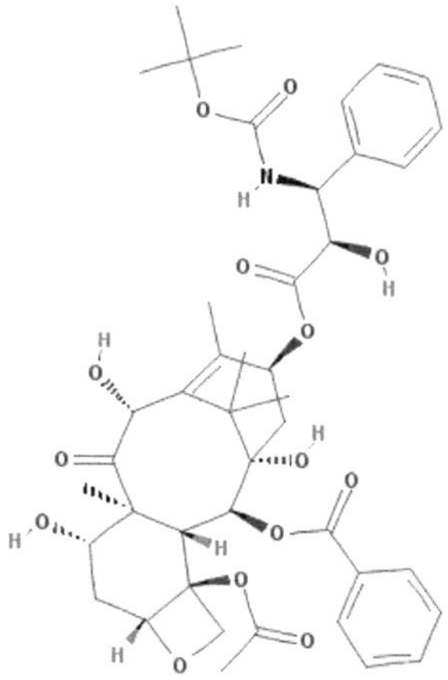	$C_{43}H_{53}NO_{14}$	807.87922
Gefitinib (123631)	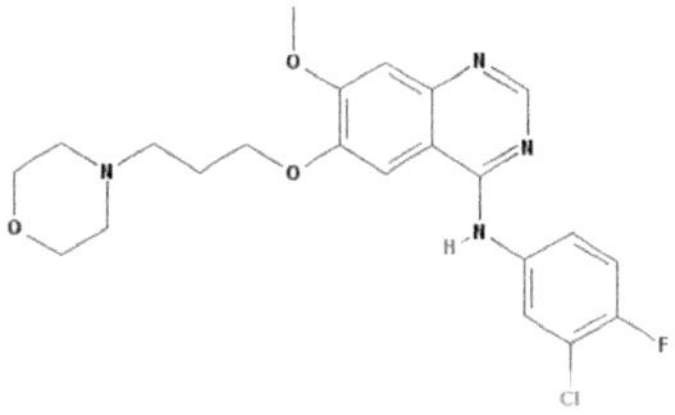	$C_{22}H_{24}C_{1}FN_{4}O_{3}$	446.902363
Andarine (9824562)	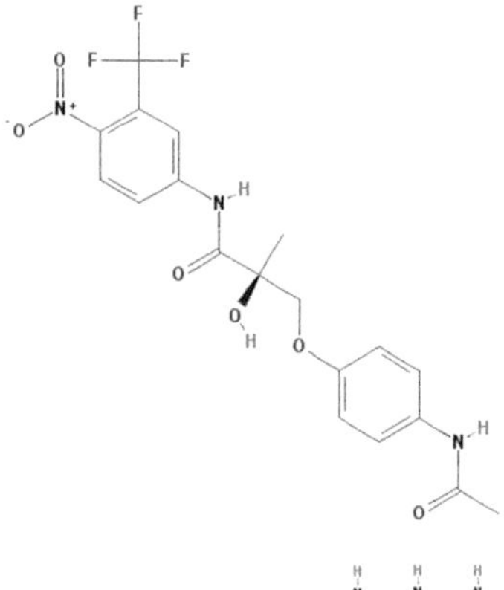	$C_{19}H_{18}F_{3}N_{3}O_{6}$	441.35793
Folinic acid (6006)	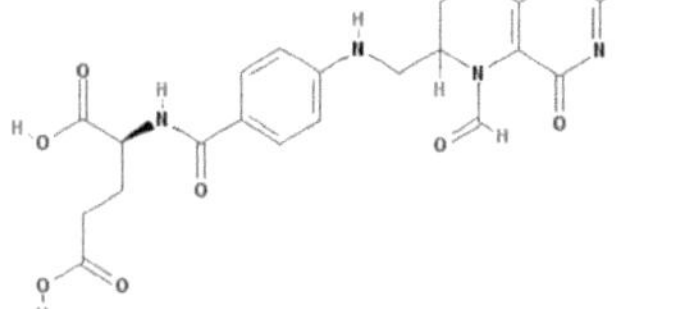	$C_{20}H_{23}N_{7}O_{7}$	473.43932

TABLE 7.3 *(Continued)*

Ligand name with PubChem ID	2D structure	Molecular formula	Molecular weight (g/mol)
Torisel (23724530)	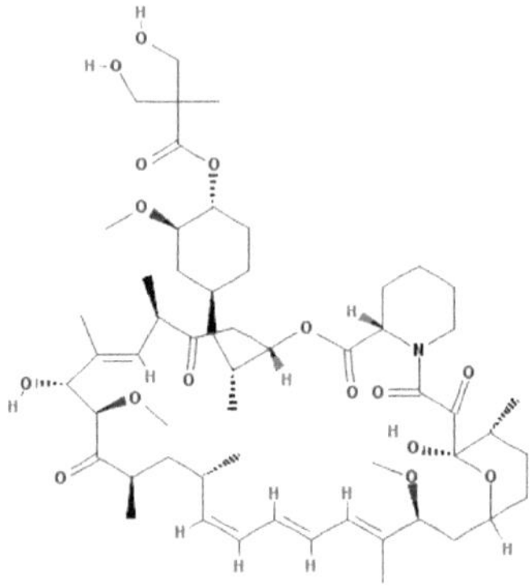	$C_{56}H_{87}NO_{16}$	1030.28708
Imiquinoid (57469)	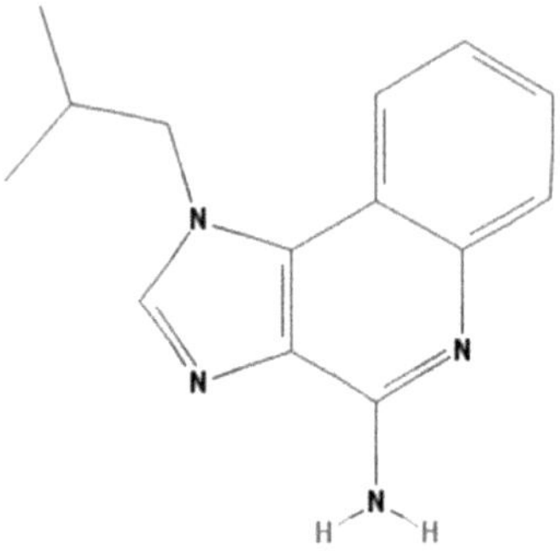	$C_{14}H_{16}N_{4}$	240.30364
Sorafenib (216239)	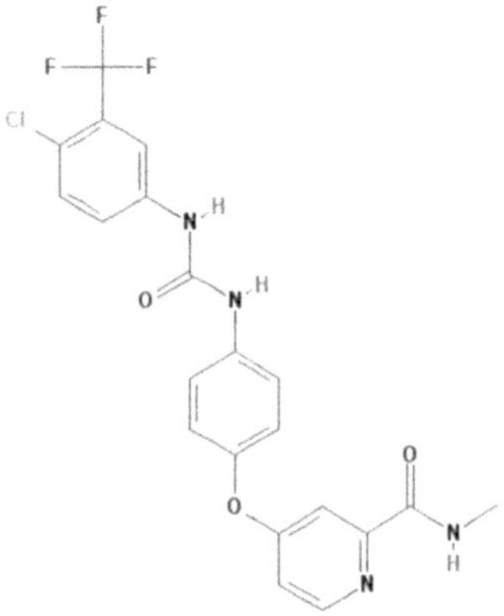	$C_{21}H_{16}C_{1}F_{3}N_{4}O_{3}$	464.82495

TABLE 7.3 *(Continued)*

Ligand name with PubChem ID	2D structure	Molecular formula	Molecular weight (g/mol)
Quercetin* (5280343)	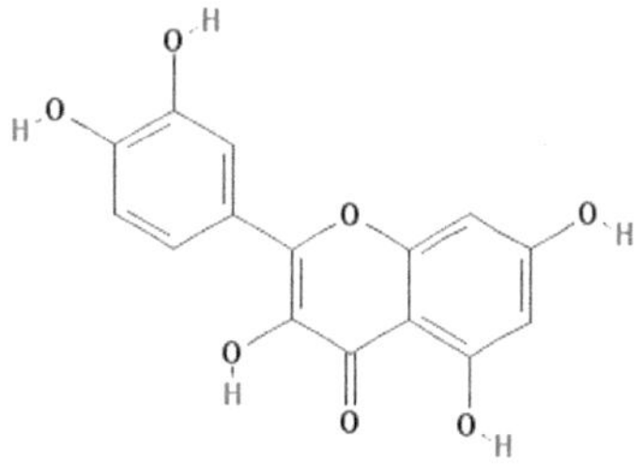	$C_{15}H_{10}O_7$	302.2357

*Represents the isolated, purified, and PubChem deposited natural substance (quercetin) from *E. littorale* was used for interaction studied against the native ligands.

7.2.1.1.3 *Active Site Prediction*

Prediction of active site for each cancer-specific protein by Q-Site finder is tabulated as Table 7.4.

7.2.1.1.4 *Molecular Dynamics*

MD calculations simulate molecular movement over time using Newton's equations of motion. Therefore, MC/SD simulation was carried out for the protein-ligand complex.

7.2.1.2 RESULTS

7.2.1.2.1 *Determination of ADME/Tox Properties*

ADME/Tox properties were analyzed for the already available and marketed synthetic ligands and compared with that of the isolated compound quercetin using Qikprop module of Schrodinger (Table 7.5). Interestingly, the phytocompound-quercetin was revealed to satisfy the LR5 (Lipinski rule of 5) in an unbiased manner was comparable to certain other synthetic ligands

TABLE 7.4 Active Site Residues of Each Targeted Proteins.

Protein name	Residues
Glucocorticoid receptor	CYS 450,HIS 451,TYR 452,GLY 453,LYS 461,LEU 507,ALA 509,ARG 510,LYS 511
Estrogen receptor	MET 315,VAL 316,SER 317,ALA 318,LEU 319,LEU 320,ASP 321,ALA 322,GLU 323,TRP 360, ALA 361,LYS 362, ARG 363,VAL 364,PRO 365,GLY 366,PHE 367,LEU 379,ILE 386,TRP 393,GLY 442,GLU 443,GLU 444,PHE 445, VAL 446,CYS 447,LEU 448,LYS 449,SER 450,LEU 453,LEU 454, VAL 478,ILE 482
Dihydrofolate reductase	ILE 7,VAL 8,ALA 9,VAL 10,ILE 16,GLY 17,LYS 18,GLY 20,ASP 21,LEU 22,TRP 24,PHE 34, LYS 55,THR 56, SER 59,VAL 115,GLY 116,GLY 117,SER 118,TYR 121
Bcl-2 protein	ASP 10,ASN 11,ARG 12,GLU 13,ILE 14,VAL 15,MET 16,LYS 17,TYR 18,ILE 19,HIS 20, TYR 21,LYS 22, LEU 23, SER 24,GLN 25,ARG 26,GLY 27,TYR 28,TRP 30,VAL 93,HIS 94, THR 96,LEU 97,ARG 98,GLN 99,ALA 100, GLY 101,ASP 102,ASP 103,PHE 104,SER 105,TRP 144, GLY 145,ARG 146,ILE 147,VAL 148,ALA 149,PHE 150, PHE 151,GLU 152,GLY 154,GLY 155, VAL 156,CYS 158,VAL 159,GLU 160,SER 161,VAL 162,SER 167, PRO 168, LEU 169,VAL 170, ASP 171,ILE 173,ALA 174,LEU 175,MET 177,THR 178,GLU 179,LEU 181,TRP 195, PHE 198,TYR 202
Epidermal growth factor	GLN 47,ARG 48,ASN 49,TYR 50,ASP 51,LEU 52,SER 53,LYS 56,THR 71,VAL 72,GLU 73, ARG 74,ILE 75,PRO 76,GLU 78,ASP 102,THR 106,GLY 107
Androgen receptor	LEU 701,LEU 704,ASN 705,LEU 707,GLY 708,AGLN 711,TRP 741,MET 742,MET 745,VAL 746, MET 749, ARG 752, PHE 764,AMET 780,MET 787,LEU 873,HIS 874,PHE 876,THR 877,LEU 880, PHE 891,MET 895,ILE 899
Thymidylate synthase	ARG 50,PHE 80,GLU 87,ILE 108,TRP 109,ASN 112,TYR 135,LEU 192,PRO 194,CYS 195,HIS 196, GLN 214, ARG 215,SER 216,GLY 217,ASP 218,LEU 221,GLY 222,VAL 223,PHE 225,ASN 226, HIS 256,TYR 258,MET 311
mTOR	TRP A2024,TRP A2028,LEU A2055,HIS A2056,ALA A2057,MET A2059,GLU A2060,GLY A2062, PRO A2063, GLU A2068,THR A2069,SER A2070,PHE A2071,ASN A2072,GLN A2073,ALA A2074, TYR A2075,GLY A2076, ARG A2077,LEU A2079,MET A2080,GLU A2081,ALA A2082,GLN A2083, GLU A2084,TRP A2085,CYS A2086, ARG A2087,TYR A2105
TLR7	TYR 556,PHE 557,LEU 558,LYS 559,GLY 560,GLU 580,VAL 581,PHE 582,LYS 583,ASP 584,LEU 585
RAF protein	SER 11,GLY 12,GLY 13,VAL 14,GLY 15,LYS 16,SER 17,ALA 18,VAL 29,GLU 30,LYS 31, TYR 32,ASP 33,THR 35,ASP 57,THR 58,ALA 59,GLY 60,THR 61,LYS 117

like prednisolone, tamoxifen, andarine, imiquimod, sorafenib, and geftinib in almost all the parameters analyzed. Among the synthetic compounds, the molecular weight, solvent accessible surface area, and number of rotatable bonds were found to be satisfied for 8–9 compounds excepting docetaxel and Torisel, which biased the specified range. However, the same parameters were predicted to lie in the preluded range in the phyto-ligand (quercetin) that showed 302.24 g/mol of molecular weight, 514.002 of SASA and 5 numbers of rotatable bonds (Table 7.5).

Similarly, the parameters such as log *P* for either hexadecane/gas or octanol/gas were also not fruitfully met by the synthetic ligands docetaxel and Torisel, which were higher or beyond the limitations of 18 and 35, respectively. Perhaps, the phyto-ligand was predicted to uphold the limitation criteria. Interestingly, all the ligands under investigation were observed to lie in the prescribed range between 4 and 45 for water/gas log *P*. However, some of the parameters like the "solute as donor for hydrogen bonding" of "percentage of oral absorption by human" were not satisfied by MTX and folinic acid, providing no reason to be ascertained as either of it influences or controls each other in a vice-versa.

Whatever the case may be, whether the ligands be if synthetic or phyto-ligand, if satisfies few parameters under consideration, could still be chosen for interactions studies in order to evaluate its performance as a good interactor, since that determines the drug likability of the molecule as it is evidenced by many researchers, here again in this perception, all the synthetic ligands in addition to the phyto-ligand was chosen to extend research on docking studies.

7.2.2 DOCKING STUDY

The cancer-specific proteins were docked with the synthetic ligands and with phytocompound quercetin and compared, in which the ligands and receptor were represented in orange and green colors, respectively, whereas the bondings were represented in blue dotted lines.

7.2.2.1 ACUTE LUEKEMIAS LYMPHOMA

The GR was docked with ligand prednisolone and quercetin compound, respectively, and compared. The analysis of GR docking with both prednisolone and quercetin was observed to interact each other by 4 hydrogen

TABLE 7.5 ADME-Tox Properties of the Synthetic Ligands of Specific Cancer Proteins Compared against the Phytocompound Quercetin.

QikProp parameters analyzed Ligands name	Normal range	Ligands evaluated										
		Prednisolone	Tamoxifen	Methotrexate	Docetaxel	Andarine	Folinic acid	Torisel	Imiquimod	Sorafenib	Gefitinib	Quercetin
Molecular weight	130.00–725.00	368.512	371.521	454.444	807.89	441.363	473.444	800.13	313.31	464.831	446.908	302.24
Total SASA	300.00–1000.00	588.329	730.6	755.46	945.183	734.198	755.505	1099.967	477.011	766.562	644.275	514.002
No. of rotatable bonds	0.0–15.00	7	9	11	13	8	10	35	3	5	8	5
Solute as donor–hydrogen bond	0.00–5.00	5	0	6	3	3	7	0	2	3	1	4
Solute as acceptor–hydrogen bond	0.00–10.00	7	2	11	16	7	14	20	3	6	7	5
Predictions for properties												
QP log *P* for hexadecane/gas	4.00–18.00	11.141	13.532	16.487	20.719	13.416	16.856	22.679	8.461	14.585	12.098	10.666
QP log *P* for octanol/gas	8.00–35.00	22.79	15.626	31.853	40.363	23.322	35.355	42.556	13.267	23.629	19.174	19.513
QP log *P* for water/gas	4.00–45.00	15.548	5.033	24.007	22.688	14.573	28.175	16.456	8.145	15.67	10.268	14.379
QP log *P* for octanol/water	< 5	1.502	6.533	−1.841	4.176	3.13	−0.844	5.766	2.48	4.097	3.494	0.362
QP log S for aqueous solubility	−6.5–0.5	−3.134	−5.937	−4.008	−5.241	−5.89	−3.61	−0.525	−3.283	−7.019	−3.05	−2.83
QP log BB for brain/blood	−3.0–1.2	−1.346	0.364	−4.798	−2.367	−1.92	−5.208	−3.28	−0.485	−0.986	0.162	−2.352
No. of primary metabolites	<7	5	3	5	10	4	4	14	0	2	5	5

TABLE 7.5 *(Continued)*

QikProp parameters analyzed Ligands name	Normal range	Ligands evaluated										
		Predniso-lone	Tamoxi-fen	Metho-trexate	Doce-taxel	Anda-rine	Folinic acid	Torisel	Imiqui-mod	Sorafenib	Gefitinib	Quercetin
Apparent Caco-2 perme-ability (nm/sec)	<25 poor, >500 great	212	2199	0	69	118	0	147	1067	312	662	19
Lipinski rule of 5 violations	4	0	1	2	2	0	2	3	0	0	0	0
Jorgensen rule of 3 violations	3	0	1	1	1	1	1	1	0	1	0	1
% Human oral absorp-tion in GI (±20%)	<25% poor	77	100	0	58	82	0	61	96	96	100	52
Qual. model for human oral absorption	>8% is high	High	High	Low	Low	Low	Low	Low	High	Low	High	Medium

bonding, however, there was a single common interaction found by LYS at 511 bonds by 2.5 and 2.6 Å. Other than this, prednisolone should get three more hydrogen interaction by hydrophobic residues and with two uncharged residues THR and LYS at position 512 and 450, respectively, by the hydrogen bond of 2.7 and 2.2 Å units (Fig. 7.1a). Interestingly, unlike that of synthetic ligand, the interaction mediated by hydrogen bonding in phyto-ligand involved the positively charged residues that are basic in nature, that is, ARG and HIS at 510 and 451 position by 2.3 and 2.0 Å distances (Fig. 7.1b). Although the synthetic ligand and quercetin binds with the active sites of the protein both binds with different residues. However, the G-score was observed to be higher (−4.36 Kcal/mol) in quercetin interacted complex than in prednisolone which showed the score of 3.77 Kcal/mol only (Table 7.6).

a) GR vs. Prednisolone **b) GR vs. Quercetin**

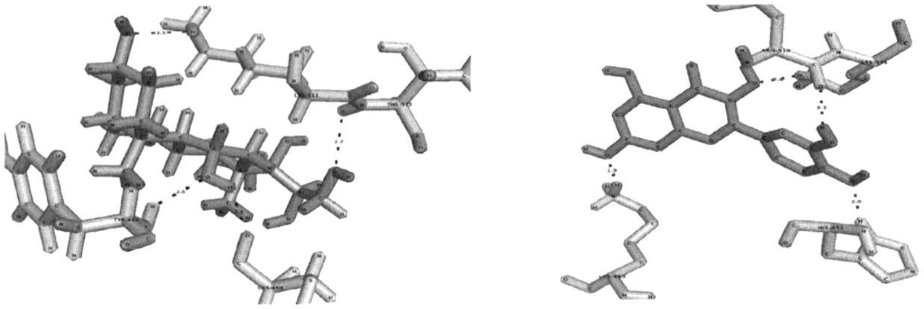

G-score: −3.77 **G-score: −4.36**

FIGURE 7.1 Snapshot of glucocorticoid receptor with prednisolone and quercetin docking for acute leukemia lymphoma.

7.2.2.2 BREAST CANCER

The ER was docked with synthetic ligand tamoxifen and quercetin compound and compared. The analysis of ER docking with tamoxifen was observed to interact with a negatively charged residue GLU 353 and a positively charged residue ARG 394 by O—H and H—O bond 2.1 and 2.2 Å distances (Fig. 7.2a). The same receptor with the purified substance quercetin interacted with a polar PRO 325, a negatively charged residue GLU 353 and hydrophobic residue TRP 393 by hydrogen bond interaction (Fig. 7.2b) with color index brown the bond length of 2.0, 1.7, and 2.1 Å, respectively. Here, both the ligands interacted with GLU at 353 but by exhibiting greater affinity with phyto-ligand by 1.7 Å distance only, which was evidenced by the higher

G-score (−4.7 Kcal/mol) unlike that of synthetic which showed only −3.37 Kcal/mol (Table 7.6).

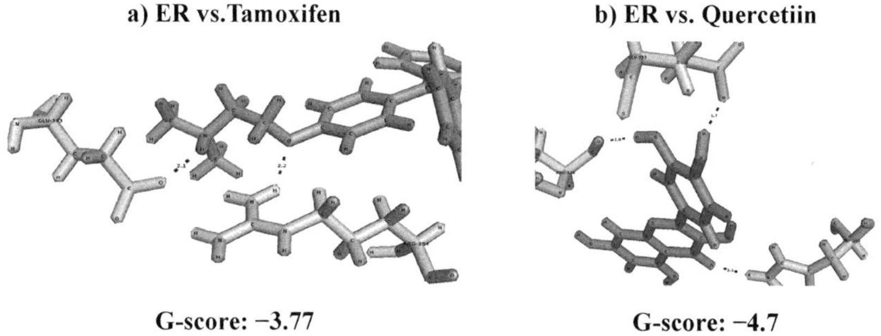

FIGURE 7.2 Snapshot of estrogen receptor with tamoxifen and quercetin docking for breast cancer.

7.2.2.3 CNS LYMPHOMA

The dihydrofolate receptor (DHFR) was docked with the synthetic ligand MTX and isolated pure compound quercetin, respectively. The interaction efficiency between them when compared in terms of G-score revealed quercetin to be higher −9.13 than MTX (−7.69 Kcal/mol) (Table 7.6) with three residues exhibiting good O—H interactions all being below 3 Å while one O—O interaction mediated by a non-polar residue VAL at 115 by extended bond length of 3.3 Å in addition to the O—H interaction. The interaction was observed with the positively charged residue ARG at 70 donating two hydrogen bonds of inter atomic distances 1.7 and 1.8 Å, however, the synthetic ligand interacted with negatively charged residue GLU 30 (2.0 Å) in addition to the hydrophobic interaction which included the residues, such as VAL 115 (2.5 Å), ILE 7 (1.9 Å), and also with uncharged amino acids GLN and ASN at position 35 and 64 forming O—H bonds of each 2.3 Å interatomic distances (Fig. 7.3a). Thus, altogether six different residues interacted with synthetic ligands, whereas only three different residues interact by phyto-ligand. Although the scoring seen to be high by phyto-ligand, could still not be considered favorable because of the fact that a specific O—O interaction, especially by respected bond length, was observed. The hydrogen bonding was observed between O—H and O—O atoms with inter atomic distances of 2.1, 3.3, 2.9, 1.5, and 1.6 Å, respectively (Fig. 7.3b).

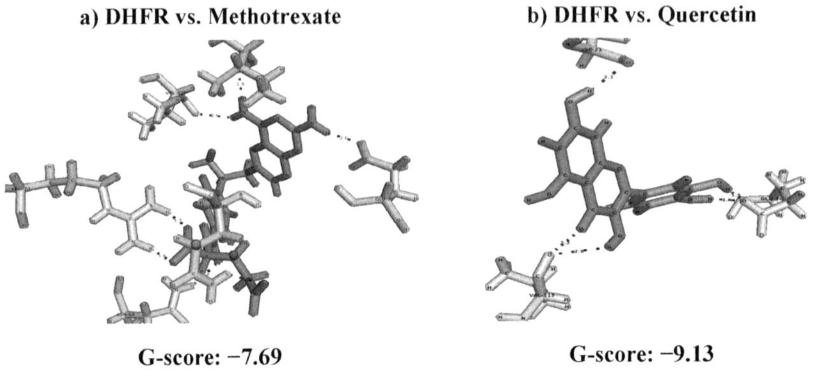

FIGURE 7.3 Snapshot of dihydrofolate receptor (DHFR) with methotrexate and quercetin docking for CNS lymphoma.

7.2.2.4 LUNG CANCER

The receptor Bcl-2 protein docked with docetaxel and isolated pure compound quercetin with the receptor and compared. The docetaxel interacted with receptor by a single positively charged residue ARG by sharing four electrons one each from positions 106 and 109 2.1 Å interatomic distance each and by 2 from position 26 by the distance of 2.2 Å interatomic distances each (Fig. 7.4a and Table 7.6). This unfavorability could also be attributed to the special kind of interaction encountered by the phyto-ligand that has remarkably involved many different polar residues, such as SER 105, TYR 108, ASP 102, and GLN 25 all being in favorable complementation, that is, below 3 Å, in addition to the ALA hydrophobic residue to lie in the same platform. SER at 105, TYR at 108 (2 bonds), ASP at 102 (2 bonds), and GLN at 25 exhibited O—H interaction by 2.5 and1.9 Å each by two different atomic co-ordinates, 2.4, 2.1, and 2.3 Å, respectively. Thus, the four polar residues alone imparted 6 O—H bonding. Moreover, a common interaction was also observed by ARG at position 26, but by longer distance (1.9 and 2.7 Å) through phyto-ligand interaction, where compound to synthetic ligand. On the whole including hydrophobic residue, a total of nine interactions was observed making the complex a very stable or strong interaction, thereby interesting to extend research on confirming the stability of the complex through molecular simulation.

a) Bcl-2 vs. Docetaxel **b) Bcl-2 vs. Quercetin**

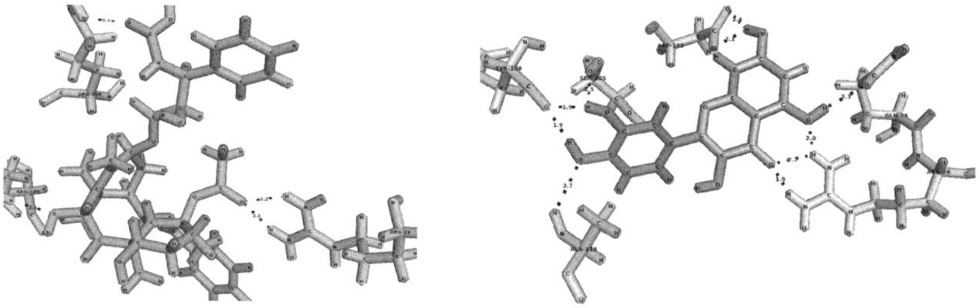

G-score: −2.97 **G-score: −5.74**

FIGURE 7.4 Snapshot of Bcl-2 protein with docetaxel and quercetin docking for lung cancer.

7.2.2.5 ORAL CANCER

The EGF was docked with synthetic ligand gefitinib and quercetin compound and were compared each other. The docking of synthetic ligand and phyto-ligand quercetin with EGFR where the receptor shares electron from the negatively charged residue GLU at 73 and ASN at 49 to both the ligands, however, it shares with N atom in the case of synthetic ligand with 2.6 and 2.3 Å (Fig. 7.5a), whereas to O atom of quercetin with 2.3 and 2.1 Å interatomic distances. Although these residues accepts electron from the H atom of the synthetic ligand forming totally four interactions, in terms of phyto-ligands forms two more interactions with polar positively charged ARG at 74, negatively charged ASP at 51, and hydrophobic LEU at 52 position of 1.8, 2.0, 2.6, and 2.5 Å where LEU at 52 position formed two hydrogen bondings between H and O atoms (Fig. 7.5b). However, the G-score was observed to be high (−5.21) in quercetin than in gefitinib which showed the G-score of −4.3 only (Table 7.6).

a) EGFR vs. Gefitinib **b) EGR vs. Quercetin**

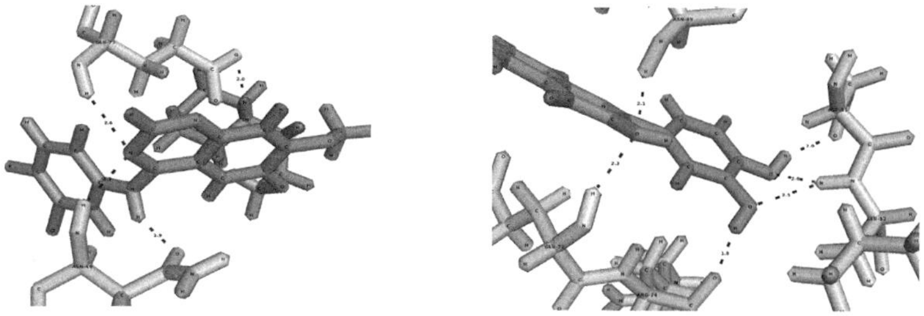

G-score: −4.3 **G-score: −5.21**

FIGURE 7.5 Snapshot of epidermal growth factor (EGF) with gefitinib and quercetin docking for oral cancer.

7.2.2.6 PROSTATE CANCER

The AR was docked with synthetic ligand andarine and quercetin and was compared, respectively. The receptor interacts both the ligands at THR 877, however, the electron sharing pair differs wherein andarine O—O atoms interact, whereas in quercetin H and O interacts of 2.8 and 2.6 interatomic distance. Apart from this interaction, andarine showed interaction with uncharged residue ASN at 705 with 2.0 Å distance (Fig. 7.6a) and quercetin interacts with negatively charged ASP at 51 (2.0 Å) and also with hydrophobic residue LEU at 704 position forming two hydrogen bonds between H and O atoms of 2.6 and 2.5 Å interatomic distance (Fig. 7.6b). In spite of O—O interaction of residue THR 877 with andarine it was observed between H and O in the case of quercetin, eventually showing more stable interactions and G-score higher in quercetin about −8.26 Kcal/mol than andarine (−3.21 Kcal/mol).

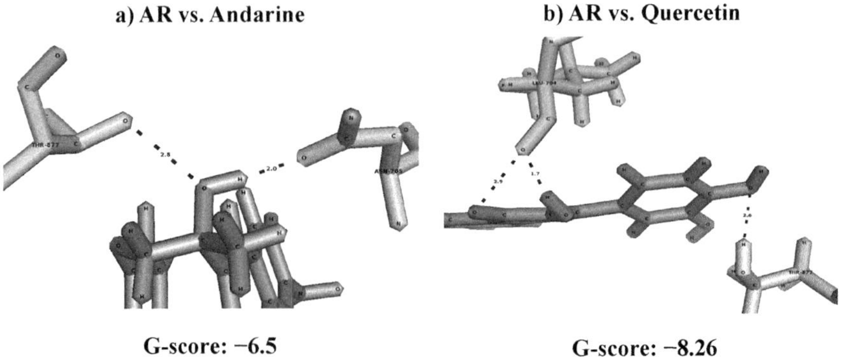

FIGURE 7.6 Snapshot of androgen receptor (AR) with andarine and quercetin docking for prostate cancer.

7.2.2.7 RECTAL CANCER

TS receptor was docked with synthetic ligand folinic acid and quercetin and interactions were compared, respectively. The G-score obtained was found higher about −10.85 Kcal/mol in TS-folinic complex than −7.33 Kcal/mol in TS-quercetin complex (Table 7.6). The TS shows six interactions with folinic acid and five with quercetin, however, two hydrophobic residues PHE at position 80 and ILE at 108 interacts with both folinic acid and quercetin, where PHE 80 shows interaction between H and O atoms of the ligands and that ILE 108 shared electrons between O and H atoms in both the cases, whereas in quercetin ILE forms one more O—O interaction of 3.3 Å distance which exceeds the beyond 3 Å (Fig. 7.7b). Other four interactions with folinic acid were observed by the uncharged residue ASN at 112 and with hydrophobic residue MET at 309 of 1.9 Å and LYS at 77 position forming two hydrogen bondings of 1.9 and 2.4 Å distances (Fig. 7.7a). Therefore, apart from the three interactions with the common residues sharing bond with both the ligands quercetin forms two bondings with polar negatively charged residue GLU at 87 of 1.8 Å and hydrophobic residue LEU at 221 of 2.3 Å where the bond length except the O—O interaction by ILE not exceeds 3 Å distance.

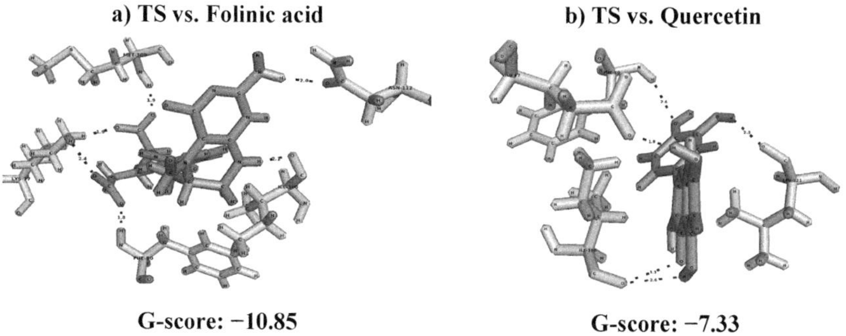

FIGURE 7.7 Snapshot of thymidylate synthase (TS) with folinic acid and quercetin docking for rectal cancer.

7.2.2.8 RENAL CANCER

The receptor mTOR was docked with synthetic ligand Torisel and quercetin and was compared, respectively. The docked complexes were observed with slightly varied G-score −3.1 Kcal/mol of mTOR-Torisel and −3.19 Kcal/mol of mTOR-quercetin complex (Table 7.6). The interaction of Torisel with mTOR showed the three electron donation of receptor by the hydrophobic residues such as VAL 2045 (2.8 Å), LYS 2046 (2.3 Å), and ALA 2059 (1.8 Å), whereas the polar negatively charged residue GLU 2053 accepts electron from Torisel and forms hydrogen bond of 2.0 Å interatomic distance (Fig. 7.8a). The mTOR interactions with quercetin showed interactions between O and H atoms with polar uncharged residues, such as ASN at 2072 (1.8 Å) and GLN at 2083 (1.9 Å) and also with polar positively charged residue HIS 2056 with 2.2 Å of interatomic distances (Fig. 7.8b).

a) mTOR vs. Torisel **b) mTOR vs. Quercetin**

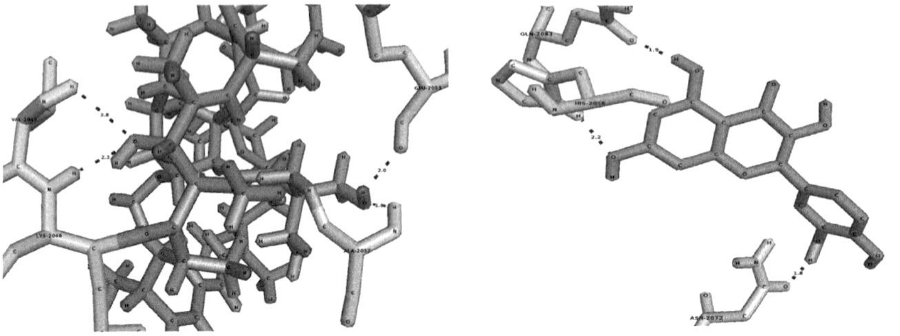

G-score: −3.1 **G-score: −3.19**

FIGURE 7.8 Snapshot of mTOR with Torisel and quercetin docking for renal cancer.

7.2.2.9 SQUAMOUS CELL CARCINOMA

TLR-7 receptor was docked with synthetic ligand imiquimod and quercetin and the interactions were compared. The interaction of imiquimod and quercetin with TLR-7 with the same residue LYS at 559 with distance of 2.1 and 2.3 Å, apart from this imiquimod showed interaction with negatively charged residue ASP at 523, whereas quercetin interacts with receptor by residues LYS 583 and shares two bonding interaction with ASP 584 of O— O interaction and between O and H atoms of distance not exceeding 3 Å (Fig. 7.9b). The G-score was observed to be higher in TLR-7 quercetin about −3.39 Kcal/mol than TLR-7 imiquimod showed −2.17 only (Table 7.6).

a) TLR-7 vs. Imiquimod **b) TLR-7 vs. Quercetin**

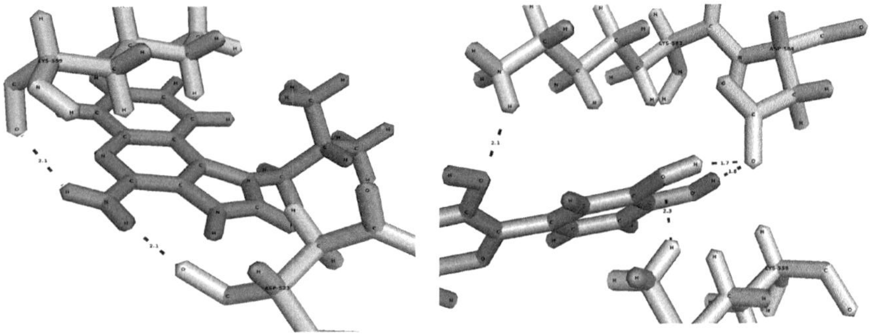

G-score: −2.17 **G-score: −3.39**

FIGURE 7.9 Snapshot of TLR-7 with imiquimoid and quercetin docking for squamous cell carcinoma.

7.2.2.10 THYROID CANCER

The receptor RAF protein was docked with synthetic ligand sorafenib and quercetin and was compared, respectively. The interactions of sorafenib and quercetin with receptor by different residues where sorafenib shows hydrogen bonding interaction with polar uncharged ASN 116 and charged ASP 33 residues with 2.4, 1.8, and 2.2 Å, respectively (Fig. 7.10a), and that ASP 523 shares two with receptor to form two hydrogen bondings (Fig. 7.10a). When comparing the interaction of RAF with sorafenib, quercetin showed interaction with hydrophobic residue LYS 31 (1.9 Å) and ALA 148 of 2.1 Å distance and also with polar uncharged residue ASP 119 (2.7 Å) (Fig. 7.10b). The G-core obtained was comparatively high in receptor with quercetin (−7.82 Kcal/mol) than synthetic ligand sorafenib showed −2.54 only (Table 7.6).

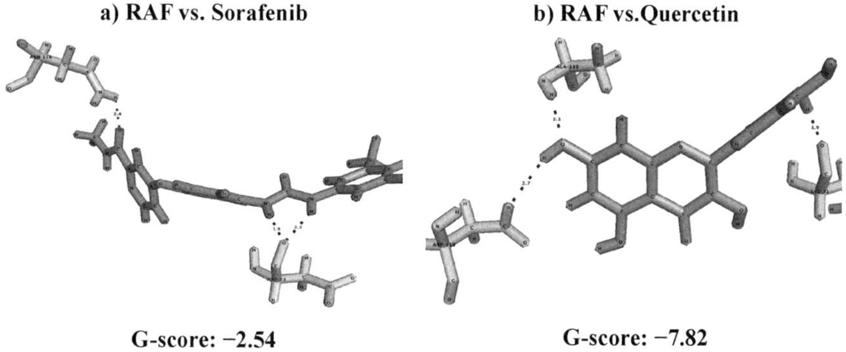

FIGURE 7.10 Snapshot of RAF protein with sorafenib and quercetin docking for thyroid cancer.

Thus, on the whole, among the ten proteins that were targeted, the level of interactions had been very excellent for the phyto-ligand (quercetin). Infact, the kind of interaction observed could be categorized into two: one which is showing more number of interactions by more or less the same residues (compared with synthetic and phto-ligand) as well with other important residues, thereby increasing the G-score by twofold or threefold; and the other could be attributable to the kind of residues (denotes the properties of the amino acid) showing additional interaction for instance O—O (e.g., In DHFR, VAL at 115th position also interacted by O—O in addition to O—H) or exhibited replaced interaction. In AR, THR at 877th position was exchanged by H—O instead of O—O which either effected or affected the docking energy G-score. For instance, the interaction between Bcl-2 protein and quercetin was observed with significant interactions (9 bonding) and best G-score

(−5.74 Kcal/mol) followed by quercetin with EGF showing six number of interactions with docking energy of −5.21 Kcal/mol. However, the complex quercetin with RAF protein showed greater G-score of −7.82 Kcal/mol could still not be considered favorable as the hydrogen bond exhibits 3.0 Å inter-atomic forces extended by this and perhaps with only few residues when compared to Bcl-2 interaction with phyto-ligand. Even though the interaction between folinic acid and TS showed six numbers of interactions with highest G-score about −10.85 Kcal/mol than quercetin still could not considered favorably as it violated the ADME-Tox properties, therefore, even on the basis of toxicity prediction and binding affinity observation the Bcl-2 protein interaction with quercetin was found significant and further taken to the dynamics analysis.

Thus from the result obtained and analyzed so far the interaction could be considered as an ideal and most favorable irrespective of the number of parameter that is met with. Even though not cent percentage properties, are met through ADME-Tox, based on the kind of the amino acid interacting and based on the distances aiding in receptor-ligand complex formation that upholds the stabilized bond length (neither less than 1 nor greater than 3 Å) would enhance the chance of considering the interaction favorable for extending research or attempting either for stable conformation or for further pharmacokinetic and dynamic studies. Thus on these basis in this study, the phyto-ligand quercetin which complex with receptor Bcl-2 was considered suitable for continuing research on molecular simulation direction. Thus it involves many different polar residues with more number of interactions in perhaps with good docking energy.

The interaction profile of cancer-specific protein with their native ligands (Phase I) and quercetin compound (Phase II) was tabulated (Table 7.6) describing the bond length formed between the atoms of receptor residues and ligands with total number of interaction for each specific protein.

7.2.3 MOLECULAR DYNAMICS

The MD simulation was carried out for the docked complex Bcl-2 vs. quercetin for lung cancer which showed significant interaction and good G-score than the other proteins interact with quercetin. Through the dynamic study, it was decided to predict the variation undergone by the complex during the simulation process. The simulation was carried out for 1 ns and to retrieve 100 samples which were observed with reduction in potential energy from −14,658.9 to −14,707.3 ê of 0–100th sample (Fig 7.11).

TABLE 7.6 A Comparative Interaction Profile of Cancer-specific Targets with Synthetic Ligands and Quercetin.

Receptor	Native ligand					Quercetin				
	Name of the ligand	**Residues interacted**	**Bond length**	**No. of bonds formed**	**G-score Kcal/mol**	**Name of the ligand**	**Residues interacted**	**Bond length**	**No. of bonds formed**	**G-score Kcal/mol**
Glucocorticoid receptor	Prednisolone	TYR 452 (H–O)	2.5	4	−3.77	Quercetin	LYS 465 (H–O)	1.9	4	−4.36
		CYS 450 (O–H)	2.21				ARG 510 (O–H)	2.3		
		THR 512 (H-O)	2.7				**LYS 511 (H–O)**	2.6		
		LYS 511 (H–O)	2.5				HIS 451 (H–O)	2.0		
Estrogen receptor	Tamoxifen	GLU 353 (O–H)	2.1	2	−3.37	Quercetin	PRO 325 (O–H)	2.0	3	−4.7
		ARG 394 (H–O)	2.2				GLU 353 (O–H)	1.7		
							TRP 393 (H–O)	2.1		
DHFR	Methotrexate	GLU 30 (O–H)	2.0	7	−7.69	Quercetin	ASP 21 (O–H)	2.1	**5**	**−9.13**
		ASN 64 (H–O)	2.3				**VAL 115 (O–O)**	3.3		
		ARG 70 (H–O)	1.7				**VAL 115 (O–H)**	2.9		
		ARG 70 (H–O)	1.8				GLU 30 (O–H)	1.5		
		VAL 115 (O–H)	2.5							
		ILE 7 (O–H)	1.9				GLU 30 (O–H)	1.6		
		GLN 35 (H–O)	2.3							
Bcl-2 Protein	Docetaxel	ARG 106 (H–O)	2.1	4	−2.97	Quercetin	SER 105 (O–H)	2.5	**9**	**−5.74**
		ARG 26 (H–O)	2.2				TYR 108 (O–H)	1.9		
							TYR 108 (O–H)	1.9		
		ARG 26 (H–O)	2.2				ALA 113 (H–O)	2.5		
							ASP 102 (O–H)	2.4		
							ASP 102 (O–H)	2.1		
							ARG 26 (H–O)	1.9		
		ARG 109 (H–O)	2.1				**ARG 26 (H–O)**	**2.7**		
							GLN 25 (O–H)	2.3		

TABLE 7.6 *(Continued)*

Receptor	Native ligand					Quercetin				
	Name of the ligand	Residues interacted	Bond length	No. of bonds formed	G-score Kcal/mol	Name of the ligand	Residues interacted	Bond length	No. of bonds formed	G-score Kcal/mol
Epidermal growth factor	Gefitinib	**GLU 73 (H—N)**	**2.6**	4	−4.3	Quercetin	**ASN 49 (H—O)**	2.1	**6**	**−5.21**
		GLU 73 (O—H)	2.0				**GLU 73 (H—O)**	2.3		
							ARG 74 (O—H)	1.8		
		ASN 49 (H—N)	2.3				ASP 51 (O—H)	2.0		
							LEU 52 (H—O)	2.6		
		ASN 49 (O—H)	1.9				LEU 52 (H—O)	2.5		
Androgen receptor	Andarine	ASN 705 (O—H)	2.0	2	−3.21	Quercetin	**THR 877 (H—O)**	**2.6**	3	
		THR 877 (O—O)	**2.8**				LEU 704 (O—H)	1.7		**−8.26**
							LEU 704 (O—O)	2.9		
Thymidylate synthase	Folinic acid	ASN 112 (O—H)	2.0	6	−10.85	**Quercetin**	**PHE 80 (H—O)**	**2.6**	5	**−7.33**
		ILE 108 (O—H)	2.1				GLU 87 (O—H)	1.8		
		PHE 80 (H—O)	1.8				**ILE 108 (O—H)**	2.6		
		MET 309 (H—O)	1.9				**ILE 108 (O—O)**	3.3		
		LYS 77 (H—O)	1.9				LEU 221 (O—H)	2.3		
		LYS 77 (H—O)	2.4							
mTOR	Torisel	VAL 2045 (H—O)	2.8	4	−3.1	Quercetin	GLN 2083 (O—H)	1.9	3	**−7.27**
		LYS 2046 (H—O)	2.3				HIS 2056 (H—O)	2.2		
		ALA 2059 (H—O)	1.8							
		GLU 2053 (O—H)	2.0				ASN 2072 (O—H)	1.8		
TLR7	Imiquimod	**LYS 559 (O—H)**	**2.1**	2	−2.17	Quercetin	LYS 583 (H—O)	2.1	4	−3.39
							ASP 584 (O—O)	1.7		
							ASP 584 (O—H)	1.8		
		ASP 523 (O—H)	2.1				**LYS 559 (H—O)**	2.3		

TABLE 7.6 *(Continued)*

Receptor	Native ligand					Quercetin				
	Name of the ligand	**Residues interacted**	**Bond length**	**No. of bonds formed**	**G-score Kcal/mol**	**Name of the ligand**	**Residues interacted**	**Bond length**	**No. of bonds formed**	**G-score Kcal/mol**
RAF protein	Sorafenib	ASN 116 (H—O)	2.4	3	**−2.54**	**Quercetin**	ALA 148 (H—O)	2.1	3	**−7.82**
		ASP 33 (O—H)	1.8				LYS 31(O—H)	1.9		
		ASP 33 (O—H)	2.2				ASP 119 (O—H)	2.7		

Note: Residues mentioned in bold were involved in interactions\ with both the synthetic ligand and quercetin.

*LipophilicEvdW–Chemscore lipophilic pair term and fraction of the total protein–Ligand Vanderwaal Energy; H bond–H bond pair Term; Electro–Electrostatic rewards; HBPenal–Penalty for Ligand with large hydrophobic contact and low H bonds score; PhobicPenal–Penalty for exposed hydrophobic Ligand groups; RotPenal–Rotatable bond penalty.

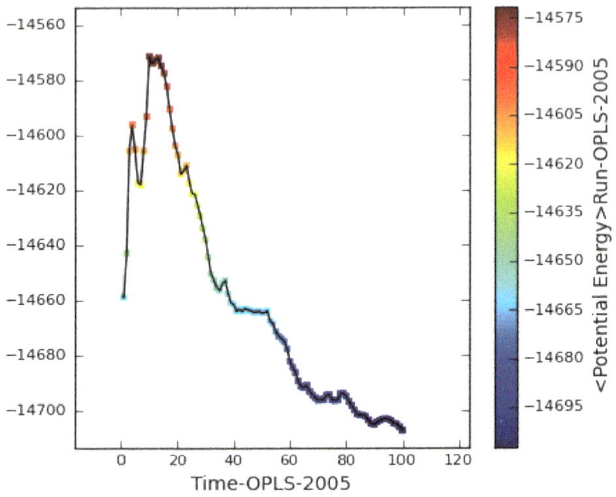

FIGURE 7.11 Dynamics plot against time vs. potential energy OPLS 2005 in run platform upto 100 iteration.

The root mean square (RMS) fluctuation of each 100 sample obtained through 1 ns simulation showed stability attained by the Bcl-2 docked with quercetin complex was plotted as graph (Fig. 7.12) where the complex at 74th sample.

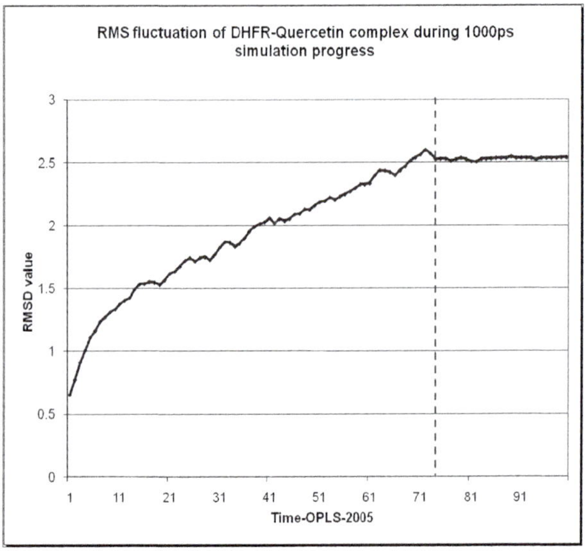

FIGURE 7.12 RMS fluctuation of Bcl-2 protein-quercetin complex.

The red line indicates the sample position attaining stability during the simulation.

7.3 DISCUSSION

In current *era*, molecular docking is one of the key tools in structural molecular biology and computer-assisted drug design. This study investigates the binding orientations and predicts binding affinities of native ligands which are currently used as synthetic drugs for cancers and also with isolated pure quercetin from *E. littorale* docking with the cancer-specific proteins. Through the docking interaction analysis, the binding mode and G-score obtained with quercetin were compared with the native ligands for each cancer specific proteins the anti-cancer efficiency of quercetin was determined. The cancer and specific proteins selected for the study are ALL, breast cancer, CNS lymphoma, lung cancer, oral cancer, prostate cancer, renal cancer, rectal cancer, SCC, and thyroid cancer and their specific proteins were GR, ER, DHFR, B-cell lymphoma-2 (Bcl-2) protein, EGFR, AR, mTOR protein, TS, TLR7, and RAF protein. The protein structures were retrieved from PDB database by concerning the resolution of the structure which should be <2.5 Å and R-value should be nearly zero and R-free value should equal to the resolution (Table 7.2). Comparative study was carried out between the native ligands and quercetin hence to understand the important structural features required to enhance the inhibitory activities and further to help producing augmented inhibitory compounds than the existing synthetic drug molecules, therefore, to avoid the side effects produced by it. Thereby Arumugam et al.[3] investigated the inhibitory potency for diabetes of plant-derived compound, namely, 1, 2 disubstituted idopyranose from *Vitex negundo* using GLIDE module of Schrodinger. He carried out docking with four targets, such as maltase glucoamylase, dipeptidyl peptidase, glycogen synthase kinase-3 and aldose reductase, and compared docking of plant compound with the docking result obtained with the known ligand.

The ligand molecules were retrieved from the PubChem database in sdf format (Table 7.3) and prepared to optimize the structure before undergoing docking studies. The prepared native ligand molecules and isolated purified quercetin were subjected for ADME/Tox for predicting the absorption, distribution, mechanism, excretion, and toxicity, therefore, to analyze the pharmacokinetics and pharmacodynamics of the ligand by accessing the drug-like properties. Through Qikprop the molecular weight, permeability through MDCK cells, log IC_{50} value for blockage of K^+ channels, gut-blood barrier, and violation of the Lipinski's rule of five was analyzed. According to the Lipinski's rule of five, quercetin does not violate the rules and the other native ligands properties were discussed as follows.

The molecular weight and other properties predicted by Qikprop for each ligand molecules were tabulated (Table 7.4) in which docetaxel (807.89 Da) and Torisel (800.13 Da) exceeded the normal range between 130 and 725 Da, whereas quercetin's molecular weight is about 302.24 Da was lesser compared with the other ligand molecules. The hydrogen donors predicted showed that MTX (6.25) and andarine (7.25) possess more than 5 hydrogen donors and prednisolones possess exactly 5 and 4 by quercetin. Moreover, the hydrogen bond acceptors eliminated the drug molecules, such as MTX (11.75), docetaxel (16.35), folinic acid (14.25), and Torisel (20.7) where quercetin exerts 5.25 of hydrogen bond donors. A partition (P) or distribution coefficient (D) is the ratio of concentrations of a compound in the two phases of a mixture of two immiscible solvents at equilibrium. The last of Lipinski rule, that is, octanol/water accessibility was observed to be about 0.362 for quercetin, whereas in other ligands such as tamoxifen (6.533) and Torisel (5.766) exceeds five. Thereby quercetin was observed to obey the rules of Lipinski and also exerts significant values for other properties, such as solvent accessible surface area, absorption in colorectal cells, and in blood–brain barrier too suggesting that quercetin is orally administered at last the calculated percentage of oral absorption showed medium for quercetin.[59]

According to the Schrodinger tool Qikprop, Jorgensen's rule of three defines the properties of drug likeliness for oral administration. The three rules are, QP log S >−5.7, QP PCaco >22 nm/s, and primary metabolites should be <7. The QP logS value determines the solubility of the small molecule in an aqueous solution which should be between −6.5 and 0.5, where all the native ligand molecules selected for the study obeys the aqueous solubility range except sorafenib about −7.019, whereas quercetin was observed with −2.83. Qikprop predicts the permeability through apparent colorectal adenocarcinoma cells (Caco) where every ligand molecules were observed above the range 25 despite quercetin has scored about 19. Thus none of the selected ligands was observed with acceptable Caco-2 cells permeability. Although the percentage of oral absorption in GI was found to be about 52 for quercetin, on the other hand, MTX and folinic acid exhibited zero values. Tamoxifen and gefitinib showed 100% of absorption in GI tract, whereas imiquimod and sorafenib found with 96% and the other native ligands, such as andarine, prednisolone, Torisel, and docetaxel observed with 82, 77, 61, and 58%.

The interaction study was carried out in two different phases where Phase I includes the docking of cancer-specific protein with synthetic ligands and Phase II with quercetin and compared each other based on the interaction profile and (G-score) XP scored during the docking process. The G-score was calculated based on the LipophilicEvdW (Chemscore lipophilic pair term and fraction of the total protein–Ligand Vanderwaal energy) HBond (H bond pair Term) Electro (Electrostatic rewards) HBPenal (Penalty for ligand with large hydrophobic contact and low H bonds score) PhobicPenal (Penalty for exposed hydrophobic ligand groups) RotPenal (Rotatable bond penalty). Comparing to the phase I study of synthetic ligands quercetin was observed with good binding affinity with the selected cancer-specific proteins except in the case of rectal cancer where TS showed higher G-score about −10.85 Kcal/mol and more number of interactions with folinic acid than quercetin (−7.33 Kcal/mol) despite folinic acid has violated ADME-Tox properties. However, the synthetic ligands prednisolone, gefitinib, andarine, and imiquimod undergo all the ADME-Tox properties these ligands were observed with minimum number of interactions compared to the phyto-ligand quercetin. The interactions of selected specific cancer proteins with quercetin based on their score and interactions number, it was found significant with Bcl-2 protein and further carried for dynamics study. Quercetin interaction with all the ten selected proteins was observed to form hydrogen bondings between O and H, O and O atoms, whereas in the case of synthetic ligands gefitinib alone showed electron sharing with N atoms of the receptor EPGR. In general, the interaction between two O atoms are said to enhance the stability of the bonding and the presence of such interaction in the active site region are due to the action of active site as a local storage sites for small molecules resulting in increased effective concentration of the ligand.[15] And quercetin was found to have strong inhibitory effects on mammalian Thioredoxin reductase (TrxRs) with IC_{50} value of 0.97 µmol/L was shown to be dose and time-dependent and attack on the reduced COOH-terminal–Cys-Sec-Gly active site of TrxR, where TrxR is one of thioredoxin system exerts a wide range of activities in cellular redox control, antioxidant function, cell viability, and proliferation which also found overexpressed in many aggressive tumors.[34]

The best G scored and effectively interacting Bcl-2 protein-quercetin was carried with MD studies to analyze the optimization of the complex so that the potential energy and to ability of the complex to attain stability were analyzed through MD study. At the initial stage of simulation (0th sample), the potential energy was observed to be −14,658.9 ê which decreases gradually and at the end of the simulation about 1000 ps it was reduced to −14,707.3 ê. The simulated protein at each 10th consecutive sample was observed for RMS deviation by and the graph plotted showed the stability attained by the protein-quercetin complex from the 74th sample (Fig. 7.12).

The Bcl-2 protein also seems to be the major inhibitor of cell death in acute myeloid leukemia where the protein was observed to be homology with Bcl-x_L and the protein consists of seven α-helices and a long loop. The proteins solubility and the biological function was studied by Petros et al.[57] which revealed that the deletion of long loop and C-termini of the protein does not interrupt the biological functions where it consists of two central, predominantly hydrophobic helices (helix 5 and 6) packed against four amphipathic α-helices. Helix 1 and 2 are oriented parallel to one another, crossing at an angle of about 45° and a long loop connects the helix 2–4 (residues 126–137) and helix 4, 5 (residues 144–163), and 6 (residues 167–192) are oriented in a nearly antiparallel fashion with a kink in helix 6 at histidine 184. An irregular turn composed of two glycine residues connects helix 6–7 (residues 195–202), which orients helix 7 orthogonal to helices 4, 5, and 6.

The protein consists of a hydrophobic groove on the surface on which the mutations in this region have been to abolish the antiapoptotic activity of Bcl-2 and block heterodimerization with other family members[84] and also reported that the hydrophobic groove includes the residues, such as leucine, isoleucine, valine, tyrosine, phenylalanine, tryptophan, aspartate, glutamate, lysine, arginine, and histidine present in α helices 3, 4, 5, and 7, even though the exact position of the above residues were not discussed. During the simulation, the protein was observed with slight structural movements, where every 10th sample of 100 samples was analyzed for its structural variation. The protein structure at 20th sample, when superimposed, showed the movement of the long loop that connects α1 and α2, loop connecting α3 and α4, and α4 helix. N and C-termini undergone greater fluctuation during the simulation progress, mainly the α3 and α4 helices were observed, therefore, it was already reported that half part of α3 contains the binding pocket of the protein. The Bcl-2 protein binding interaction with quercetin was observed involving the residues, such as aspartic acid, serine, tyrosine, alanine, arginine, and glutamine in which three active residues aspartic acid alone involved in the interaction in this study.

KEYWORDS

- **cancer targets**
- **drugs**
- **quercetin**
- **molecular docking**
- **dynamic simulation**

REFERENCES

1. Allen, M. P. *Introduction to Molecular Dynamics Simulation*; John von Neumann Institute for Computing: Jülich, Germany, 2004; Vol. 23, pp 1–28.
2. Ania, D. L. N.; Rolando, R. Current Methodology for the Assessment of ADME-Tox Properties on Drug Candidate Molecules. *Biotecnol. Apl.* **2008,** *25*, 97–110.
3. Arumugam, S.; Manikandan, R.; Chinnasamy, A. Molecular Docking Studies of 1, 2 Disustituted Idopyranose from *Vitex negundo* with Anti-diabetic Activity of Type 2 Diabetes. *Int. J. Pharma. Bio. Sci.* **2011,** *2* (1), B68–B83.
4. Asnaghi, L.; Bruno, P.; Priulla, M.; Nicolin, A. mTOR: A Protein Kinase Switching Between Life and Death. *Pharmacol. Res.* **2004,** *50* (6), 545–549.
5. Barnham, K. J.; Torres, A. M.; Alewood, D.; Alewood, P. F.; Domagala, T.; Nice, E. C.; Norton, R. S. Role of the 6–20 Disulfide Bridge in the Structure and Activity of Epidermal Growth Factor. *Protein Sci.* **1998,** *7* (8), 1738–1749.
6. Buchholz, T. A. Radiation Therapy for Early-stage Breast Cancer after Breast-conserving Surgery. *N. Engl. J. Med.* **2009,** *360* (1), 63–70.
7. Cancer Facts and Figures, 2010.
8. Carmona, R. H. A Report of the Surgeon General. *The Health Consequences of Involuntary Exposure to Tobacco Smoke;* US Department of Health and Human Services: Washington, DC, 2006.
9. Chen, M. J.; Shimada, T.; Moulton, A. D.; Cline, A.; Humphries, R. K.; Maizel, J.; Nienhuis, A. W. The Functional Human Dihydrofolate Reductase Gene. *J. Biol. Chem.* **1984,** *259* (6), 3933–3943.
10. Chen, M. J.; Shimada, T.; Moulton, A. D.; Harrison, M.; Nienhuis, A. W. Intronless Human Dihydrofolate Reductase Genes are Derived from Processed RNA Molecules. *Proc. Natl. Acad. Sci. USA.* **1982,** *79* (23), 7435–7439.
11. Clarke, S. J.; Rivory, L. P. Clinical Pharmacokinetics of Docetaxel. *Clin. Pharmacokinet.* **1999,** *36* (2), 99–114.
12. Clauditz, T. S.; Reiff, M.; Gravert, L.; Gnoss, A.; Tsourlakis, M. C.; Sauter, G.; Bokemeyer, C.; Knecht, R.; Wilczak, W. Human Epidermal Growth Factor Receptor 2 (HER2) in Salivary Gland Carcinomas. *Pathology* **2011,** *43* (5), 459–464.
13. Corn, B. W.; Marcus, S. M.; Topham, A.; Hauck, W.; Curran, W. J. Will Primary Central Nervous System Lymphoma be the Most Frequent Brain Tumor Diagnosed in the Year 2000? *Cancer* **1997,** *79* (12), 2409–2413.

14. Dahlman-Wright, K.; Cavailles, V.; Fuqua, S. A.; Jordan, V. C.; Katzenellenbogen, J. A.; Korach, K. S.; Maggi, A.; Muramatsu, M.; Parker, M. G.; Gustafsson, J. A. International Union of Pharmacology. LXIV. Estrogen Receptors. *Pharmacol. Rev.* **2006,** *58* (4), 773–781.
15. Daskalakis, V.; Varotsis, C. Binding and Docking Interactions of NO, CO and O_2 in Heme Proteins as Probed by Density Functional Theory. *Int. J. Mol. Sci.* **2009,** *10* (9), 4137–4156.
16. Davis, M.; Williams, R.; Chakraborty, J. Prednisone or Prednisolone for the Treatment of Chronic Active Hepatitis? A Comparison of Plasma Availability. *Br. J. Clin. Pharmacol.* **1978,** *5* (6), 501–505.
17. Djulbegovic, M.; Beyth, R. J.; Neuberger, M. M. Screening for Prostate Cancer: Systematic Review and Meta-analysis of Randomised Controlled Trials. *BMJ* **2010,** *341*, c4543.
18. Eby, N. L.; Grufferman, S.; Flannelly, C. M.; Schold, S. C.; Vogel, F. S.; Burger, P. C. Increasing Incidence of Primary Brain Lymphoma in the US. *Cancer* **1988,** *62* (11), 2461–2465.
19. FDA Approves New Drug for Advanced Kidney Cancer. May 30, 2007.
20. Fine, H. A.; Mayer, R. J. Primary Central Nervous System Lymphoma. *Ann. Intern. Med.* **1993,** *119* (11), 1093–1104.
21. Florescu, A.; Amir, E.; Bouganim, N.; Clemons, M. Immune Therapy for Breast Cancer in 2010—Hype or Hope? *Curr. Oncol.* **2011,** *18* (1), 9–18.
22. Frank, G. R. Role of Estrogen and Androgen in Pubertal Skeletal Physiology. *Med. Pediatr. Oncol.* **2003,** *41* (3), 217–221.
23. Funanage, V. L.; Myoda, T. T.; Moses, P. A.; Cowell, H. R. Assignment of the Human Dihydrofolate Reductase Gene to the q11----q22 Region of Chromosome 5. *Mol. Cell. Biol.* **1984,** *4* (10), 2010–2016.
24. Gerullis, H.; Ecke, T. H.; Imer, C.; Heuck, C. J.; Otto, T. mTOR-inhibition in Metastatic Renal Cell Carcinoma. Focus on Temsirolimus: A Review. *Minerva Urol. Nefrol.* **2010,** *62* (4), 411–423.
25. Giordano, S. H.; Cohen, D. S.; Buzdar, A. U.; Perkins, G.; Hortobagyi, G. N. Breast Carcinoma in Men: A Population-based Study. *Cancer* **2004,** *101* (1), 51–57.
26. Gorlova, O. Y.; Weng, S. F.; Zhang, Y. Aggregation of Cancer among Relatives of Never-Smoking Lung Cancer Patients. *Int. J. Cancer.* **2007,** *121* (1), 111–118.
27. Gupta, S.; Hastak, K.; Ahmad, N.; Lewin, J. S.; Mukhtar, H. Inhibition of Prostate Carcinogenesis in TRAMP Mice by Oral Infusion of Green Tea Polyphenols. *Proc. Natl. Acad. Sci. USA.* **2001,** *98* (18), 10350–10355.
28. Hardy, L. W.; Stroud, R. M.; Santi, D. V.; Montfort, W. R.; Jones, M. O.; Finer-Moore, J. S. Atomic Structure of Thymidylate Synthase: Target for Rational Drug Design. *Science* **1987,** *235* (4787), 448–455.
29. Herbst, R. S. Review of Epidermal Growth Factor Receptor Biology. *Int. J. Radiat. Oncol. Biol. Phys.* **2004,** *59* (2), 21–26. doi:10.1016/j.ijrobp.2003.11.041. PMID 15142631.
30. Herrlinger, U.; Schabet, M.; Bitzer, M.; Petersen, D.; Krauseneck, P. Primary Central Nervous System Lymphoma: From Clinical Presentation to Diagnosis. *J. Neurosurg.* **2000,** *92,* 261–266.
31. Hu, M. I.; Vassilopoulou-Sellin, R.; Lustig, R.; Lamont, J. P. Thyroid and Parathyroid Cancers. In *Cancer Management: A Multidisciplinary Approach,* 11th ed.; Pazdur, R.,

Wagman, L. D., Camphausen, K. A., Hoskins, W. J., Eds.; Cmp United Business Media: Banglore, 2008.

32. Izbicka, E.; Campos, D.; Carrizales, G.; Tolcher, A. Biomarkers for Sensitivity to Docetaxel and Paclitaxel in Human Tumor Cell Lines *In Vitro*. *Cancer Genom. Proteom.* **2005,** *2*, 219–222.
33. Jemal, A.; Elizabeth, W. Cancer Statistics. *Cancer J. Clin.* **2010,** *60* (5), 277–300.
34. Jun, L.; Laura, V. P.; Jianguo, F.; Salvador, R. N.; Boris, Z.; Arne, H. Inhibition of Mammalian Thioredoxin Reductase by Some Flavonoids: Implications for Myricetin and Quercetin Anticancer Activity. *Cancer Res.* **2006,** *66* (8), 4410–4418.
35. Kaneda, S.; Gotoh, O.; Shimizu, K.; Nalbantoglu, J.; Takeishi, K.; Seno, T.; Ayusawa, D. Structural and Functional Analysis of the Human Thymidylate Synthase Gene. *J. Biol. Chem.* **1990,** *265* (33), 20277–20284.
36. Konstantopoulou, M.; Lord, M. G.; Macfarlane, A. W. Treatment of Invasive Squamous Cell Carcinoma with 50-percent Imiquimod Cream. *Dermatol. Online J.* **2006,** *12* (3), 10.
37. Kumar, P.; Clark, M. *Clinical Medicine,* 6th ed.; Elsevier: London, 2005; pp 683–648.
38. Lacroix, M. Significance, Detection and Markers of Disseminated Breast Cancer Cells. *Endocr. Relat. Cancer* **2006,** *13* (4), 1033–1067.
39. Lambrou, G. I.; Vlahopoulos, S.; Papathanasiou, C.; Papanikolaou, M.; Karpusas, M.; Zoumakis, E.; Tzortzatou-Stathopoulou, F. Prednisolone Exerts Late Mitogenic and Biphasic Effects on Resistant Acute Lymphoblastic Leukemia Cells: Relation to Early Gene Expression. *Leuk Res.* **2009,** *33*, 1684–1695.
40. Leo, A.; Hansch, C.; Elkins, D. Partition Coefficients and their Uses. *Chem. Rev.* **1971,** *71* (6), 525–616.
41. Levin, E. R. Integration of the Extranuclear and Nuclear Actions of Estrogen. *Mol. Endocrinol.* **2005,** *19* (8), 1951–1959.
42. Li, P.; Wood, K.; Mamon, H.; Haser, W.; Roberts, T. Raf-1: A Kinase Currently Without a Cause but Not Lacking in Effects. *Cell* **1991,** *64* (3), 479–482.
43. Lien, E.; Ingalls, R. R. Toll-like Receptors. *Crit. Care Med.* **2002,** *30* (1), 1–11.
44. Lipinski, C. A.; Lombardo, F.; Dominy, B. W.; Feeney, P. J. Experimental and Computational Approaches to Estimate Solubility and Permeability in Drug Discovery and Development Settings. *Adv. Drug Del. Rev.* **2001,** *46*, 3–26.
45. Lu, N. Z.; Wardell, S. E.; Burnstein, K. L.; Defranco, D.; Fuller, P. J.; Giguere, V.; Hochberg, R. B.; McKay, L.; Renoir, J. M.; Weigel, N. L.; Wilson, E. M.; McDonnell, D. P.; Cidlowski, J. A. International Union of Pharmacology. LXV. The Pharmacology and Classification of the Nuclear Receptor Superfamily: Glucocorticoid, Mineralocorticoid, Progesterone, and Androgen Receptors. *Pharmacol. Rev.* **2006,** *58* (4), 782–797.
46. Luscombe, N. M.; Greenbaum, D.; Gerstein, M. What is Bioinformatics? An Introduction and Overview. *Yearb. Med. Inform.* **2001,** *1*, 83–99.
47. Margaret, H. R.; Neil, A. F.; Leigh, A. S.; Frank, L. G. Histologic Variants of Squamous Cell Carcinoma of the Skin. *Cancer Control* **2001,** *8* (4), 354–363.
48. Martin, K.; Andrew, J. M. Molecular Dynamics Simulations of Biomolecules. *Nat. Struct. Biol.* **2002,** *9*, 646–652.
49. Massarweh, S.; Osborne, C. K.; Creighton, C. J.; Qin, L.; Tsimelzon, A.; Huang, S.; Weiss, H.; Rimawi, M.; Schiff, R. Tamoxifen Resistance in Breast Tumors is Driven by Growth Factor Receptor Signaling with Repression of Classic Estrogen Receptor Genomic Function. *Cancer Res.* **2008,** *68* (3), 826–383.

50. Michael, L.; Susanna, M.; Annika, J.; Matthias, S.; Alexander, S.; Katjana, O.; Axel, G.; Christopher, B.; Marlies, V.; Horst, U.; Ingo, G. H.; Hendrik, P.; Uwe, S. Association of Genetic Variants of Methionine Metabolism with Methotrexate-induced CNS White Matter Changes in Patients with Primary CNS Lymphoma. *Neuro Oncol.* **2009,** *11*, 2–8.
51. Miller, D. C.; Hafez, K. S.; Stewart, A.; Montie, J. E.; Wei, J. T. Prostate Carcinoma Presentation, Diagnosis, and Staging: An Update from the National Cancer Data Base. *Cancer* **2003,** *98* (6), 1169–1178.
52. Minna, J. D.; Schiller, J. H. *Harrison's Principles of Internal Medicine,* 17th ed.; McGraw-Hill: New York, NY, 2008; pp 551–562.
53. Mok, T. S. Gefitinib or Carboplatin-paclitaxel in Pulmonary Adenocarcinoma. *N. Eng. J. Med.* **2009,** *361*, 947–957.
54. Mol, F.; Mol, B. W.; Ankum, W. M.; Van Der Veen, F.; Hajenius, P. J. Current Evidence on Surgery, Systemic Methotrexate and Expectant Management in the Treatment of Tubal Ectopic Pregnancy: A Systematic Review and Meta-analysis. *Hum. Reprod. Update* **2008,** *14* (4), 309–319.
55. Mulders, P. F.; Brouwers, A. H.; Hulsbergen-van der Kaa, C. A.; van Lin, E. N.; Osanto, S.; de Mulder, P. H. Guideline 'Renal Cell Carcinoma (in Dutch; Flemish)'. *Ned. Tijdschr. Geneeskd.* **2008,** *152* (7), 376–380.
56. Pao, W.; Miller, V.; Zakowski, M. EGF Receptor Gene Mutations are Common in Lung Cancers from "Never Smokers" and are Associated with Sensitivity of Tumors to Gefitinib and Erlotinib. *Proc. Natl. Acad. Sci. USA* **2004,** *101* (36), 13306–13311.
57. Petros, A. M.; Medek, A.; David, G. N.; Daniel, H. K.; Yoon, H. S.; Kerry, S.; Edmund, D. M.; Tilman, O.; Stephen, W. F. Solution Structure of the Antiapoptotic Protein Bcl-2. *Proc. Natl. Acad. Sci. USA* **2001,** *98* (6), 3012–3017.
58. Ravi Rajagopalan, P. T.; Zhang, Z.; McCourt, L.; Dwyer, M.; Benkovic, S. J.; Hammes, G. G. Interaction of Dihydrofolate Reductase with Methotrexate: Ensemble and Single-molecule Kinetics. *Proc. Natl. Acad. Sci. USA* **2002,** *99* (21), 13481–13486.
59. Ranelletti, F. O.; Maggiano, N.; Serra, F. G. Quercetin Inhibits p21-RAS Expression in Human Colon Cancer Cell Lines and in Primary Colorectal Tumors. *Int. J. Cancer* **2000,** *85*, 438–445.
60. Rang, H. P.; Dale, M. M.; Ritter, J. M.; Moore, P. K. The Pituitary and the Adrenal Cortex. In *Pharmacology,* 5th ed.; Hunter, L., Ed.; Churchill Livingstone: London, 2003; pp 413–415.
61. Raudrant, D.; Rabe, T. Progestogens with Antiandrogenic Properties. *Drugs* **2003,** *63* (5), 463–492.
62. Rhen, T.; Cidlowski, J. A. Antiinflammatory Action of Glucocorticoids: New Mechanisms for Old Drugs. *N. Engl. J. Med.* **2005,** *353* (16), 1711–1723.
63. Rini, B. I.; Rathmell, W. K.; Godley, P. Renal Cell Carcinoma. *Curr. Opin. Oncol.* **2008,** *20* (3), 300–306.
64. Saman, W.; Seppo, P.; Toru, N.; Victor, R. P.; Markku, P.; Heidi, K.; Onni, N. Demonstration of Ethanol-induced Protein Adducts in Oral Leukoplakia (Pre-cancer) and Cancer. *J. Oral Pathol. Med.* **2008,** *37* (3), 157–165.
65. Sariego, J. Breast Cancer in the Young Patient. *Am. Surg.* **2010,** *76* (12), 1397–1400.
66. Saurabh, S.; Sanjaykumar, C.; Prashant, S.; Gomase, V. S. Computational Approach towards the B-Cell Lymphoma-2 Protein: A Noticeable Target for Cancer Proteomics. *J. Pharmacol. Res.* **2010,** *1* (1), 1–8.

67. Scheinfeld, N. Three Cases of Toxic Skin Eruptions Associated with Methotrexate and a Compilation of Methotrexate-induced Skin Eruptions. *Dermatol. Online J.* **2006,** *12* (7), 15.
68. Shintani, S.; Li, C.; Mihara, M.; Nakashiro, K.; Hamakawa, H. Gefitinib ('Iressa'), an Epidermal Growth Factor Receptor Tyrosine Kinase Inhibitor, Mediates the Inhibition of Lymph Node Metastasis in Oral Cancer Cells. *Cancer Lett.* **2003,** *201* (2), 149–155.
69. Smith, M. A. Secondary Leukemia or Myelodysplastic Syndrome After Treatment with Epipodophyllotoxins. *J. Clin. Oncol.* **2001,** *17* (2), 569–577.
70. Sordella, R.; Bell, D. W.; Haber, D. A.; Settleman, J. Gefitinib-sensitizing EGFR Mutations in Lung Cancer Activate Anti-apoptotic Pathways. *Science* **2004,** *305* (5687), 1163–1167.
71. Sridhar, S. S.; Hedley, D.; Siu, L. L. Raf Kinase as a Target for Anticancer Therapeutics. *Mol. Cancer Ther.* **2005,** *4* (4), 677–685.
72. Srivastava, V.; Gupta, S. P.; Siddiqi, M. I.; Mishra, B. N. Molecular Docking Studies on Quinazoline Antifolate Derivatives as Human Thymidylate Synthase Inhibitors. *Bioinformation* **2010,** *4* (8), 357–365.
73. Stams, W. A.; den Boer, M. L.; Beverloo, H. B. Expression Levels of TEL, AML1, and the Fusion Products TEL-AML1 and AML1-TEL Versus Drug Sensitivity and Clinical Outcome in t(12;21)-Positive Pediatric Acute Lymphoblastic Leukemia. *Clin. Cancer Res.* **2005,** *11* (8), 2974–2980.
74. Susan, J. H.; Dirkjan, H.; George, F. M.; Thomas, S. K.; Adam, W. C.; Ilse, G. M.; Carl, F. S.; Danielle, M. M.; Chrysalyne, S.; Rachael, A. C. Imiquimod Enhances IFN-γ Production and Effector Function of T Cells Infiltrating Human Squamous Cell Carcinomas of the Skin. *J. Invest. Dermatol.* **2009,** *129* (11), 2676–2685.
75. Takimoto, C. H.; Calvo, E.; Pazdur, R.; Wagman, L. D.; Camphausen, K. A.; Hoskins, W. J. Principles of Oncologic Pharmacotherapy. In *Cancer Management: A Multidisciplinary Approach;* 11th ed. UBM Medica: London, UK, 2008, pp 42–58.
76. Thompson, I. M.; Pauler, D. K.; Goodman, P. J. Prevalence of Prostate Cancer among Men with a Prostate-specific Antigen Level < or = 4.0 ng Per Milliliter. *N. Engl. J. Med.* **2004,** *350* (22), 2239–2246.
77. Thun, M. J.; Hannan, L. M.; Adams-Campbell, L. L.; Adami Hans, O. Lung Cancer Occurrence in Never-Smokers: An Analysis of 13 Cohorts and 22 Cancer Registry Studies. *PLoS Med.* **2008,** *5* (9), e185.
78. Venturi, S.; Venturi, M. Iodine in Evolution of Salivary Glands and in Oral Health. *Nutr. Health* **2009,** *20* (2), 119–134.
79. Wan, X.; Shen, N.; Mendoza, A.; Khanna, C.; Helman, L. J. CCI-779 Inhibits Rhabdomyosarcoma Xenograft Growth by an Antiangiogenic Mechanism Linked to the Targeting of mTOR/HIF-1alpha/VEGF Signaling. *Neoplasia* **2006,** *8* (5), 394–401.
80. Werning John, W. *Diagnosis, Management, and Rehabilitation: Oral Cancer;* Thieme Publishers: New York, NY, 2007. ISBN 978-1588903099.
81. Wilhelm, S. M.; Adnane, L.; Newell, P.; Villanueva, A.; Llovet, J. M.; Lynch, M. Preclinical Overview of Sorafenib, a Multikinase Inhibitor that Targets Both Raf and VEGF and PDGF Receptor Tyrosine Kinase Signaling. *Mole. Cancer Ther.* **2008,** *7* (10), 3129–3140.
82. Yager, J. D.; Davidson, N. E. Estrogen Carcinogenesis in Breast Cancer. *N. Engl. J. Med.* **2006,** *354* (3), 270–282.

83. Yao, V.; Berkman, C. E.; Choi, J. K.; O'Keefe, D. S.; Bacich, D. J. Expression of Prostate-Specific Membrane Antigen (PSMA), Increases Cell Folate Uptake and Proliferation and Suggests a Novel Role for PSMA in the Uptake of the Non-polyglutamated Folate, Folic Acid. *Prostate* **2010,** *70* (3), 305–316.
84. Yin, X. M.; Oltvai, Z. N.; Korsmeyer, S. J. BH1 and BH2 Domains of Bcl–2 are required for Inhibition of Apoptosis and Heterodimerization with Bax. *Nature* **1994,** *369*, 321–323.
85. Yu, H.; Adedoyin, A. ADME-Tox Drug Discovery: Integration of Experimental and Computational Technologies. *Drug Discov. Today* **2003,** *8*, 852–861.

CHAPTER 8

THE TRENDS OF INDONESIAN ETHANOL PRODUCTION: PALM PLANTATION BIOMASS WASTE

TEUKU BEUNA BARDANT[1], HERU SUSANTO[1,2*], and INA WINARNI[3]

[1]*Indonesian Institute of Science, Jakarta, Indonesia*

[2]*Department of Information Management, College of Management, Tunghai University, Taichung City, Taiwan, ROC*

[3]*Forest Products Research and Development Center, Bogor, Indonesia*

[*]*Corresponding author. E-mail: heru.susanto@lipi.go.id*

CONTENTS

ABSTRACT

The emergence of innovative chemistry informatics (cheminformatics) is threatening the sustainability of bioethanol production from palm oil empty fruit bunch in Indonesia. With the friendliness and convenience offered by cheminformatics, many bioethanol production processes nowadays prefer to determining optimum condition and forecasting the results computationally through response surface methodology (RSM). RSM is one of cheminformatics[1] that explores the relationships between several explanatory variables and one or more response variables. Indonesia had established bioethanol production pilot plant by using fully automatic-computerized-system supported by sufficient ICT emerging technology and database. Bioethanol production from cellulose-based materials is one of the popular bioprocess engineering research field in Indonesia in the last 5 years since palm plantation biomass waste became the most concern potency for bioethanol raw material since Indonesia is the world's leading producer of palm oil. The purpose of this chapter is to assess the impact of RSM as a practical cheminformatics to support bioprocess technology and bioethanol technology processes by optimizing the operations variables and easily adopted in automatic-computerized-system ICT emerging technology.

8.1 INTRODUCTION

The term cheminformatics was born referring to the combination of chemistry and informatics. Initially, this field of chemistry did not have a name until 1998, when the term cheminformatics was coined by Frank K. Brown. He defined it as "...mixing of those information resources to transform data into information and information into knowledge for the intended purpose of making better decisions faster in the area of drug lead identification and optimization." The recent development of cheminformatics makes it possible to be applied to data analysis for various industries like paper and pulp, dyes, and other allied industries.

Response surface methodology (RSM-CI) is one of cheminformatics that explores the relationships between several explanatory variables and one or more response variables. RSM was introduced by George. E. P Box and K. B. Wilson with the main objective of obtaining an optimal response.

[1]Called by RSM-CI; Response Surface Methodology-Chemistry Informatics.

Box and Wilson suggested a second order polynomial equation as the model template. RSM had already familiar to researchers in bioprocess engineering which usually deals with long-time microbial growth and required to consider many operations' variables. RSM will significantly increase the research efficiency for determining optimum condition and forecasting the results. Bioethanol production from cellulose-based materials is one of popular bioprocess engineering research field in Indonesia in the last 5 years. Palm plantation biomass waste became the most important concern potency for bioethanol raw material, as Indonesia is the world's leading producer of palm oil.

Currently, Indonesia had established ethanol production pilot plant by using palm oil empty fruit bunch (EFB) as a raw material in a fully automatic-computerized-system. Introducing cheminformatics approach by presenting optimum condition in the mathematical equation will give a significant advantage. The mathematical model can easily be applied in the automatic-computerized-system which is supported by sufficient database.

This study emphasizes the application of RSM as a practical cheminformatics to support bioprocess technology by optimizing the operations variables in terms that are easily adopted in automatic-computerized-system, mathematical model. Ethanol production process from palm plantation biomass waste in Indonesia is an interesting case study of how this cheminformatics is applied. In Section 8.2, a brief explanation about the potency of Indonesian palm oil plantation biomass waste is given. In Section 8.3, a brief explanation of bioethanol production process from cellulose-based materials is delivered. The crucial variables to be optimized were substrate loading, enzyme loading, and surfactant addition dose. The importance of surfactant additions, the third variable that makes RSM application required, will be explained in Section 8.4. Practical examples of optimization using RSM in bioethanol production from palm oil EFB will be described in Section 8.5. The mathematical model modification from EFB enzymatic hydrolysis will be applied to palm oil trunk enzymatic hydrolysis. The modification process and its application will be discussed Section 8.6. Quoting George E. P. Box statement, "essentially, all models are wrong, but some are useful." Thus, the discussion will be concluded by a summary of some current limitations of the mathematical model obtained by using cheminformatics and future prospects of cheminformatics in dealing with limitations and the application possibilities in broader fields, other than drug discoveries, and its related topics.

8.2 INDONESIAN POTENCY IN PALM PLANTATION BIOMASS

Today, Indonesia is the largest palm oil producer in the world. Palm oil plantations have been developed since 1968 from only 79,209 Ha that belonged to the Indonesian government and the other 40,451 Ha that belonged to the private companies. Small holder industrialists started to contribute in producing palm oil in 1979, mainly as the impact of transmigration program created by the government which triggered agro-industries outside Java island. Along with the plantation expansion as shown in Figure 8.1, the production kept on increasing. In 2015, Indonesian total palm oil production was estimated to be 30,948.931 tons in which only 2.49% came from the government plantations (Directorate General of Estate 2014).

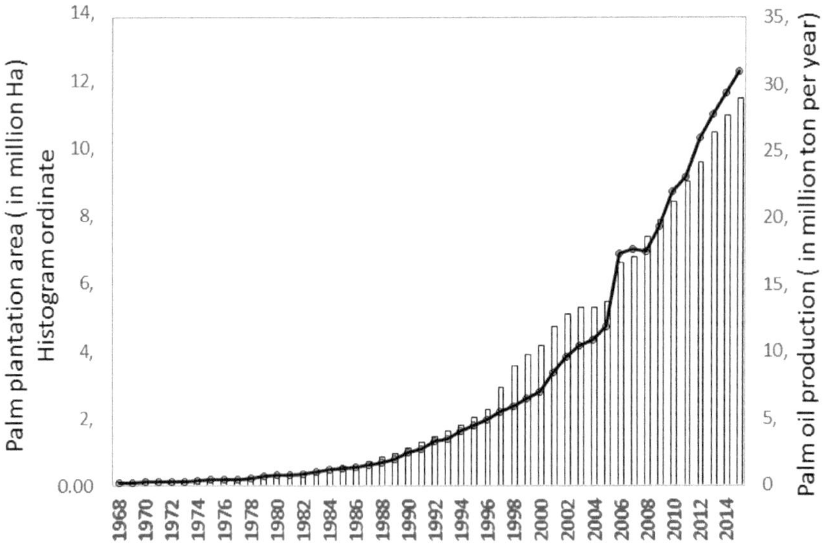

FIGURE 8.1 Plantation expansion and production level.

The EFB potency can be extrapolated from palm oil production, as the EFB composition in fresh fruit bunch is similar to palm oil, 22–23% w.[6] This huge potency was waiting to be explored for the sake of Indonesian prosperity. The most efficient way to detach palm fruit from its bunch is by steam cooking the fresh bunch in an autoclave. This process caused the EFB to be available in wet conditions which became the main challenge for utilizing it as a construction material or solid fuel by direct combustion just like palm kernel. Recent utilization of EFB was as in situ composting materials which consume transportation cost from extraction plant to the plantations. The utilization is usually limited to the plantations that have the same owner with the extraction plant. Bioethanol production from EFB next to palm oil extraction plant gave a promising profitable solution. EFB as the wet raw material will not be a problem since the pulping process will boil it in soda solution or through other thermomechanical processes. Energy produced from kernel and fiber combustion exceeds the requirement of extraction plants in which the latest technology is installed. This excess energy can be used by bioethanol production in pulping and distillation unit. Recently, some extraction plants use the excess energy for generating power plant and supply electricity to the adjacent dwellings. This utilization was limited by the wired infrastructure for distributions since the plant and plantation are usually placed in a relatively remote area. Converting the excess energy in the form of bioethanol will give cheaper infestations in distribution infrastructure by using the trucking system.

The other potential palm plantation biomass waste was the palm trunk. To keep the plantation productivity at its best, the palm tree cannot be older than 20 years; thus, regeneration program needs to be planned for the sake of sustainability. As can be seen in Figure 8.1, by assuming there will be no relocation in the existing plantation area, then theoretically there will be 224,528 Ha of palm oil plantation in regeneration program in all around Indonesia in 2016, similar with the number of plantations developed in 1996. Palm oil plantations have 136–143 trees per Ha with the average weight of tree being 330 kg.[6] Thus, the potency of palm trunk as biomass stock is 10–10.6 million ton in all over Indonesia in 2016. Since the expansion keeps on happening since 1997, the palm trunk potency also climbed up—not to mention the new generation of short-frond palm oil trees that gave the possibility to plant more palm trees per hectare. The main obstacle of palm trunk utilization is its low mechanical strength that makes it not suitable for timber and furniture industries, except for particle board. Challenge for particle board industries in Indonesia is the lack of resin availability in the domestic market. Developing new resin factories in Indonesia is still not feasible due to raw material unavailability and environmental law.

Utilization of palm plantation biomass waste as raw material for ethanol production is still feasibly competitive and strongly supported by the Indonesian government. One of the significant supports is to develop pilot plants in PP Kimia LIPI. Reports about the pilot plant performance were already delivered in a previous study.[4] In the next section, the whole production process of bioethanol from cellulosic biomass as it was applied in the pilot plant is briefly described.

8.3 ETHANOL PRODUCTION FROM CELLULOSIC BIOMASS WASTE: THE BASICS

Basically, the process consists of four steps: pretreatment which converts biomass waste to pulp; enzymatic hydrolysis which converts pulp to glucose; fermentation which converts glucose to ethanol; and distillation to purify the ethanol to meet fuel grade. Figure 8.2 gives a brief description of how these four steps are connected. For converting biomass waste, the most common material is wood-based materials; converting it into pulp is an ancient technology that defines civilization. Since Tsai Lun invented paper, the production technology has really evolved and been established. This is a promising support for ethanol production from biomass. Room for improvement lies in increasing process efficiency by utilizing heat and chemicals as least as possible to meet the next steps' requirements.

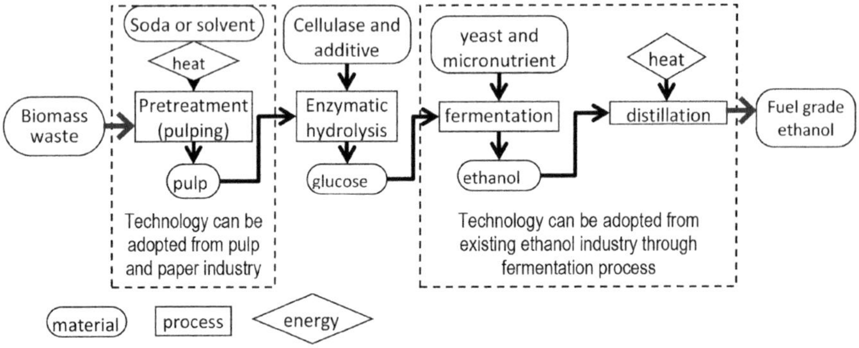

FIGURE 8.2 Flowchart of bioethanol production from biomass waste.

Fermentation and distillation technologies were also acknowledged since the history of humankind. Record and evidence of it can be found in every ancient civilization, mainly in Egypt and China. This long-lived technology had constantly improved so that nowadays, many industries are offering fermentation and distillation licenses. In terms to meet fuel grade requirement, ethanol–water mixture needs to be separated above their azeotrope point. Mostly, newly developed technologies create improvement in this part of the process to make the separation more efficient. One of the popular approaches was to use mineral-based adsorbent to withdraw water.

Thus, enzymatic hydrolysis is the unique and crucial step in the whole process which determines the economic feasibility. The hydrolysis rate decreases during the process which leads to decreased yields and prolonged times lead to higher production cost. (; A close relative to be benchmark was starch-based bioethanol production. In 1998, the portion of enzyme cost in ethanol production from starch in Brazil was up to 40% of the total cost, higher than raw material, utilities, and other expenses which were 20%, 28%, and 12%, respectively. Through continuous research, the portion of enzyme cost became less than 8% although there were contributions of increasing raw material prices and fuel which make their portion 32% and 35%, respectively. Research in developing high activity of amylase had reached the efficiency of 1 g of enzyme (protein equivalent) for 1 gallon of ethanol. Recently, cellulase still lies in 100 g for a gallon of ethanol.

Novozymes became the leading industry in the race for providing reliable cellulase in commercial scale. They claimed that their cellulase can be sold in US$ 2 per gallon which makes ethanol production from corncob as feasible as if it was produced from cassava (Novozymes press release February 15, 2010). Danisco Inc. as the competitor also introduced their latest cellulase in a similar period of time. As quoted by Reuters, Danisco Spokesman, Rene Tronborg, claimed that by using their cellulase, bioethanol production from hayes has feasibilities comparable with the production process using corn. This statement was announced in *Renewable Fuels Association's 15th Annual National Ethanol Conference* in Orlando, Florida. This event was a follow-up action of Barack Obama's statement to reduce dependency on fossil fuel at the early of 2010.

Another approach can be used in developing effective cellulose-to-ethanol process for liquid fuel application. To improve yields and reduce reduction time and yet reducing the production cost, several technical approaches were studied. Conducting simultaneous hydrolysis and fermentation process for bioethanol production had already been done by using pulp of palm oil EFB.[1,3] Kinetic study of enzymatic hydrolysis process had already conducted to synchronize the hydrolysis reaction rate with the fermentation rate.[2] This study had intensively developed to a pilot scale unit.[4] Conducting hydrolysis and fermentation simultaneously gave significant advantages compared to hydrolysis and fermentation in series.[5]

The sense of chemical engineering designing economically feasible process needs to be involved in each approach, which is seldom overlooked by chemists and biologists while developing enzymes. There is a huge opportunity to adapt existing pulping, fermentation, and distillation technology, as shown in Figure 8.2, which will contribute to lower investment cost and maintenance. Thus, it is easier to calculate depreciation which leads to lower overall production cost. To adopt existing-commercially-reliable distillation unit for ethanol purification, hydrolysis, and fermentation need to be designed for producing fermentation broth with suitable specification. One of Indonesia's well-known ethanol producer, PT. Madubaru Yogyakarta, produces ethanol from molasses setting 8–9% v ethanol concentration in its feed specification for the distillation unit operating normally. The distillation unit itself was designed to purify fermentation broth which contains 5% v of ethanol at its lowest. Feed with less ethanol concentration cannot be purified by using this similar unit. That is why in this study, 8% v was chosen as a benchmark in model validation and will be described later. The most logical approach for increasing ethanol concentration in fermentation broth was by loading more substrates in the hydrolysis process. The theoretical conversion of glucose to ethanol through fermentation using *Saccharomyces cerevisiae* is 51%. Thus, by including cellulose content in pulp into account, the pulp loading in hydrolysis process need to be over 20% w of reaction system.

Performing cellulose hydrolysis experiments with high loading substrate is very difficult by only relying on usual laboratory apparatus, such as orbital shaker or magnetic stirrer as reported by previous study. Despite of this technical challenge, high loading substrate also needs to consider substrate inhibitions. This is some unique phenomena in reaction system that assisted by enzyme which is more reactant does not always mean it will result in more products. In case of cellulose hydrolysis, the enzyme can be trapped within cellulose fibers if the fibers are not dispersed adequately, which has happened in reaction with high loading cellulose substrate. Not to mention the more negative impact caused by lignin impurities since more lignin is in the reaction system.

8.4 SURFACTANT ADDITION FOR ENHANCING CELLULOSE HYDROLYSIS PERFORMANCE: THE THIRD VARIABLE

Performing enzymatic hydrolysis of cellulosic biomass with high loading substrate implicated complex variable considerations since viscosity and substrate inhibitions will affect the end results. High enzyme concentrations are needed to reach high cellulose conversion, and enzyme recycling is difficult due to adsorption of enzymes to residual lignocelluloses. Enhancement of cellulose hydrolysis performance by adding surfactants to the hydrolysis mixture has been reported by several authors. Increased hydrolysis yield by addition of surfactants has been reported for delignified steam-exploded wood, bagasse, and corn stover. Appearance of lignin impurities also affects the process since lignin can deactivate cellulase, thus reduce the performance of overall enzymatic hydrolysis. Surfactant addition to hydrolysis process can reduce the destructive effect of lignin impurities. The effect of surfactant on enzymatic hydrolysis of steam-pretreated spruce (SPS), with relatively high lignin impurities, was the object of intense research for use in an ethanol producing process.

Surfactant addition in the hydrolysis process with high loading substrates using two naturally different raw materials, water hyacinth pulp, and EFB pulp, were observed. As it can be seen in Figure 8.3 for the cellulose conversion of water hyacinth pulp, addition of TWEEN 20, and SPAN 85 gave significant boost for cellulase. In cellulose hydrolysis conducted by 15 FPU/g-pulp cellulose, water hyacinth pulp produced by only boiling it in 1 M NaOH in atmospheric pressure gave 45.48% w cellulose conversion which can be increased to 52.29% w by producing the pulp in autoclave (2 Atm pressure). Better results can be obtained by adding 1% v of TWEEN 20 to the water hyacinth boiled-pulp which gave 56.39% w. TWEEN 20 not only saves production cost by reducing energy for pulping process, but also reduces the enzyme loading. Addition of 1% v TWEEN 20 and SPAN 85 to water hyacinth boiled-pulp hydrolysis process gave 51% and 53.79% w cellulose conversion, respectively, by only using 10 FPU/g-pulp. Thus, all studies agreed that surfactant additions would improve the performance of hydrolysis step. The next issue then will be how the surfactants will affect the next step, fermentation process.

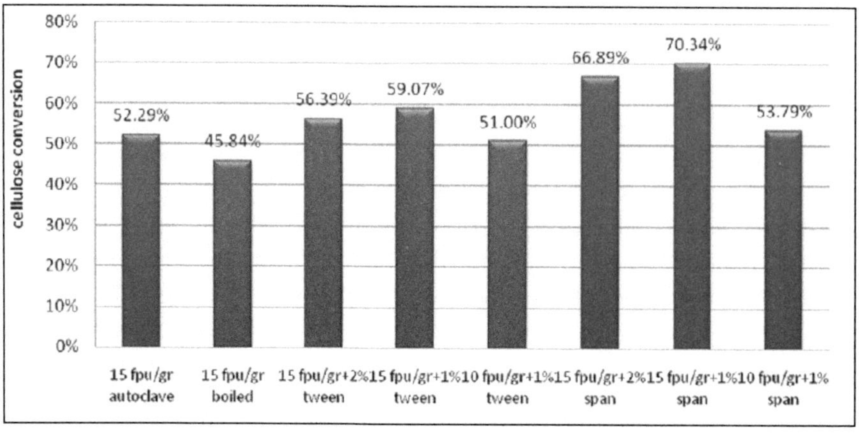

FIGURE 8.3 Comparison of TWEEN 20 and SPAN 85 addition in increasing cellulose conversion of enzymatic hydrolysis reaction to the reaction that is using pulp produced by autoclave. Surfactant additions not only save production cost by reducing energy for pulping process but also reduce the enzyme loading.

Surfactants were commonly introduced in fermentation systems to avoid foam generation in the systems which can interfere with the monitoring and measurement tools, and assist contamination by plugging the ventilation system. There are three types of surfactants; anionic, cationic, and non-ionic surfactants. Most of the microbe membrane surfaces have electronegative charge and will be affected by anions and cationic surfactants. The toxicity of surfactant to microbial cell is attributed to an extraction of lipids in the cell membrane. That is why these surfactants are also usually used as bactericides, such as quaternary ammonium compounds. There was a rule of thumb proposed to predict surfactant toxicity based on its hydrophilic–lipophilic balance (HLB). Generally, increasing HLB tends to increase surfactant toxicity. As shown in previous study in fermentation of butanol production, surfactants with HLB value higher than 12 were toxic to microbes while surfactant with HLB in range 1–7 was not toxic. Nonionic surfactants, such as linear alcohol ethoxylates, are biodegradable and show low toxicity in an aquatic environment. On the other hand, a fortunate coincidence has been proven that the most effective surfactant that has been used as addition in cellulose enzymatic hydrolysis is nonionic surfactant. Thus, addition of nonionic surfactant for enhancing ligno-cellulose enzymatic hydrolysis followed by fermentation in a system with no prior sterilization is preferable. The surfactant effect is higher at low cellulase concentration.

The mechanism of surfactant effects in ligno-cellulose hydrolysis was extensively studied. It was concluded that nonionic surfactants could enhance the cellulase performance by overcoming the negative effect of cellulose crystalline. Different mechanisms have been proposed for explaining positive effect of surfactant addition to an enzymatic hydrolysis of cellulose. The surfactant could change the nature of the substrate, for example, by increasing the available cellulose surface. Surfactant effects on enzyme–substrate interaction have been proposed by preventing it from inactivation enzymes that adsorbed into cellulose fiber. Surfactant addition facilitates desorption of enzymes from substrate. Even previous studies showed that surfactant could overcome the effect of cellulose crystalline; it should be noted that all the previous results were obtained from pure cellulose.

The effect of lignin impurities in pulp is still beyond these explanations which have important role in utilizing pulp as raw material for bioethanol production process. Considering the effect of lignin impurities takes the effectiveness of pretreatment into account. Lignin-free pulp required high energy and/or chemical loading in the pretreatment, thus, high production cost. Defining optimum surfactant addition in the hydrolysis process can reduce the requirement in pretreatment process and leads to reduce production cost.

Substrate loading and surfactant additions have some similarities. Adding too much substrate will cause substrate inhibition in hydrolysis process and thus, lower the conversion but adding a few substrates will cause higher energy required in the purification. Adding too much surfactant will give lethal effect for the yeast in the next process but adding a few surfactants cannot counter the negative impact of lignin impurities. Thus, defining optimum condition of surfactant doses will be as important as defining feasible substrate loading. Surfactant doses become the third variable to be considered besides substrate loading and enzyme loading.

8.5 RSM-CI APPLICATION IN DEFINING OPTIMUM CONDITION OF EFB ENZYMATIC HYDROLYSIS PROCESS

The appearance of surfactant doses as the third variable makes classical method in determining optimum condition no longer efficient. The classical method of optimization involves varying one parameter at a time and keeping the other constant. In two-variable systems, whether or not there were interrelations between variables, classical method will give same optimum condition, regardless of which variable was firstly set as constant. In system with three variables or more, classical method become inefficient and fails to explain relationship between the variables.

RSM was used in several optimizations in enzymatic reactions. This methodology was recommended due to its ability to consider multivariable problems. In this study, ester sorbitan (mixture of 1,4-anhydrosorbitol, 1,5-anhydrosorbitol, and 1,4,3,6-dianhydrosorbitol) also known as SPAN 85 was the chosen commercial surfactant based on one-point-test as shown in Figure 8.4. SPAN 85 has HLB value only 1.8. These results support the application of SPAN 85 as a nonionic surfactant in cellulase hydrolysis. The concerned hydrolysis process is the one that was followed by fermentation without any treatment for removing or reducing the effect of surfactant and cellulase prior fermentation.

In the hydrolysis system, which uses 10 FPU/g-substrates, addition of 1% v of SPAN 85 could increase the cellulose conversion from 38.55% to 87.30% and clearly more effective than TWEEN 20. This result is comparable with the hydrolysis system, which used 15 FPU/g-substrates. Thus, based on one-point-test, surfactant additions can save enzyme's consumption up to 33%. Since the enzyme cost still become a big part of overall

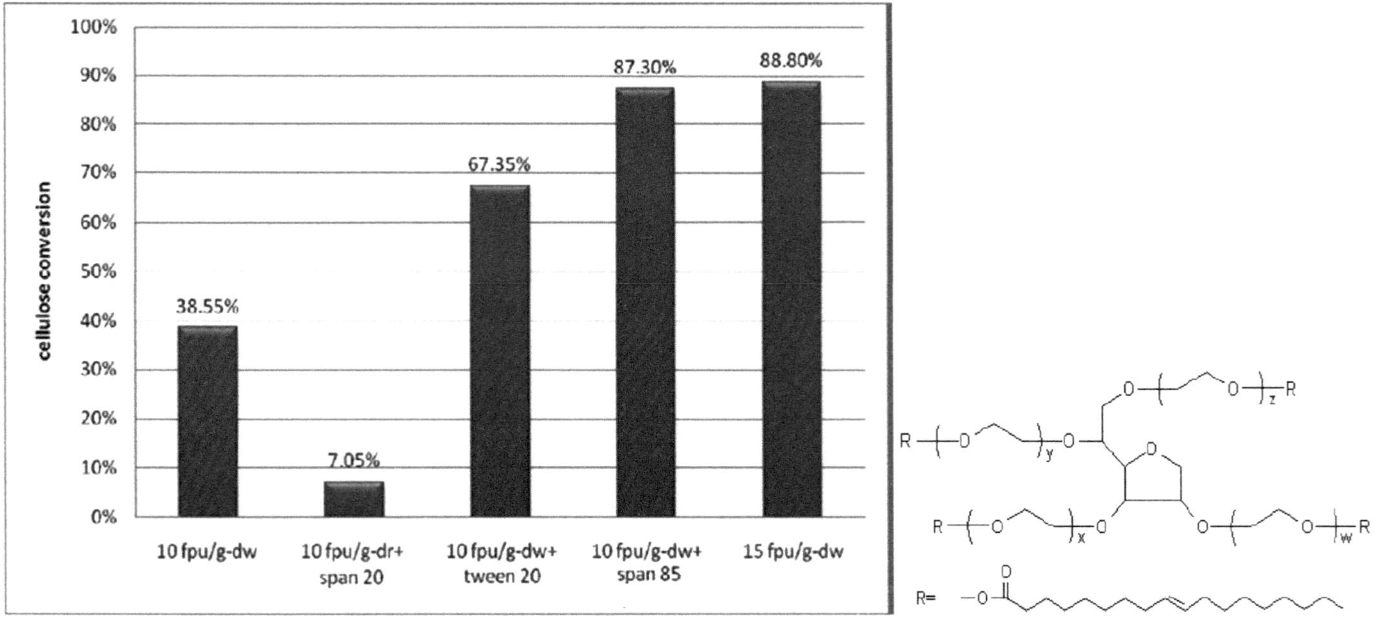

FIGURE 8.4 Left: The effect of TWEEN 20 and SPAN 85 addition in increasing cellulose conversion of enzymatic hydrolysis reaction. Right: The molecular structure of SPAN 85, surfactant that was used in this study.

production cost, optimization of surfactant addition in cellulase hydrolysis reaction will be observed by using RSM method with the main objective to save more enzyme.

The application of RSM method was started by creating a three-level-three-factor central composite rotatable design (CCRD). The variables and its levels selected for the cellulase hydrolysis were SPAN 85 concentration (0–2% v), cellulase concentration (10–15 FPU/g substrate-dw), and substrate loading (20–30% dw). Substrate was palm oil EFB pulp made from EFB from Malimping, Indonesia by using Kraft pulping with lignin impurities 9.34% w. Commercial cellulase Novozymes was used and SPAN 85 was purchased from Merck. Fermentation processes were conducted in same condition, by adding 1% w of powdered yeast, 0.3% w of urea, and 0.1% w of NPK. Other affecting variables, such as operating temperature was kept constant at Indonesian room temperature (25–28°C) and pH solution was set at 4.8 by using 50 mM sodium acetate buffer. All process was conducted in a system with a total volume of 100 mL. Duration for hydrolysis and fermentation was set at 48 and 60 h, respectively. In this study, the observed variables were arranged as tabulated in Table 8.1, which use the least number of required experiments.

The data obtained from experiments were resulted reducing sugar, ethanol concentration, and ethanol conversion. The experimental results, thus called actual value, for respective variables are also shown in Table 8.1. Resulted reducing sugar is the measured reducing sugar concentration which was obtained after the enzymatic hydrolysis process, got completed, as the feed for fermentation process. The resulted ethanol concentration is the ethanol concentration in the fermentation broth that was measured after the fermentation process got completed. The ethanol conversion was calculated as percentage weight by comparing the resulted ethanol weight to the respective pulp weight as the process feed.

All of these dependent variables went through model-fitting process to a second-order polynomial equation by using software for statistic SPSS 14. The basic principle of model fitting process was creating a mathematical model then use it to recalculate all dependent variables that were tested in experiments. Dependent variables data obtained from recalculations, thus called predicted value, will be compared to the dependent variables data obtained from experiments by using a parameter called coefficient of determination. The accepted mathematical model was the one that gave coefficient of determination above 0.95 in the range 0–1.

The accepted mathematical models were shown in term of dependent variables, Y_1 for resulted reducing sugar, Y_2 for resulted ethanol concentration,

TABLE 8.1 Central Composite Rotatable Quadratic Polynomial Model, Experimental Data, and Actual Predicted Values for Three-level-three-factor Response Surface Analysis.

Independent variables			Dependent variables					
Substrate loading (% dw)	**Enzyme (FPU/ gr substrate)**	**Span 85 concentration (% vol)**	**Resulted reducing sugar (% dw)**		**Resulted ethanol concentration**		**Ethanol conversion (%)**	
X_1	X_2	X_3	**Actual**	**Predicted**	**Actual**	**Predicted**	**Actual**	**Predicted**
20 (−1)	10 (−1)	0 (−1)	11.98	11.94	4.39	4.69	21.95	23.10
25 (0)	15 (1)	1 (0)	21.76	20.49	6.97	6.46	27.88	26.18
30 (1)	12.5 (0)	2 (1)	22.74	25.17	5.6	6.10	18.67	20.05
20 (−1)	15 (1)	2 (1)	10.65	10.65	5.59	5.59	27.95	27.95
30 (1)	15 (1)	0 (−1)	23.53	25.93	7.1	7.90	23.67	26.20
25 (0)	10 (−1)	2 (1)	24.82	22.39	6.93	6.05	27.72	26.34
25 (0)	12.5 (0)	0 (−1)	22.93	20.57	7.16	6.43	28.64	24.96
30 (1)	10 (−1)	1 (0)	22.03	23.48	4.89	5.13	16.30	16.68
20 (−1)	12.5 (0)	1 (0)	8.92	9.08	6.34	6.67	31.70	33.16
Coefficient determination (R^2)			0.9906		0.9623		0.9839	

and Y_3 for ethanol conversion. The independent variables were stated as X_1 for substrate loading, X_2 for enzyme loading, and X_3 for the SPAN 85 doses. Based on their coefficient of determinations, all values were satisfyingly above 0.95. It can be concluded that these statistical models will give good predictions for ethanol production process by using EFB pulp. By displaying the model prediction in 3D graph as shown in Figure 8.2–8.4, the optimum condition was easier to be determined. These graphs were created by using CAD software and become strong statements for the existing cheminformatics. This is wonderful evidence that information technology gives strong support for chemistry and chemical engineering works.

$$\begin{aligned} &Y_1 = -55.3491 + 7.5018X_1 - 5.6530X_2 - 2.0763X_3 - 0.1443X_1^2 + 0.1185X_2^2 \\ &+ 1.1450X_3^2 + 0.0983X_1X_2 - 0.0519X_1X_3 + 0.1327X_2X_3Y_2 = 4.1471 + 0.0434X_1 \\ &- 0.4240X_2 - 10.8375X_3 - 0.04732X_1^2 - 0.2798X_2^2 - 1.5456X_3^2 + 0.2541 \\ &X_1X_2 - 0.6038X_1X_3 + 0.7078X_2X_3Y_3 = 9.4757 - 0.6878X_1 + 3.2328X_2 \\ &+ 44.9815X_3 - 0.1651X_1^2 - 1.1526X_2^2 - 6.0872X_3^2 + 0.8819 \\ &X_1X_2 - 2.2216X_1X_3 + 2.2607X_2X_3 \end{aligned} \tag{8.1}$$

However, additional analyses were required for these models to make sure they are aligned with existing biochemical theories. First analysis was focused on resulted reducing sugar that gave different trend compared to resulted ethanol concentration and ethanol conversion. Analyzing resulted reducing sugar was mainly about the enzyme performance and its kinetics. Based on basic theory of reaction kinetics, more substrate concentration will give more products. This theory was suitable with the model prediction that gave minimum value, and the value increases aligned with substrate loading. Even though there is substrate inhibition theory in enzymatic reactions, but it is still not happening in our substrate loading scope 20–30% dw (Fig. 8.5).

Surfactant addition was applied for reducing the negative effect of lignin impurities as aforementioned. Increasing substrate loading will require higher dose of surfactant addition. Thus, it is very reasonable that increasing in resulted reducing sugar will be obtained as surfactant doses increased. Small discrepancies were observed in the minimum value which were not exactly at the point of origin (0,0).

In analyzing resulted ethanol concentration and ethanol conversion, *S. cerevisiae* or yeast performance was included. The resulted mathematical model gave predictions that aligned with theory about surfactant effect to yeast by showing an optimum value. SPAN 85 doses lower than optimum value was indicating that the amount of surfactant could not overcome the negative effect of lignin impurities, thus lowering the cellulase performance. On the other hand, SPAN 85 doses higher than optimum values caused harmful effect to yeast by its increasing toxicity as previously explained. The effect that is mentioned at last cannot be observed if the observed process has only hydrolysis step based on resulted reducing sugar. Of course, this is only happening if the process were conducted in similar duration and temperature with the one that is used in this study, 48 h hydrolysis and 60 h fermentation at 25–28°C.

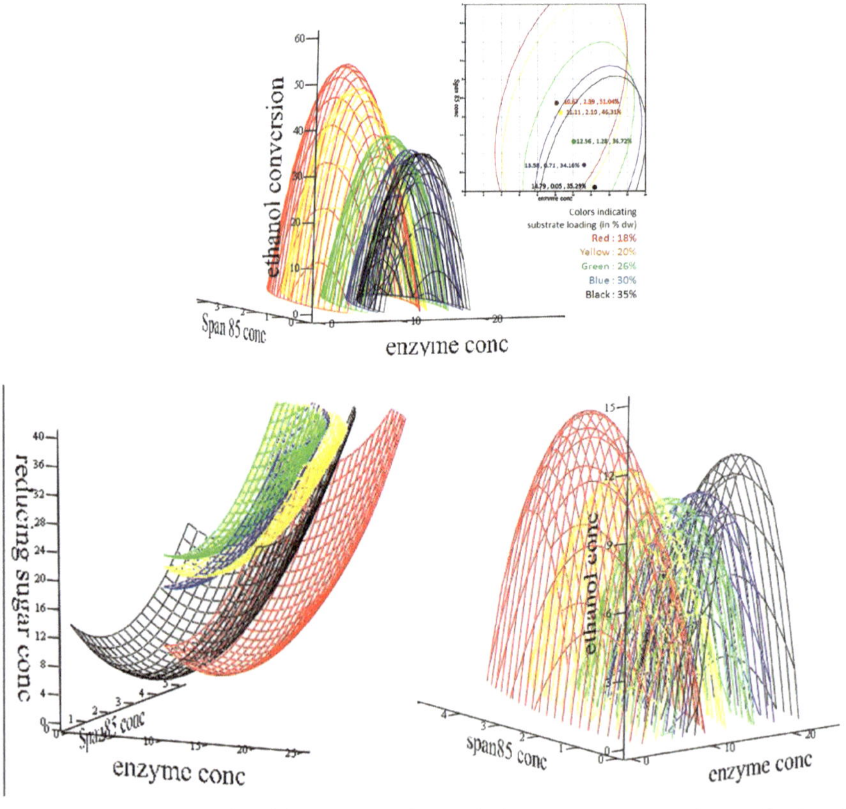

FIGURE 8.5 Profile of cellulose conversion, reducing sugar concentration, and ethanol concentration that was predicted by the obtained mathematical model.

Interesting alignment between statistical conclusion with bioprocess theory and facts gave confidence for conducting model verification. In verification step, several random conditions that calculated to give ethanol concentration in fermentation broth over 8% v were used. This 8% v limit was adjusted to the lowest accepted ethanol concentration in fermentation broth to be directly sent to distillation unit in one of existing Indonesian ethanol production factory. The calculated results of depended variables were further called predicted value. List of tested random conditions and their respective predicted values are shown in Table 8.2. These random conditions were then sent to experimental test in laboratory to obtain the, so called, actual values. As it can be observed in Table 8.2, the obtained actual values were very close with the predicted ones. Thus, the mathematical model had satisfyingly proved as a reliable and useful prediction tool. As it can be seen in the series of tested variables, there are several conditions that can be used for obtaining ethanol concentration over 8% v which adapts to substrate loading and enzyme loading simultaneously. Suitable combination of substrate and enzyme loading was required and this can be fulfilled by RSM analysis. The mathematical model involving the term X_1X_2 in the equation represents the mutual effect of substrate loading and enzyme loading. By involving this term, the sugar concentrations as feed for fermentation were adjusted to obtain ethanol concentration in fermentation broth over 8% v. As pulp substrate loading increased, the enzyme loading was adjusted at lowest possible concentration. This adaptation condition will be very useful in large scale/commercial applications as a precaution if substrate or enzyme supply chain were interrupted. These results cannot be obtained from conventional analysis approach, which observes the effect of one variable by keeping the other variables at constant value.

As can be seen in this case study, there will be a certain amount of doubt if the statistical results do not align with background facts and theories. In worse case, the model also cannot give satisfying predictions. In this problem, second order polynomial equation may not be the suitable empirical model for the observed phenomena. The options for still using the same experimental data is very limited since no variables were set constant to be analyzed in conventional way. For avoiding waste of time and experimental resources, adequate reference studies of the similar or related field were required. In other problem, the optimum condition cannot be found. The most recommended way for searching it was to add more experiments, with broader variable range, which is also designed in CCRD and repeat the RSM analysis sequences. Because, usually the problem came from unsuitable selection of variable range with optimum value which does not lie in the range. Since RSM analyzing process is only dealing with empirical approach, it will not be easy to create solid correlation between the results with scientific theory and fundamentals and thus, difficult to plan scientific improvement for the observed process.

TABLE 8.2 List of Random Conditions and Their Respective Predicted Values that Tested in Laboratory Experiments. The Results from Laboratory Were Called Actual Values.

Independent variables			Dependent variables					
Substrate loading (% dw)	**Enzyme (FPU/ gr substrate)**	**Span 85 concentration (% vol)**	**Resulted reducing sugar (% dw)**		**Resulted ethanol concentration (%v)**		**Ethanol conversion (%)**	
X_1	X_2	X_3	**Actual**	**Predicted**	**Actual**	**Predicted**	**Actual**	**Predicted**
36.5	14.75	1	21.76	20.66	10.48	10.07	28.71	27.19
23	12.5	1	15.63	16.50	8.23	8.41	35.78	36.20
21.25	11	1	12.10	14.05	8.96	8.89	41.67	39.00
20	10.25	1	10.51	11.01	7.77	8.91	38.85	40.12
31	12	1	24.75	21.39	8.81	9.07	33.88	35.32
26	13.25	1	21.98	23.37	8.25	8.24	26.61	27.46
Coefficient determination (R^2)			0.8880		0.9189		0.9163	

8.6 MODIFICATION OF EFB MATHEMATICAL MODEL TO BE USED IN PALM TRUNK ENZYMATIC HYDROLYSIS PROCESS

RSM had been demonstrated giving significant contribution for biochemical engineer in determining optimum conditions. However, obtaining optimum condition from laboratory experiments is not an easy task, especially in bioprocess like bioethanol from cellulose. It had been mentioned that for completing one experiment, 108 h were required and not including preparations. Possibility of using obtained equation for predicting similar or closely related raw material seems very worthy to be examined. This study continued by implementing the equation from hydrolysis and fermentation of palm oil EFB pulp to the process that uses palm oil trunk.

As mentioned earlier, palm oil trunks become a potential raw material due to regeneration of palm oil plantation which actually happened in these recent years in Indonesia. Cross-sectional profile of palm oil trunk can be seen in Figure 8.6. As a part of monocotyledon plants, palm oil trunk does not have cambium with less-structured arrangement of its fibers. By excluding the barks, palm oil trunk structure can be classified as parenchyma that lies inside the circle and vascular bundle at the outside.

FIGURE 8.6 Right: Cross-sectional area of palm oil trunk. The part that inside the circle is parenchyma and the one outside the circle is vascular bundle.

Vascular bundles have more structured fibers than parenchyma. Botanists believe this is due to its function for trunk's outer protection. Parenchyma, on the other hand, is having softer structure and high water adsorption capacity that are very useful in transporting nutrient. It is also having a function as nutrient storage, thus small amount of starch and plant sap were found in parenchyma. Based on this structure and chemical composition, an attempt to direct utilization of parenchyma as feed for enzymatic hydrolysis process, without prior pulping were conducted, but gave unsatisfied results. In this study, palm oil trunk from Malingping, Indonesia were chopped and milled to 50 meshes. Some of the resulted material went to mechanical separation for obtaining vascular bundle and parenchyma part. The other part was not separated and both non-separated and vascular bundle part went through Kraft pulping process that has similar condition with EFB. Pulp of whole trunks had 6.96% w lignin impurities and the vascular bundle had it 13.70% w. Later, both pulps were separately used as raw material for bioethanol production process.

Similar condition in hydrolysis and fermentation process for EFB pulp were applied to both of pulps, 48 h hydrolysis followed by 60 h fermentation at 25–28°C. Fermentation processes were conducted in same condition, by adding 1% w of powdered yeast, 0.3% w of urea, and 0.1% w of NPK. Observed variables followed the selected condition for EFB as shown in Table 8.3 with the similar goal of obtaining ethanol concentration in fermentation broth over 8% v. The resulted reducing sugar concentrations from hydrolysis of whole trunk pulp and its comparison with hydrolysis of EFB pulp is shown in Figure 8.7.

Data was performed in histogram with selected condition number as abscise. The numbers were referred to the similar optimum condition as mentioned in Table 8.3. The reducing sugar from palm oil empty bunch hydrolysis showed better results than resulted reducing sugar from palm oil trunks. The result of correlative test between trunks and EFB in high-loading substrate enzymatic hydrolysis based on their resulted reducing sugar is also shown in Figure 8.7.

TABLE 8.3 Variables for Hydrolysis Whole Trunk Pulp and Vascular Bundle Pulp (Left).

Optimum condition number	Substrate loading (% dw)	Enzyme (FPU/g substrate)	Span 85 Conc. (% vol)
1	20	10.25	1
2	21.25	11	1
3	23	12	1
4	26	12.5	1
5	31	13.25	1
6	36.5	14.75	1

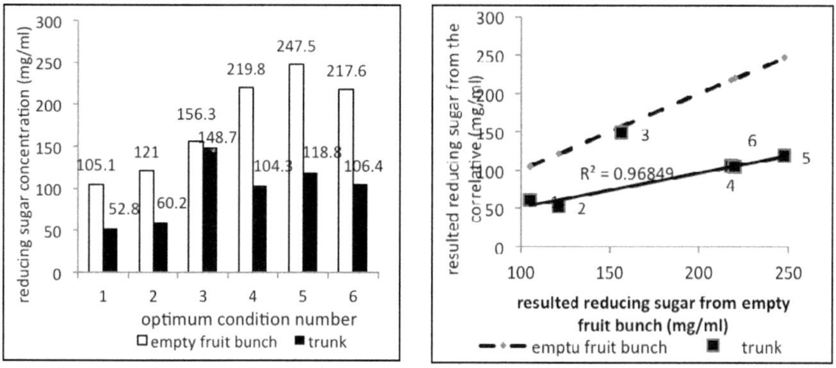

FIGURE 8.7 Resulted reducing sugar obtained from palm oil trunks hydrolysis.

Perfect correlation was simulated by correlating the EFB results to itself as showed in the graph as dashed line. Square dots were representing the resulted reducing sugar from palm oil trunks hydrolysis arranged correlatively to the ones obtained from palm oil EFB in the same optimum operation condition. By excluding data obtained from condition number 3, calculated correlative coefficient was 0.9993. This result supported the conclusion that optimum conditions for high-loading-substrate enzymatic hydrolysis of EFB were similar with optimum conditions for trunk pulp.

Reported data in Figure 8.8 are the comparison of resulted ethanol concentration between palm oil EFB and palm oil trunk. Since all operating conditions of fermentation were set similar, the difference solely caused by the difference of hydrolysis process and outside disturbances. No sterilization prior fermentation is needed in order to reduce operational cost if this similar process is applied in commercial scale. Unsterilized fermentations also allowed the enzymatic hydrolysis to continue during the fermentation, which is also called simultaneous saccharification fermentation (SSF). However, contamination possibilities need to be considered as outside disturbance.

It is shown in Figure 8.6 that the resulted ethanol from trunks was lower than those obtained from EFB. This is aligned with the resulted reducing sugar reported above. In EFB pulp, 1% v of SPAN 85 could effectively reduce the negative effect of lignin residue up to 9.34% w. As previously described, the hydrophobic part of the surfactant binds through hydrophobic interactions to lignin on the lignocellulose fibers and the hydrophilic head group of the surfactant prevents unproductive binding of cellulase to lignin. The content of lignin residue in palm oil trunk pulp was only 6.96% w. Thus, lower cellulase performance in trunk pulp, which was indicated by lower resulted ethanol concentration compared to EFB pulp, was not caused by lignin impurities. It could be concluded that the reason for lower resulted ethanol concentration was more dominantly caused by the higher crystalline cellulose content in whole trunk pulp. Later it will be discussed, that there were two factors that affect cellulase performance, which are lignin residue in pulp, and cellulose crystalline in pulp.

The result of correlative test between trunks and EFB in high-loading substrate enzymatic hydrolysis based on their resulted ethanol concentration is also shown in Figure 8.8. Unfortunately, none of the tested conditions gave ethanol concentration over 8% v in fermentation broth. However, no data was excluded and the best correlative coefficient was calculated as 0.808. Aligning this conclusion with correlative test using sugar concentration it is acceptable to assume that optimum conditions for high-loading-substrate enzymatic hydrolysis for EFB and those for trunk pulp were similar.

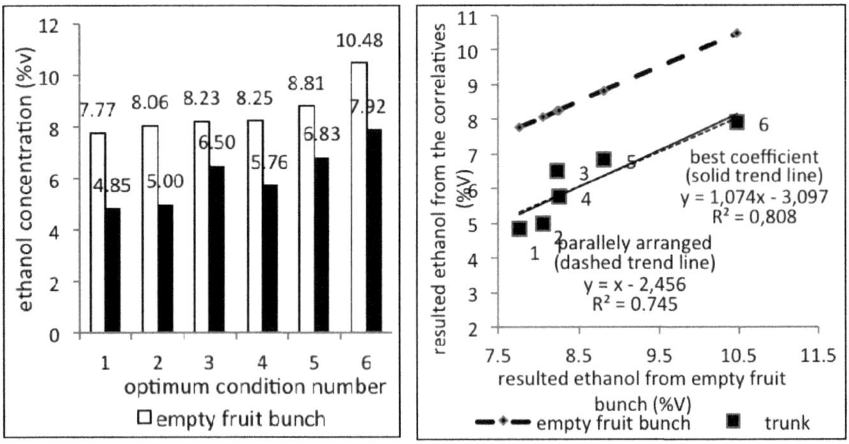

FIGURE 8.8 Ethanol obtained from palm oil trunks hydrolysis and continued by fermentation.

The trend line's slope for trunks was 1.074, which is very close to 1. It means the trunks trend line can be considered as parallel to the dashed line for EFB. Arranged trend line to meet the slope equal to 1 reduced the correlative coefficient to 0.7451 and is still considered acceptable. From the arranged trend line equation, there was a constant value obtained −2.456% v, then mathematical equation for predicting resulted ethanol concentration from palm oil trunks can be derived from eq 8.1.

$$\begin{aligned}Y_2 &= (4.1471-2.456)+0.0434X_1-0.4240X_2-10.8375X_3-0.04732X_1^2\\&-0.2798X_2^2-1.5456X_3^2-0.2541X_1X_2-0.6030X_1X_3+0.7070X_2X_3\\Y_2 &= 1.6904+0.0434X_1-0.4240X_2+10.8375X_3-0.04732X_1^2-0.2798X_2^2\\&-1.5456X_3^2+0.2541X_1X_2-0.6038X_1X_3+0.7078X_2X_3\end{aligned} \tag{8.2}$$

The predicting results of eq 8.2 are shown in Table 8.4. The coefficient of determination for eq 8.2 was much lower than prediction made by eq 8.1 for EFB but still usable in brief prediction. Thus, using RSM model that was obtained from a certain substrate was very promising to be applied to similar substrates. Increasing the experimental number more will hopefully give better correlations. Similar correlation test was conducted for vascular bundle pulp. Unfortunately, the correlation of EFB optimum conditions to vascular bundle was not as satisfying as the correlation between EFB and whole trunks. Resulted reducing sugar of vascular bundle pulp gave acceptable correlation with sugars obtained from EFB pulp from respective optimum conditions as shown in Figure 8.9. However, the correlation of resulted ethanol concentration between EFB and vascular bundle gave unacceptable results. The correlative coefficient is only 0.219, as shown in Figure 8.10, which is much lower than the previous study which used whole trunk, 0.7451.

TABLE 8.4 Comparison of Obtained Experimental Ethanol Concentration from Trunks Pulp and EFB Pulp to Their Correlative Prediction by Mathematical Model.

Optimum condition number	Substrate loading (% dw)	Enzyme (FPU/g substrate)	Resulted ethanol concentration from trunks (% v)		Resulted ethanol concentration from EFB (% v)	
			Obtained	Predicted (eq 8.2)	Obtained	Predicted (eq 8.1)
1	20	10.25	4.85	5.55	7.77	8.01
2	21.25	11	5.00	5.63	8.06	8.09
3	23	12	6.50	5.65	8.23	8.11
4	26	12.5	5.76	5.78	8.25	8.24
5	31	13.25	6.83	6.61	8.81	9.07
6	36.5	14.75	7.92	7.61	10.48	10.07
Coefficient of determination			0.7451		0.9189	

In a previous study, parts of palm oil trunks were separated mechanically to its vascular bundle and parenchyma, then each part was measured for its hemicelluloses content and cellulose crystallinity index. The results show that vascular bundle and parenchyma had hemicellulose content 25.47% and 35.57% w, respectively, and cellulose crystallinity index 76.78 and 69.70, respectively. Cellulose crystalline index of palm oil EFB was not found from previous studies. However, in another previous study in *Miscanthus*, it was shown the correlation between hemicelluloses content and crystalline cellulose in its natural existence. The higher the hemicellulose content in one observed part of the plant, the lower the cellulose crystallinity index. These previous studies in palm oil trunk and *Miscanthus* were supporting a conclusion about correlation between hemicelluloses content and celluloses crystallinity index.

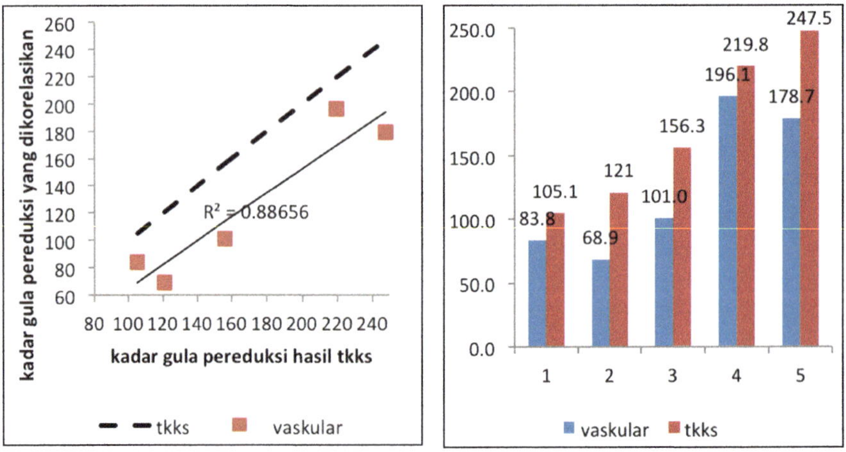

FIGURE 8.9 Ethanol obtained from palm oil vascular bundle hydrolysis and continued by fermentation.

In another previous study, effects of non-cellulose sugar in natural plants to cellulose crystallinity were specified only to galactoglucuronoxylan. Three monocotyledons which are Italian ryegrass, pineapple, and onion were measured for its galactoglucuronoxylan content and correlated the results to cellulose crystallinity index. The results showed that higher galactoglucuronoxylan content will lead to higher crystallinity index when the cellulose is naturally biosynthesized. Galactoglucuronoxylan has similar properties with starch due to its water solubility and branched molecular structure. Starch was found in palm oil trunks but not in EFB. This fact could strengthen our previous conclusion that palm oil trunks have higher cellulose crystallinity index than its EFB. Not only because it had lower hemicellulose content but also because of the existence of starch in palm oil trunks.

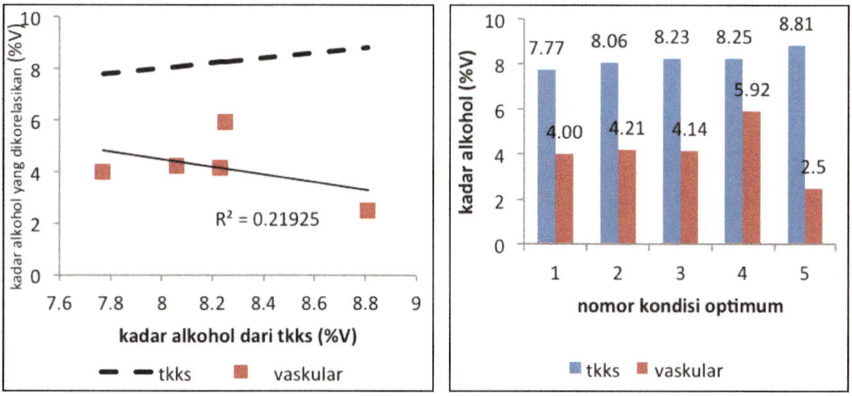

(a) (b)

FIGURE 8.10 (a) Kurva kolerasi dan (b) kadar alkohol tkks dengan vascular bundle.

Hemicelluloses content in parenchyma, EFB, whole trunks, and vascular bundle were 35.57, 35.3, 31.8, and 25.47% w, respectively. Cellulose crystallinity index for vascular bundle and parenchyma were 76.78 and 69.70, respectively. It will be reasonable to assume that cellulose crystallinity index in EFB was close to parenchyma, 69.70, due to no significant difference between hemicelluloses content in EFB and parenchyma and the index for whole trunk would be between 69.70 and 76.70. Ordering from the highest cellulose crystallinity index to the lowest in its natural existence would be vascular bundle (76.78) > whole trunk > EFB = parenchyma (69.70). Since both raw materials went through Kraft pulping process with similar operation conditions, the cellulose crystallinity index in resulted pulp gave similar tendencies. These facts support the experimental finding that the order of enzymatic hydrolysis performance from the lowest was vascular bundle < whole trunk < EFB.

These results showed the limitation of RSM in searching for general model that can be applied to similar or closely related conditions. In case of bioethanol production from palm plantation biomass waste, general model cannot be obtained, even the biomass came from one specific species. However, scientific and experimental facts were aligned with the RSM model which opens windows of opportunities for using RSM in this field. One promising possibility was establishing clear raw material characteristics, including lignin impurities and cellulose crystallinity index, as the term for the existing RSM model works satisfyingly.

8.7 CONCLUSIONS AND FUTURE RESEARCH

8.7.1 CONCLUSIONS

The RSM-CI is a medium for bioethanol process to extend its activities beyond traditional boundaries that were characterized by space and time constraints, leading to accessible reach of processes directly anytime, anywhere.

RSM-CI systems are one of cheminformatics that explores the relationships between several explanatory variables and one or more response variables that allow users to access bioethanol processing systems directly by clicks of a mouse or touch of a finger.

The adoption of RSM-CI systems by bioethanol production, enhancing cellulose hydrolysis, and optimum enzymatic condition of EFB enzymatic are high, threatening software as the systems allow users to interact directly. Adoption of cheminformatics systems by users, especially during bioethanol production processes, in Indonesia is also high, threatening the competitiveness of determining optimum condition and forecasting the results.

However, in Indonesia, with the rapid advancement of technology, there is no doubt that traditional stages of bioethanol processes will face more challenges in the future. In general, RSM-CI promises more efficiency and effectiveness stages and forecasting to grab optimum and economical scalability of product rather than traditional processes.

8.7.2 FUTURE RESEARCH DIRECTION

This study incorporates two main issues of RSM-CI features for enabling bioethanol production from palm oil of EFB in Indonesia; determining optimum condition, and forecasting the result of production. With the promising features and user acceptance of RSM-CI performance, quality, reliability, and usability, it is highly recommended for bioethanol producer to consider adopting RSM-CI methodology as it has been valued as an effective, efficient tool to assist plant to get efficient-effective condition and helps to set a clear strategy to get optimum result confidently.

The study triggers a future research direction in bioethanol production strategy which would focus on enhancing bioethanol plant, forecasting cellulase performance, EFB enzymatic hydrolysis process, change management associated with the implementation of RSM-CI, RSM-CI software release management, ICT service continuity management, and configuration management to handle bioethanol production.

The future direction of this study can also accommodate and customize RSM-CI to fit with other bioprocesses; such as other material of bioethanol production, bioethanol supply chain management, and after production handling care. RSM-CI was designed as an adaptable framework that can be implemented and extended to processes by customizing its control.

KEYWORDS

- **chemistry informatics**
- **response surface methodology**
- **RSM-CI**
- **ICT emerging technology**
- **EFB**

REFERENCES

1. Triwahyuni, E.; Sudiyani, Y.; Muryanto; Abimanyu, H. The Effect of Substrate Loading on Simultaneous Saccharification and Fermentation Process for Bioethanol Production from Oil Palm Empty Fruit Bunches. *Energy Procedia* **2015,** *68*, 138–146.
2. Barlianti, V.; Dahnum, D.; Triwahyuni, E.; Aristiawan, Y.; Sudiyani, Y.; Muryanto. Enzymatic Hydrolysis of Oil Palm Empty Fruit Bunch to Produce Reducing Sugar and its Kinetic. *J. Menara Perkebunan* **2015,** *83* (1), 37–43.
3. Triwahyuni, E.; Muryanto; Fitria, I.; Abimanyu, H. Pemanfaatan Limbah Tandan Kosong Kelapa Sawit untuk Produksi Bioetanol dengan Optimasi Proses Sakarifikasi dan Fermentasi. *J. Energi dan Lingkungan (Enerlink)* **2014,** *10*, 27–32.
4. Sudiyani, Y.; Styarini, D.; Triwahyuni, E.; Sembiring, K.; Sudiyarmanto; Aristiawan, Y.; Han, M.; Abimanyu, H.Utilization of Biomass Waste Empty Fruit Bunch Fiber of Palm Oil for Bioethanol Production Using Pilot–Scale Unit. *Energy Procedia* **2013,** *32*, 31–38.
5. Dahnum, D.; Tasum, S. O.; Triwahyuni, E.; Nurdin, M.; Abimanyu, H. Comparison of SHF and SSF Processes Using Enzyme and Dry Yeast for Optimization of Bioethanol Production from Empty Fruit Bunch. *Energy Procedia* **2015,** *68*, 107–116.
6. Pahan, I. *Panduan Lengkap Kelapa Sawit: Manajemen Agribisnis dari Hulu sampai Ilir;* Niaga Swadaya: Bogor, Indonesia, 2006.

CHAPTER 9

CURRENT TRENDS IN BIOTECHNOLOGICAL PRODUCTION OF FRUCTOOLIGOSACCHARIDES

ORLANDO DE LA ROSA[1], DIANA B. MUÑIZ-MÁRQUEZ[2], JORGE E. WONG-PAZ[2], RAÚL RODRÍGUEZ-HERRERA[1], ROSA M. RODRÍGUEZ-JASSO[1], JUAN C. CONTRERAS-ESQUIVEL[1], and CRISTÓBAL N. AGUILAR[1*]

[1]*Food Research Department, School of Chemistry, University Autonomous of Coahuila, Saltillo CP 25280, Coahuila, Mexico*

[2]*Engineering Department, Technological Institute of Ciudad Valles, National Technological of Mexico, Ciudad Valles 79010, San Luis Potosí, Mexico*

[*]*Corresponding author. E-mail: cristobal.aguilar@uadec.edu.mx*

CONTENTS

ABSTRACT

Nutritional and therapeutic benefits of prebiotics have captured the interest of consumers and the food industry for use as food ingredients in order to create functional foods. Fructooligosaccharides (FOS) are alternative sweeteners, consisting of 1-kestose, 1-nystose, and 1β-fructofuranosilnystose produced from sucrose by the action of fructosyltransferase (2.4.1.9) and β-fructofuranosidase (3.2.1.26) from plant bacteria, yeasts, and fungi. FOS are low caloric, are non-cariogenic, and aid in the absorption of minerals such as calcium and magnesium in the gut, reduce levels of cholesterol, triglycerides, and phospholipids and stimulate the development of the intestinal and colon microflora. This chapter is focused on reviewing the functional properties of FOS, biotechnological production, and recent trends.

9.1 INTRODUCTION

Nowadays, there is growing interest in people to improve their health through good nutrition. Nutraceutical ingredients and functional foods have attracted special attention in the development of new products due to the human health benefits observed.[1]

Prebiotics are considered as nutraceutical ingredients and they are used in the functional foods processing (falta cita). Prebiotics are indigestible ingredients by humans and have a positive influence on the body of the host by selectively stimulating the growth and/or activity of bacteria or a limited number of bacterial species in the colon because they are substrates for growth and metabolism of probiotic bacteria. It is great interest to the general public because it provides a better balance in the intestinal ecosystem and improves host health.[2] Great efforts in developing strategies in the daily diet for modulating the composition and activity of the microbiota, using prebiotics, probiotics, and a combination of both (symbiotic) has been explored.[3–6]

There are three essential criteria for a food ingredient to be classified as a prebiotic:

1. must not be hydrolyzed or absorbed in the upper gastrointestinal tract,
2. must be a selective substrate for one or a limited number of probiotics, and
3. must be able to alter the colonies of microflora to a better and healthier composition.[2,7]

Among prebiotics, fructooligosaccharides (FOS) in addition to meeting the above requirements they attract attention due to its properties and its great economic potential for the sugar industry.[8] Having sweetness of 0.4–0.6 times compared with sucrose, these being used in the pharmaceutical industry as artificial sweeteners.[9]

The health benefits and applications of FOS in nutrition have been well documented and these include activation of the immune system and resistance to infection, FOS are low caloric because they are rarely hydrolyzed by digestive enzymes and are not used as an energy source in the body so they can be safe for inclusion in products for diabetics, are non-cariogenic by which they could be used in chewing gums and dental products, and playing an important role in reducing cholesterol, triglycerides, and phospholipids, as well as help to improve the absorption of minerals such as calcium and magnesium in the gut.[10,11,7]

9.2 PREBIOTICS

A prebiotic is a food ingredient that beneficially affects the host by selectively stimulating the growth and/or activity of one or a limited number of "probiotics" bacteria and thus improves host health. Prebiotics are short-chain carbohydrates which are not digested by human digestive enzymes and selectively stimulate the activity of certain groups of beneficial bacteria for the body.[13] In the intestine, prebiotics are fermented by beneficial bacteria to produce short-chain fatty acid (SCFA). Prebiotics also have many other health benefits, such as reduce risk of suffering cancer of the large intestine and increase absorption of calcium and magnesium. Among the best known prebiotics are carbohydrates particularly oligosaccharides such as galactooligosaccharides (GOS), maltooligosaccharides, FOS, xylooligosaccharides, inulin, and hydrolysates.[12]

Prebiotics are found in many vegetables and fruits and, also are considered components of functional foods which have significant technological advances. Their addition to these foods improves the sensory characteristics such as taste and texture, which also enhances the stability of foams and emulsions.

According to the definition of a functional food, it is one that is part of the human diet and is shown to provide additional health benefits and reduce the risk of chronic diseases through its additional benefits.

A functional food can be classified if it meets one of the following:

1. foods with natural bioactive substances (e.g., dietary fiber),
2. food supplemented with bioactive substances (e.g., probiotics and antioxidants), and
3. food ingredients derived and introduced to conventional foods (e.g., prebiotics).[14]

Carbohydrate prebiotics are short-chain non-digestible by human digestive enzymes and are called short-chain carbohydrates resistant. They are also called nondigestible oligosaccharides which are soluble in 80% ethanol. A prebiotic is a non-active constituent of food that reaches the colon and is selectively fermented. The benefit to the host is mediated by selectively stimulating the growth and/or activity of one or a limited number of bacteria.[15]

Prebiotics pass through the small intestine to the colon and become accessible for probiotic bacteria without having been exploited by intestinal bacteria. Lactulose, GOS[16], FOS, inulin, and its hydrolysates, maltooligosaccharides, resistant starch, and prebiotic are normally used in the human diet, including FOS are the most studied and widely marketed[14] (Table 9.1).

TABLE 9.1 Nondigestible Oligosaccharides with Bifidogenic Properties Available in the Market.

Compound	Molecular structure
Cyclodextrins	(Gu)n
Fructooligosaccharides	(Fr)n–Gu
Galactooligosaccharides	(Ga)n–Gu
Gentiooligosaccharides	(Gu)n
Glycosylsucrose	(Gu)n–Fr
Isomaltooligosaccharides	(Gu)n
Isomaltulose (or palatinose)	(Gu–Fr)n
Lactosucrose	Ga–Gu–Fr
Lactulose	Ga–Fr
Maltooligosaccharides	(Gu)n
Raffinose	Ga–Gu–Fr
Soybean oligosaccharides	(Ga)n–Gu–Fr
Xylooligosaccharides	(Xy)n

Ga, galactose; Gu, glucose; Fr, fructose; Xy, xylose.

9.3 FRUCTOOLIGOSACCHARIDES

FOS are also known as oligofructose and usually are called as oligosaccharides derived from inulin, mainly composed of 1-kestose (GF2), 1-nystose (GF3), and 1-β–fructofuranosylnystose (GF4), in which fructosyl units are linked in the β-(2–1) position of a sucrose molecule.[17,18] The formula GFn which indicates the degree of polymerization by the number of fructose molecules which are presents and linked to glucose (Fig. 9.1).

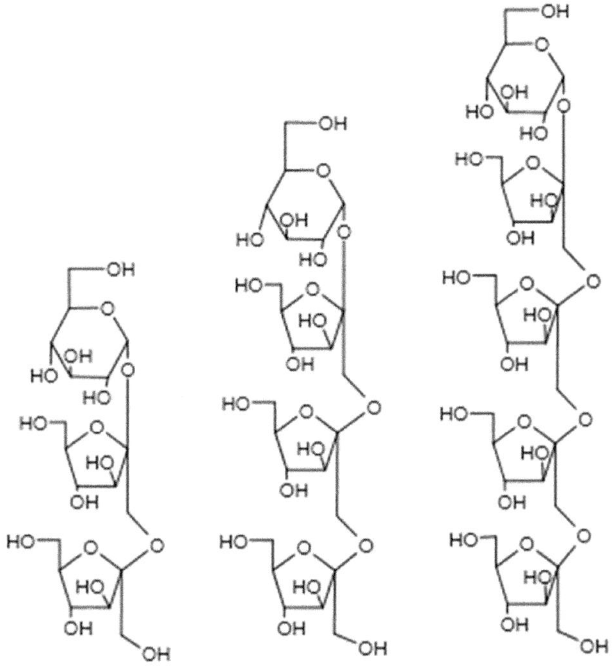

FIGURE 9.1 FOS structures: 1-kestose (GF2, left), 1-nystose (GF3, middle), and 1-fructofuranosyl nystose (GF4, right).

9.3.1 FUNCTIONAL PROPERTIES

Among the functional properties of the FOS, the ability to reduce the risk of chronic diseases such as colon cancer, ulcerative colitis, intestinal cancer, cardiovascular disease, and obesity,[18] is due in large part digestibility of these prebiotics in the body, because they are hardly digested by digestive enzymes and gastric juice, that enables the most prebiotics to arrive intact the intestine and colon where they are susceptible to be fermented by probiotic bacteria and beneficial intestinal microflora which produce metabolites such as SCFA such as butyrate and propionate between that stand out because of its anti-cancer effect.[19] Butyrate can be a source of energy for colon epithelial cells that also thought to promote proliferation and differentiation of cells in the intestine, it has effects of inhibiting cancer cells, colonic adenomas and carcinomas, and induction of apoptosis, preventing the tumor formation.[14,20–23] These effects are regulated by the expression of differentiation markers, alkaline phosphatase, glutathione-S-transferase, and other response genes as well as the suppression of expression of 2-cyclooxygenase and also alters the epigenome through inhibition deacetylases of histone. Propionate may have the anti-inflammatory ability over cancer cells in the colon.[23–25]

Among other properties, the improvement and regulation of the immune system are highlighted. This is due to the activation of proliferation and differentiation of intestinal epithelial cells and colon promoted by the formation of butyrate, development, and growth of the epithelial barrier which provides protection against pathogens because it hinders their attachment to the gut, in addition to the development and regulation of gut-associated lymphoid tissue which forms the largest area of immune tissue in the human body, comprising the innate immunity with important roles from NOD and toll-like receptors.[23,26]

Among other benefits for its prebiotic nature and as soluble fiber, FOS has the ability to have an impact on obesity. It is now emerging an overview of how the consumption of FOS can make an impact in reducing the consumption of food and energy for people obese leading to weight loss and improving health. This is because they are non-digestible by human digestive enzymes by binding type possessing β(2–1). It is thought that these prebiotics fermentation by the beneficial microflora achieves immune regulation with anti-inflammatory effects improving intestinal permeability and improved metabolism. In a study,[27] 52 women with type-2 diabetes were given a dose of 10 g oligofructose B+ inulin per day reduction decreased glucose levels in plasma was observed and a decrease in glycosylated hemoglobin levels. In a similar study,[28] other beneficial health effects of oligosaccharides were observed and the reduction of the levels of cholesterol and triglycerides was observed after 12 weeks.[29] This is thought to be due to inhibition of lipogenic enzyme in the liver, resulting from the action of propionate produced by the fermentation of prebiotics. This reaches the liver via the portal vein and inhibits the pathways of cholesterol by inhibition HMG-CoA reductase.[30]

FOS also have a beneficial impact on mineral absorption, in studies by Al-Sheraji et al. and Roberfroid[14,31] with the administration of 15 g/day oligofructose or inulin 40 g/day an increase in apparent calcium absorption was observed. Also, it has been observed the increase in magnesium absorption caused by the ingestion of FOS.[14,32]

Prebiotics such as FOS are considered safe for inclusion in traditional diets because of their presence as natural ingredients in food and plants. According to data from the United States Department of Agriculture, it is estimated that the average daily consumption of FOS from chicory ranges between 1 and 4 g/day.[14,33] A study showed that FOS from chicory has no toxicity to organs and that these compounds are not mutagenic, carcinogenic, or teratogenic.[34] Other results show that these fructans are well tolerated in amounts up to 20 g/day, can trigger diarrhea if doses of 30 g/day or more.[14,35]

9.3.2 FOS NATURAL SOURCES

FOS are naturally found in vegetables such as onions and garlic and, also found in tomatoes and honey, accompanied by the inulin from chicory roots (*Cichorium intybus L.*), bulbs in Jerusalem artichoke (*Helianthus tuberosus*), some monocots such as rye, barley, rice, wheat, and banana,[36,37] but they are found in very small amounts and their presence depends the season. Because of this, currently, it has drawn attention to the production of FOS through biotechnology.[38,39]

FOS can be produced from sucrose by the action of enzymes with the transfructosylating enzyme fructosyltransferase (FTase) (EC 2.4.1.9) and the β-fructofuranosidase (FFase) (EC 3.2.1.26) derived from plants and microorganisms (Fig. 9.2). It has been found that microorganisms of genera *Aureobasidium* spp., *Penicillium* spp., *Aspergillus* spp., and *Fusarium* spp. have the ability to produce these enzymes.[12,37,40–44]

9.4 FOS PRODUCTION

For the production of FOS, high initial substrate concentration is required for efficient tranfructosylation.[45] Other products of this reaction are fructose and glucose which has been found to be an inhibitor of the reaction of transfructosylation when it accumulates in the media.

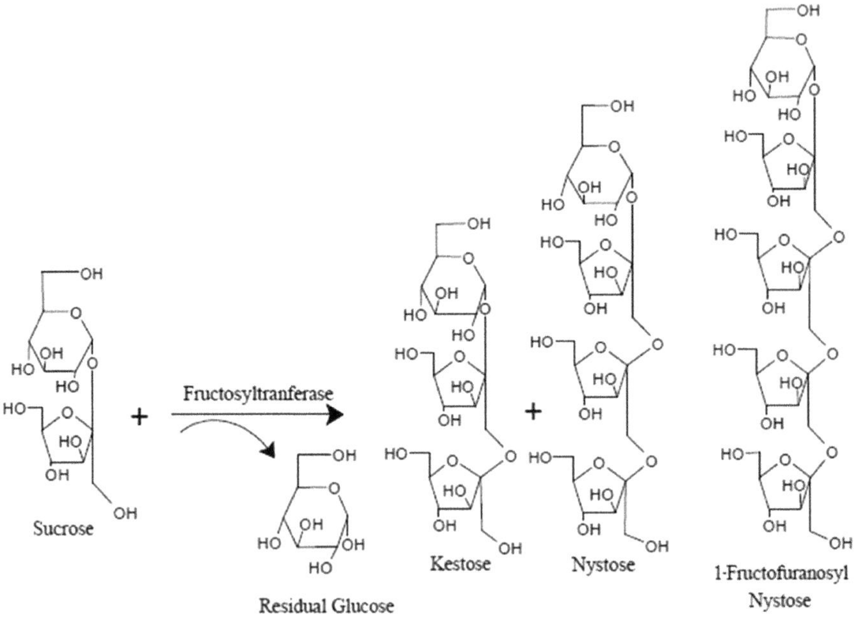

FIGURE 9.2 FOS (kestose, nystose, and 1-fructofuranosyl nystose) synthesis as from sucrose by the action of the enzyme fructosyltranferase.

Transferase activity acts on sucrose breaking the β-(1,2) link and transferring fructosyl group to an acceptor molecule such as sucrose, releasing glucose. This reaction produces FOS which units are linked by a β-(2, 1) bond in position of sucrose.[6] Due to the enzymes that catalyze the above reaction, there is a difference in the opinion regarding nomenclature by different authors referred to both FOS producing enzyme FFase (EC 3.2.1.26)[46,47,42] and FTase (EC 2.4.1.9).[37,38,40,48–50] Both enzymes have been reported in literature as FOS producing enzymes and have also been shown that these have both activities, hydrolyzing activity (U_H) and transfructosylating activity (U_T). Their activities vary greatly (ratio, U_T/U_H) depending on the nature of these enzymes, if they are isolated from plants, bacteria, yeasts and fungi, the genus, species, and strain. The manner in which the enzyme activity is defined varies according to the author, the activity units are defined as the amount of enzyme that transfers them 1 μmol of fructose per minute,[37,51,52] or as the amount of enzyme which liberates 1 μmol of glucose per minute,[53,54] or by 1 μmol nitrophenol liberated per minute per p-nitrophenol-α-D-glucopyranoside[55] or as the amount of enzyme which produces 1 μmol of kestose per minute.[37,56,57]

Besides these there are more different analytical methods that are used to determine the production of FOS by this enzyme so it is difficult to establish a specific definition for comparison.[37] Activity units may also vary greatly if these enzymes are intracellular or extracellular. Nguyen et al.[46] reported the production of β-FFase FOS by intracellular and extracellular *Aspergillus niger* IMI 303 386 showing the best results transfructosylation intracellular β-FFase.

Hidaka et al.[45] evaluated the ratio (U_T/U_H) for producing different microorganisms showing FOS resulted in *A. niger* ATCC 20611 strain with high productivity having a transfer activity much greater than its hydrolytic activity (U_T/U_H) = 14.2 after 1 day incubation, 12.2 after 3 days.

Antošová and Polakovic[50] showed different characteristics of different enzymes with transferase activity from different sources, both microorganisms and plants had different characteristics of these oligosaccharides.

The molecular mass of fungal FTase is in a range from 18.000 to 600.000 and homopolymer formed from 2 to 6 monomeric units. Many papers have defined temperature and pH optimum for activity of these enzymes between 50 and 60°C, and a pH of 4.5–6.5, respectively.[6,37] FTases from plants have different pH range and temperature as FTase from Jerusalem artichoke which optimal pH is between 3.5 and 5, and the optimum temperature between 20°C and 25°C in the case of 1-SST and pH 5.5–7 and temperature 25–35°C for 1-FFT.[50]

Yoshikawa et al.[42] reported in a work the production of at least five types of β-FFase in the cell wall of *Aureobasidium pullulans* DSM2404 catalyzing the reaction of transfructosylation.

In this test, the FFaseI was predominant in the period of formation of FOS while FFase levels II–V increased in the period of degradation of FOS. Ratios (U_T/U_H) of FFases IV were 14.3, 12.1, 11.7, 1.28, and 8.11, respectively, where FFaseI proved to have the best (U_T/U_H) ratio further was the only enzyme showed activity with glucose in the medium the other enzymes were strongly inhibited by the presence of glucose.

Figure 9.3 shows the action of both activities hydrolyzing (U_H) and tranfructosylating (U_T).

There are mainly two methods that can be used for the production of these enzymes with transfuctosylation activity and FOS production by fermentation, submerged fermentation (SmF), and solid-state fermentation (SSF) which will be described below.

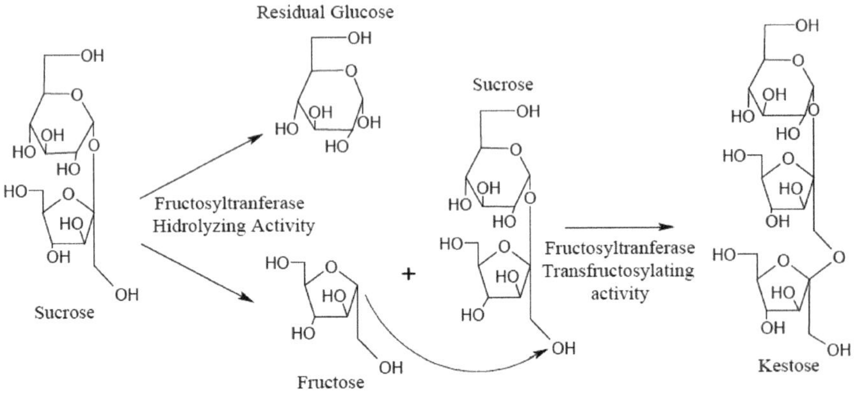

FIGURE 9.3 Kestose formation by the action of fructosyltranferase enzyme hydrolyzing activity (U_H), and transfructosylating activity (U_T), hydrolyzing a sucrose molecule and transferring a fructosyl group to an another acceptor molecule of sucrose, respectively.

9.4.1 FOS PRODUCTION BY SMF

The most common and studied method for production of FOS is the transfructosylation of sucrose in two steps, in the first step by SmF enzyme is produced, and in the second step the enzyme is reacted with the carbon source for production of FOS under controlled conditions.[12]

The variables studied to define generally the best operating conditions for the production of the enzyme are the source of carbon and nitrogen its concentration, time of cultivation, agitation, and aeration. Other important factors are the addition of various minerals, small amounts of amino acids, polymers, and surfactants.[6,37]

Silva et al.[57] evaluated at three different levels important factors for the production of FOS, carbon and nitrogen sources, percentage of sucrose, and yeast extract, respectively, inoculum percentage, pH, temperature, agitation, concentration of urea, and the average concentration of various mineral salts K_2HPO_4 $(NH_4)_2SO_4$, $MgSO_4$, $ZnSO_4$, and $MnSO_4$. The sucrose concentration proved to be a positive parameter for the formation of FOS because the enzymes catalyze the reaction of transfructosylation at high substrate concentrations. Higher productivity conversion was 54.7% and 223 g/L total FOS with an initial concentration of 400 g/L sucrose. As mineral salts, $MnSO_4$ proved to be the only mineral presented a significant effect stimulant for production of FOS. The K_2HPO_4 is described as a source of micronutrient for cell growth as well as being a buffer solution. Its optimal concentration varies between 4 and 5 g/L.[53,54,58–60]

The effect of pH on the average production of FTase and microorganism growth has been reported. A pH of 5.5 has been found as the best initial value for the production of *Aspergillus oryzae* FTase CFR202,[54,59] *Aspergillus japonicus* JN19,[61] and *Penicillium purpurogenum*.[62]

As for the 1-step process, Dominguez et al.[12] evaluated a one-step system for the production of FOS in which obtained under optimal production conditions 64.1 g FOS/g sucrose, being the temperature the most significant parameter in this trial.

9.4.2 FOS PRODUCTION BY SSF

The SSF presents a growing interest and high potential for small-scale units. Some advantages of this process are the simplicity of operation, high volumetric productivity, product concentration, an initial investment inexpensive, requires low power in addition to that there is less risk of contamination due to its high concentration of inoculum and low humidity in the reactor, and a simpler separation process.[12,17,59,63]

The disadvantages of the SSF are the difficult measurement of parameters such as pH and aeration, the type of sampling is destructive, and many complications for fermentations in larger scale arise due to problems of heat transfer and oxygen in the media and a homogeneous diffusion of the substrate.

Food, agriculture, and forestry industry produces large volumes of waste that can be utilized as materials for SSF. Examples include potato waste, corn cob, tapioca bagasse, sugar cane bagasse, and wheat grain waste, among others.[64,65] This makes the process costs are low compared to SmF and an alternative to the disposal of industrial wastes.

Various agro-industrial by-products of wheat cereal products, corn, sugarcane bagasse, and by-products of the processing of coffee and tea have been used as substrates for the production of FTase in SSF by *A. oryzae* CFR 202.[41]

Mussatto et al.[17] studied the ability to colonize different synthetic materials (polyurethane foam, stainless steel sponge, vegetable fiber, pumice, zeolites, and glass fiber) from *A. japonicus* ATCC 20236 to produce FOS from sucrose (165 g/L).

Dietary fiber was the best support for the growth of *A. japonicus* (1.25 g/L carrier) producing 116.3 g/L FOS (56.3 g/L 1-kestose, 46.9 g/L 1-nystose, and 13.1 g/L 1-β-fructofuranosyl nystose) with 69% of yield (78% based only on the amount of sucrose consumed), reporting high activity of the enzyme β-FFase (42.9 U/mL).

Mussatto and Teixeira[63] assessed various media for the production of FOS looking to reduce process costs studying the use of supplemented and non-supplemented media, and seeking an increase in the production yield of FOS resulting in a high production of FOS 128.7 g/L of β-FFase activity (71.3 U/mL), using as support coffee silverskin showing similar results in trials with the supplemented and unsupplemented support.

It has been sought to increase the performance and productivity of the FOS in the SSF assessing conditions that affect this process, in one study[66] conducted fermentations with coffee husk evaluating different moisture levels 60, 70, and 80% sucrose solution with 240 g/L, and a solution of spores of *A. japonicus* 2 × 10 ^ 5, 2 × 10 ^ 6, or 2 × 10 ^ 7 spores/g dry support, and different temperature tests 26, 30, and 34°C for 20 h, showing that the humidity did not influence the production of FOS or FFase enzyme, the temperature being between 26 and 30°C, and the inoculum of 2 × 10 ^ 7 esp/g material which maximized the production of FOS to 208.8 g/L FOS and a productivity of 10.44 g/L* h FFasa 64.12 U/mL and with a productivity of 4 U/mL* h.

9.5 IMPROVED PRODUCTION YIELDS OF FOS

A considerable disadvantage for the production of large-scale FOS is that the resulting mixture of the bioreactor consists of different carbohydrates: monosaccharides, unreacted disaccharides, and oligosaccharides. The incomplete conversion creates great challenges to producers of FOS because a purer product would increase their value and their utility in other food and pharmaceuticals. To generate a purer mixture, several studies had tried to remove the digestible carbohydrates from FOS mix, in which they have used different bioengineering strategies for accomplishing this. In such strategies, it has been applied different separation techniques using different technologies as well as additional steps of using enzymes and selective bioconversion fermentations. Production yields generally range from 55 to 60%, based on the initial sucrose concentration due to inhibition of the reaction by-products generated during the reaction.

One of the factors which directly influence the production of FOS by the reaction of transfructosylation, is the accumulation of residual glucose in the medium during the reaction which strongly inhibits production of FOS.[6,10,42,47]

Because of this, different authors have sought to remove the glucose produced during the reaction by means of systems with mixed-enzymes, in such systems, the goal is to eliminate the reaction by-product which is an inhibitor, and thus the use of the substrate can be maximized. Tanriseven and Gokmen[49] proposed an interesting system in which producing FOS from sucrose using an enzyme commercial preparation Pectinex Ultra SP-L (Novozymes A/S, Denmark) after sugar mixture was processed using Leuconostoc mesenteroides B-512 FM dextransucrase to convert all remaining unreacted sugars to isomaltooligosaccharides which also increase the activity of bifidobacteria[47] used a preparation of commercial glucose oxidase to convert glucose generated to gluconic acid and was then precipitated as calcium gluconate using calcium carbonate for pH control of the reaction. This proved to be effective and the system occurred more than 90% (w/w) of FOS dry basis, the remainder being glucose, sucrose, and a small amount of calcium gluconate.

Yoshikawa et al.[67] produced FOS from sucrose with various enzyme preparations of FFase obtaining a yield of 62% with preparation FFaseI, then the reaction was just using glucose isomerase (GI) added in a ratio of activity 1:2 and FFase:GI obtaining a maximum yield of 69% FOS.

Various studies have used strains of *Saccharomyces cerevisiae*, *Zimomonas mobilis*, and *Pichia heimii* to remove glucose and fructose accumulated during fermentation,[7,68–70] where these were completely fermented and produced ethanol, carbon dioxide, and a small amount of sorbitol as a by-product. A high content of FOS (98%) was obtained in the mixture after efficient removal of the released glucose and unreacted sucrose in the medium.[54]

Whereby the microbial treatment proved to be a good alternative to increase the percentage of FOS in the reaction mixture by removal of mono and disaccharides, this being suitable process for the enzymatic production of FOS, but otherwise the implementation of this methodology is relatively new and still need to be developed to get good yields in addition to using this type of process would involve a further step in the purification to remove biomass and other metabolites formed during fermentation to obtain FOS with less contaminants which would increase the cost of production.

Other techniques used to remove sugars from the medium are nanofiltration and microfiltration systems which have been proposed in different jobs to remove low molecular weight carbohydrates of the oligosaccharide mixture.[71–74] These techniques have reached high blends of FOS above 90% g FOS/gsucrose[71] and above 80%.[74] These systems have demonstrated good production yields of FOS but unfortunately these make the process cost rises.

Different studies on the production of FOS have reported the use of enzymes immobilized in calcium alginate beads, methacrylamide polymeric beads, epoxy acrylic—activated beads (Eupergit C), epoxy—activated polymethacrylate (Sepabeads EC-EP5), glass porous ion exchange resin (Amberlite IRA 900 CI), and various polymeric, and ceramic filter membrane,[11,75–80] where it is recognized that the immobilization of enzymes has proved to be an effective tool to retain enzymes in the reactors as well as providing greater stability to the enzyme in pH and temperature changes, allowing continuous operation system. Among the disadvantages of these systems are microbial contamination, can occur adsorption of feed components and hard pipe columns.

Also other techniques have been implemented as enzyme engineering[22] discloses a process for the production of 6-kestose in which uses a modified invertase expressed by Saccharomyces which exhibits improved activity of transfructosylation wherein the 6-kestose was produced with high specificity representing 95% total FOS which could make interesting use of genetic engineering as a tool to enhance the activity of the enzymes involved in the synthesis of FOS and improve their production.

The market of prebiotics is increasing, whereby, to meet a growing world demand is necessary the implementation of different techniques and bioprocess strategies which opens new possibilities and trends for production of FOS.

9.6 FOS MARKET

There is an increasing trend toward the production of prebiotic ingredient-based food products to provide innovative solutions to the consumers. Consumer demands are changing and are highly influenced by the increasing consumption of health products.[81]

Manufacturers are focusing on the development of new products to provide a wide range of end applications to answer the new demands of the consumer for new food products with nutritional and health added value.[81] As seen previous in this review, FOS belong to the prebiotics group and has been used as a sweetener over the past few years primarily in the food and beverage industry.

Transparency Market Research[82] forecasts that the global prebiotic ingredients market will improve, it states that the global prebiotic ingredients market is anticipated to reach USD4.5 billion by the end of 2018 and according to study by Grand View Research Inc.,[83] is projected to reach USD 5.75 billion by 2020. In terms of volume Global prebiotics market was

581.0 kilotons in 2013 and is expected to reach 1084.7 kilotons by 2020, growing at a Compound Annual Growth Rate of 9.3% from 2014 to 2020.[83]

Prebiotics major application sectors are in food products such as dairy, bakery, cereals, meat, and sports drinks are also driving the growth of prebiotic ingredients market.[81]

Food and beverage was the largest application segment with market volume of 488.1 kilotons in 2013.[83] Other applications include animal feeding and pharmaceuticals. Animal feed has not yet been fully developed and explored and hence, the main focus of manufacturers in animal feed is in research and development of sustainable products. Prebiotic ingredient demand for animal feed is expected to reach USD 429.3 million in 2018 and is expected to have a positive impact on the FOS market.[82]

Increasing demand for low calorie or fat-free food which provides added health benefits is expected to have a positive impact on the market.[84] The inulin and sucrose FOSs are the novel dietary fibers that fulfill these considerations.[84] The United States, Germany, Japan, and China dominate the functional food market and expected to witness growth owing to increased demand for dietary supplements. In East Asia, North America, and Europe, the FOS products are primarily employed as dietary fibers, thus growth of dietary fiber sector will increase FOS demand.[84]

Development of symbiotic combining FOS with targeted probiotic strains coupled with introduction of new products such as drinking yogurts, low-fat reduction creams, chocolates, and bakery products are expected to be key drivers for the industry over the forecast period. The product is expected to substitute numerous sweeteners in the food and beverage industry including aspartame, sucralose, and xylitol on account of their superior properties and cost effectiveness.[84]

Among the FOS products commercially available are Beneshine™ [P-type powder and liquid (>95% purity)] from sucrose by the enzyme FTase is conducted by Shenzhen Victory Biology Engineering Co., Ltd., China.[84] Meioligo manufactured by a Key player in the FOS industry Meiji Seika Kaisha Ltd. in Japan. FortiFeed® P-95, it is another product of which consists of a prebiotic soluble fiber in the mixture which contains about 95% short-chain FOS dry basis (Corn Products International, Inc.). Actilight® FOS product (Beghin-Meiji Industries, France), is a highly bifidogenic product, produced from sucrose. Another commercially available product is NutraFlora® US GNC Nutrition. Orafti Active Food Ingredients in USA produces Raftilose, an inulin that contains FOSs in addition to polysaccharides. Among the FOS products available on the market from inulin,

OLIFRUCTINE-SP® manufactured by Company Nutriagaves of Mexico S.A. de C.V., a mixture of fructan extracted from the juice of the agave plant Tequilana Weber blue variety. Other manufacturers include Jarrow Formulas, Cheil Foods and Chemicals Inc.

The following Table 9.2 summarizes the companies that handle products of FOS and the brand name of these products is done.

TABLE 9.2 Commercially Available Food-grade FOS.

Substrate	Manufacturer	Trademark
Sucrose	Beghin-Meiji Industries, Francia	Actilight®
	Cheil Foods and Chemicals Inc., Korea	Oligo-sugar
	GTC Nutrition, EU	NutraFlora®
	Meiji Seika Kaisha Ltd., Japan	Meioligo®
	Victory Biology Engineering Co., Ltd., China	Beneshine™ P-type
Inulin	Beneo-Orafti, Bélgica	Orafti®
	Cosucra Groupe Warcoing, Bélgica	Fibrulose®
	Sensus, Holanda	Frutalose®
	Nutriagaves de Mexico S.A. de C.V., Mexico	OLIFRUCTINE-SP®

In addition to the marketing of food grade FOS are available in analytical grade market with purities of 80–99%. Companies that include more supply of FOS 1-kestose (GF2), nystose (GF3), and 1F-fructofuranosyl nystose β-(GF4) are Sigma-Aldrich, Megazyme, and Wako Chemicals GmbH.[6,85]

9.7 CONCLUSIONS

The prebiotics market is gaining strength and is growing greatly in recent years due to the increasing interest in people for these products because it has raised awareness for a healthier life. This document has been reviewed many positive effects on health and in reducing the risk of disease by FOS, in addition to its many physicochemical and physiological properties and its wide application in the food industry. We reviewed the various challenges that are in the production of FOS, noting that there has been a breakthrough in the techniques and technologies in bioprocesses to improve yields, quality, and purity of the final product, however, due to the increasing demand evaluating more FOS producing strains with greater efficiency in shorter periods of time, the use of genetic engineering to create strains with better transfructosylation activity, and improve production processes where industrial and agro-industrial wastes are used in order to establish processes allow low-cost production with good yields of FOS.

KEYWORDS

- **fructooligosaccharides**
- **prebiotics**
- **nutraceutical**
- **functional foods**
- **solid-state fermentation**
- **fructosyltransferase**
- **β-fructofuranosidase**

REFERENCES

1. Bitzios, M.; Fraser, I.; Haddock-Fraser, J. Functional Ingredients and Food Choice: Results from a Dual-mode Study Employing Means-end-chain Analysis and a Choice Experiment. *Food Policy* **2011,** *36* (5), 715–725.
2. Kovács, Z.; Benjamins, E.; Grau, K.; Ur Rehman, A.; Ebrahimi, M.; Czermak, P. Recent Developments in Manufacturing Oligosaccharides with Prebiotic Functions. *Adv. Biochem. Eng. Biotechnol.* **2013,** *143*, 257–295. Retrieved from http://www.ncbi.nlm.nih.gov/pubmed/23942834
3. Howlett, J., Ed.; *Functional Foods—from Science to Health and Claims;* ILSI Europe—Concise Monograph Series: Brussels, Belgium, 2008.
4. Szajewska, H. Probiotics and Prebiotics in Preterminfants: Where are We? Where are We Going? *Early Hum. Dev.* **2010,** *86*, S81–S86.
5. De Preter, V.; Hamer, H. M.; Windey, K.; Verbeke, K. The Impact of Pre and/or Probiotics on Human Colonicmetabolism: Does it Affect Human Health? *Mol. Nutr. Food Res.* **2011,** *55*, 46–57.
6. Dominguez, A. L.; Rodrigues, L. R.; Lima, N. M.; Teixeira, J. A. An Overview of the Recent Developments on Fructooligosaccharide Production and Applications. *Food Bioprocess Technol.* **2013,** *7*, 1–14.
7. Mutanda, T.; Mokoena, M. P.; Olaniran, A. O.; Wilhelmi, B. S.; Whiteley, C. G. Microbial Enzymatic Production and Applications of Short-chain Fructooligosaccharides and Inulooligosaccharides: Recent Advances and Current Perspectives. *J. Ind. Microbiol. Biotechnol.* **2014,** *41*, 893–906.
8. Godshall, M. A. Future Directions for the Sugar Industry. *Int. Sugar J.* **2001,** *103* (1233), 378-384.

9. Biedrzycka, E.; Bielecka, M. Fructooligosaccharides Production from Sucrose by *Aspergillus* sp. N74 Immobilized in Calcium Alginate. *Trends Food Sci. Technol.* **2004,** *15*, 170–175.
10. Yun, J. W. Fructooligosaccharides Occurrence, Preparation, and Application. *Enzyme Microb. Technol.* **1996,** *19* (2), 107–117.
11. Tanriseven, A.; Aslan, Y. Immobilization of Pectinex Ultra SP-L to Produce Fructooligosaccharides. *Enzyme Microb. Technol.* **2005,** *36*, 550–554.
12. Dominguez, A.; Nobre, C.; Rodrigues, L. R.; Peres, A. M.; Torres, D.; Rocha, I.; Lima, N.; Teixeira, J. A. New Improved Method for Fructooligosaccharides Production by *Aureobasidium pullullans*. *Carbohydr. Polym.* **2012,** *89*, 1174–1179.
13. Quigley, M. E.; Hudson, G. J.; Englyst, H. N. Determination of Resistant Short-chain Carbohydrates (Non-digestible Oligosaccharides) Using Gas-liquid Chromatography. *Food Chem.* **1999,** *65*, 381–390.
14. Al-Sheraji, S. H.; Ismail, A.; Manap, M. Y.; Mustafa, S.; Yusof, R. M.; Hassan, F. A. Prebiotics as Functional Foods: A Review. *J. Funct. Foods* **2013,** *5*, 1542–1553.
15. Gibson, G. R.; Roberfroid, M. B. Dietary Modulation of the Human Colonic Microbiota. Introducing the Concept of Prebiotics. *J. Nutr.* **1995,** *125*, 1401–1412.
16. Sako, T.; Matsumoto, K.; Tanaka, R. Recent Progress on Research and Applications of Non-digestible Galacto-oligosaccharides. *Int. Dairy J.* **1999,** *9*, 69–80.
17. Mussatto, S. I.; Aguilar, C. N.; Rodrigues, L. R.; Teixeira, J. A. Colonization of *Aspergillus japonicus* on Synthetic Materials and Application to the Production of Fructooligosaccharides. *Carbohydr. Res.* **2009,** *344* (6), 795–800.
18. Scheid, M. M. A.; Moreno, Y. M. F.; Maróstica Junior, M. R.; Pastore, G. M. Effect of Prebiotics on the Health of the Elderly. *Food Res. Int.* **2013,** *53*, 426–432.
19. Vitali, B.; Ndagijimana, M.; Maccaferri, S.; Biagi, E.; Guerzoni, M. E.; Brigidi, P. An In Vitro Evaluation of the Effect of Probiotics and Prebiotics on the Metabolic Profile of Human Microbiota. *Anaerobe* **2012,** *18* (4), 386–391.
20. Duncan, S. H.; Louis, P.; Flint, H. J. Lactate-utilizing Bacteria, Isolated from Human Feces, that Produce Butyrate as a Major Fermentation Product. *Appl. Environ. Microbiol.* **2004,** *70*, 5810–5817.
21. Gomes, A. C.; Bueno, A. A.; Souza, R. G. M. de.; Mota, J. F. Gut Microbiota, Probiotics and Diabetes. *Nutr. J.* **2014,** *13*, 60. Retrieved from http://www.pubmedcentral.nih.gov/articlerender.fcgi?artid=4078018&tool=pmcentrez&rendertype=abstract
22. Marín-navarro, J.; Talens-perales, D.; Polaina, J. One-pot Production of Fructooligosaccharides by a *Saccharomyces cerevisiae* Strain Expressing an Engineered Invertase. *Appl. Microbiol. Biotechnol.* **2015,** *99*, 2549–2555.
23. Serban, D. E. Gastrointestinal Cancers: Influence of Gut Microbiota, Probiotics and Prebiotics. *Cancer Lett.* **2014,** *345*, 258–270.
24. Domokos, M.; Jakus, J.; Szeker, K.; Csizinszky, R.; Csiko, G.; Neogrady, Z.; Csordas, A.; Galfi, P. Butyrate-induced Cell Death and Differentiation are Associated with Distinct Patterns of ROS in HT29-derived Human Colon Cancer Cells. *Dig. Dis. Sci.* **2010,** *55*, 920–930.
25. Scharlau, D.; Borowicki, A.; Habermann, N.; Hofmann, T.; Klenow, S.; Miene, C.; Munjal, U.; Stein, K.; Glei, M. Mechanisms of Primary Cancer Prevention by Butyrate and Other Products Formed during Gut Flora-mediated Fermentation of Dietary Fiber. *Mutat. Res.* **2009,** *682*, 39–53.
26. O'Hara, A. M.; Shanahan, F. The Gut Flora as a Forgotten Organ. *EMBO Rep.* **2006,** *7*, 688–693.

27. Dehghan, P.; Gargari, B. P.; Jafar-abadi, M. A. Oligofructose-enriched Inulin Improves some Inflammatory Markers and Metabolic Endotoxemia in Women with Type 2 Diabetes Mellitus: A Randomized Controlled Clinical Trial. *Nutrition* **2013,** *30,* 418–423.
28. Depeint, F.; Tzortzis, G.; Vulevic, J.; I'anson, K.; Gibson, G. R. Prebiotic Evaluation of a Novel Galactooligosaccharide Mixture Produced by the Enzymatic Activity of *Bifidobacterium bifidum* NCIMB 41171, in Healthy Humans: A Randomized, Double-blind, Crossover, Placebo-controlled Intervention Study. *Am. J. Clin. Nutr.* **2008,** *87,* 785–791.
29. Rastall, R. A.; Gibson, G. R. Recent Developments in Prebiotics to Selectively Impact Beneficial Microbes and Promote Intestinal Health. *Curr. Opin. Biotechnol.* **2015,** *32,* 42–46. Retrieved from http://dx.doi.org/10.1016/j.copbio.2014.11.002
30. Levrat, M. A.; Favier, M. L.; Moundras, C.; Remesy, C.; Demigne, C.; Morand, C. Role of Dietary Propionic Acid and Bile Acid Excretion in the Hypocholesterolemic Effects of Oligosaccharides in Rats. *J. Nutr.* **1994,** *124,* 531–538.
31. Roberfroid, M. Functional Food Concept and Its Application to Prebiotics. *Dig. Liver Dis.* **2002,** *34,* S105–S110.
32. Bornet, F. R. J.; Brouns, F.; Tashiro, Y.; Duviller, V. Nutritional Aspect of Short-chain Fructooligosaccharieds: Natural Occurrence, Chemistry, Physiology and Health Implications. *Dig. Liver Dis.* **2002,** *34,* S111–S120.
33. Moshfegh, A. J.; Friday, J. E.; Goldman, J. P.; Ahuja, J. K. Presence of Inulin, and Oligofructose in the Diets of Americans. *J. Nutr.* **1999,** *129,* 1407S–1411S.
34. Carabin, I. G.; Flamm, W. G. Evaluation of Safety of Inulin and Oligofructose as Dietary Fibre. *Regul. Toxicol. Pharmacol.* **1999,** *30,* 268–282.
35. Den Hond, E.; Geypens, B.; Ghoos, Y. Effect if High Performance Chicory Inulin on Constipation. *Nutr. Res.* **2000,** *20,* 731–736.
36. Król, B.; Grzelak, K. Qualitative and Quantitative Composition of Fructooligosaccharides in Bread. *Eur. Food Res. Technol.* **2006,** *223* (6), 755–758.
37. Maiorano, A. E.; Piccoli, R. M.; Da Silva, E. S.; De Andrade Rodrigues, M. F. Microbial Production of Fructosyltransferases for Synthesis of Pre-biotics. *Biotechnol. Lett.* **2008,** *30,* 1867.
38. Sangeetha, P. T.; Ramesh, M. N.; Prapulla, S. G. Production of Fructosyl Transferase by *Aspergillus oryzae* CFR 202 in Solid-state Fermentation Using Agricultural By-products. *Appl. Microbiol. Biotechnol.* **2004,** *65* (5), 530–537.
39. Nobre, C.; Teixeira, J. A.; Rodrigues, L. R. Fructo-oligosaccharides Purificationfrom a Fermentative Broth Using an Activated Charcoal Column. *N. Biotechnol.* **2012,** *29,* 395–401.
40. Yun, J. W.; Kang, S. C.; Song, S. K. Continuous Production of Fructooligosaccharides from Sucrose by Immobilized Fructosyltransferase. *Biotechnol. Tech.* **1995,** *9,* 805–808.
41. Sangeetha, P. T.; Ramesh, M. N.; Prapulla, S. G. Production of Fructosyltransferase by *Aspergillus oryzae* CFR 202 in Solid-state Fermentation Using Agricultural By-products. *Appl. Microbiol. Biotechnol.* **2004,** *65,* 530–537.
42. Yoshikawa, J.; Amachi, S.; Shinoyama, H.; Fujii, T. Multiple B-fructofuranosidases by *Aureobasidium pullulans* DSM2404 and Their Roles in Fructooligosaccharide Production. *FEMS Microbiol. Lett.* **2006,** *265,* 159–163.
43. Jung, K. H.; Bang, S. H.; Oh, T. K.; Park, H. J. Industrial Production of Fructooligosaccharides by Immobilized Cells of *Aureobasidium pullulans* in a Packed Bed Reactor. *Biotechnol. Lett.* **2011,** *33,* 1621–1624.

44. Mussatto, S. I.; Ballesteros L. F.; Martins, S.; Maltos, D. A. F.; Aguilar, C. N.; Teixeira, J. A. Maximization of Fructooligosaccharides and β-Fructofuranosidase Production by *Aspergillus japonicas* under Solid-state Fermentation Conditions. *Food Bioprocess. Technol.* **2013,** *6*, 2128–2134.
45. Hidaka, H.; Hirayama, M.; Sumi, N. A Fructooligosaccharide-producing Enzyme from *Aspergillus niger* ATCC 20611. *Agric. Biol. Chem.* **1988,** *52*, 1181–1187.
46. Nguyen, Q. D.; Mattes, F.; Hoschke, A.; Rezessy-Szabo, J.; Bhat, M. K. Production, Purification and Identification of Fructooligosaccharides Produced by B-fructofuranosidase from *Aspergillus niger* IMI 303386. *Biotechnol. Lett.* **1999,** *21* (3), 183–186.
47. Sheu, D. C.; Lio, P. J.; Chen, S. T.; Lin, C. T.; Duan, K. J. Production of Fructooligosaccharides in High Yield Using a Mixed Enzyme System of Fructofuranosidase and Glucose Oxidase. *Biotechnol. Lett.* **2001,** *23* (18), 1499–1503.
48. Patel, V.; Saunders, G.; Bucke, C. Production of Fructooligosaccharides by *Fusarium oxysporum. Biotechnol. Lett.* **1994,** *16* (11), 1139–1144.
49. Tanriseven, A.; Gokmen, F. Novel Method for the Production of a Mixture Containing Fructooligosaccharides and Isomaltooligosaccharides. *Biotechnol. Tech.* **1999,** *13* (3), 207–210.
50. Antosová, M.; Polakovic, M. Fructosyltrasnferases: The Enzymes Catalyzing Production of Fructooligosaccharides. *Chem. Pap.* **2001,** *55* (6), 350–358. Retrieved from http://www.chempap.org/?id=7&paper=422
51. Chen, W. C.; Liu, C. H. Production of B-fructofuranosidase by *Aspergillus japonicus. Enzyme Microb. Technol.* **1996,** *18*, 153–160.
52. Dorta, C.; Cruz, R.; Neto, P. O.; Moura, D. J. C. Sugarcane Molasses and Yeast Powder Used in the Fructooligosaccharides Production by *Aspergillus japonicus*-FCL 119T and *Aspergillus niger* ATCC 20611. *J. Ind. Microbiol. Biotechnol.* **2006,** *33*, 1003–1009.
53. Park, J. P.; Oh, T. K.; Yun, J. W. Purification and Characterization of a Novel Transfructosylating Enzyme from *Bacillus macerans* EG-6. *Process Biochem.* **2001,** *37*, 471–476.
54. Sangeetha, P. T.; Ramesh, M. N.; Prapulla, S. G. Fructooli-gosaccharide Production Using Fructosyl Transferase Obtained from Recycling Culture of *Aspergillus oryzae* CFR 202. *Process Biochem.* **2005,** *40*, 1085–1088.
55. Wang, X. D.; Rakshit, S. K. Improved Extracellular Transferase Enzyme Production by *Aspergillus foetidus* for Synthesis of Isooligosaccharides. *Bioprocess Eng.* **1999,** *20*, 429–434.
56. L'Hocine, L.; Jiang, Z.; Jiang, B.; Xu, S. Purification and Partial Characterization of Fructosyltransferase and Invertase from *Aspergillus niger* AS0023. *J. Biotechnol.* **2000,** *81*, 73–84.
57. da Silva, J. B.; Fai, A. E. C.; dos Santos, R.; Basso, L. C.; Pastore, G. M. Parameters Evaluation of Fructooligosaccharides Production by Sucrose Biotransformation Using an Osmophilic *Aureobasium pullulans* Strain. *Procedia Food Sci.* **2011,** *1*, 1547–1552.
58. Vandáková, M.; Platková, M.; Antošová, M.; Báles, V.; Polakovič, M. Optimization of Cultivation Conditions for Production of Fructosyltransferase by *Aureobasidium pullulans. Chem. Pap.* **2004,** *58* (1), 15–22.
59. Sangeetha, P. T.; Ramesh, M. N.; Prapulla, S. G. Recent Trends in the Microbial Production, Analysis and Application of Fructooligosaccharides. *Trends Food Sci. Technol.* **2005,** *16*, 442–457.
60. Shin, H. T.; Baig, S. Y.; Lee, S. W.; Suh, D. S.; Kwon, S. T.; Lin, Y. B.; Lee, J. H. Production of Fructo-oligosaccharides from Molasses by *Aureobasidium pullulans* Cells. *Bioresour. Technol.* **2004,** *93*, 59–62.

61. Wang, L. M.; Zhou, H. M. Isolation and Identification of a Novel *A. japonicus* JN19 Producing β-fructofuranosidase and Characterization of the Enzyme. *J. Food Biochem.* **2006,** *30*, 641–658.
62. Dhake, A. B.; Patil, M. B. Effect of Substrate Feeding on Production of Fructosyltransferase by *Penicillium purpurogenum. Braz. J. Microbiol.* **2007,** *38*, 194–199.
63. Mussatto, S. I.; Teixeira, J. A. Increase in the Fructooligosaccharides Yield and Productivity by Solid-state Fermentation with *Aspergillus japonicus* Using Agro-industrial Residues as Support and Nutrient Source. *Biochem. Eng. J.* **2010,** *53*, 154–157.
64. Couto, S. R.; Sanromán, M. A. Application of Solid-state Fermentation to Food Industry–A Review. *J. Food Eng.* **2006,** *76* (3), 291–302.
65. Orzua, M. C.; Mussatto, S. I.; Contreras-Esquivel, J. C.; Rodriguez, R.; de la Garza, H.; Teixeira, J. A.; Aguilar, C. N. Exploitation of Agro Industrial Wastes as Immobilization Carrier for Solid-state Fermentation. *Ind. Crops Prod.* **2009,** *30* (1), 24–27.
66. Mussatto, S. I.; Ballesteros, L. F.; Martins, S.; Maltos, D. A. F.; Aguilar, C. N.; Teixeira, J. A. Maximization of Fructooligosaccharides and β-fructofuranosidase Production by *Aspergillus japonicus* under Solid-state Fermentation Conditions. *Food Bioprocess Technol.* **2013,** *6* (8), 2128–2134.
67. Yoshikawa, J.; Amachi, S.; Shinoyama, H.; Fujii, T. Production of Fructooligosaccharides by Crude Enzyme Preparations of Beta-fructofuranosidase from *Aureobasidium pullulans. Biotechnol. Lett.* **2008,** *30* (3), 535–539.
68. Crittenden, R. J.; Playne, M. J. Purification of Food Grade Oligosaccharides Using Immobilised Cells of *Zymomonas mobilis. Appl. Microbiol. Biotechnol.* **2002,** *58*, 297–302.
69. Yoon, E. J.; Yoo, S. H.; Chac, J.; Lee, H. G. Effect of Levan's Branching Structure on Antitumor Activity. *Int. J. Biol. Macromol.* **2004,** *34*, 191–194.
70. Sheu, D. C.; Chang, J. Y.; Wang, C. Y.; Wu, C. T.; Huang, C. J. Continuous Production of High-purity Fructooligosaccharides and Ethanol by Immobilized *Aspergillus japonicus* and *Pichia heimii. Bioprocess Biosyst. Eng.* **2013,** *36* (11), 1745–1751.
71. Nishizawa, K.; Nakajima, M.; Nabetani, H. Kinetic Study on Transfructosylation by β-fructofuranosidase from *Aspergillus niger* ATCC 20611 and Availability of a Membrane Reactor for Fructooligosaccharide Production. *Food Sci. Technol. Res.* **2001,** *7*, 39–44.
72. Goulas, A.; Tzortzis, G.; Gibson, G. R. Development of a Process for the Production and Purification of α-and β-galactooligosaccharides from *Bifobacterium bifidum* NCIMB 41171. *Int. Dairy J.* **2007,** *17*, 648–656.
73. Goulas, A. K.; Kapasakalidis, P. G.; Sinclair, H. R.; Rastall, R. A.; Grandison, A. S. *J. Membr. Sci.* **2002,** *209* (1), 321.
74. Sheu, D. C.; Duan, K. J.; Cheng, C. Y.; Bi, J. L.; Chen, J. Y. *Biotechnol. Prog.* **2002,** *18* (6), 1282.
75. Yun, J. W.; Jung, K. H.; Oh, J. W.; Lee, J. H. *Appl. Biochem. Biotechnol.* **1990,** *24/25*, 299–308.
76. Hayashi, S.; Nonoguchi, M.; Takasaki, Y.; Ueno, H.; Imada, K. Purification and Properties of B-fructofuranosidase from *Aureobasidium* sp. ATCC 20524. *J. Ind. Microbiol.* **1991,** *7*, 251–256.
77. Hayashi, S.; Tubouchi, M.; Takasaki, Y.; Imada, K. Long-term Continuous Reaction of Immobilized β-fructofuranosidase. *Biotechnol. Lett.* **1994,** *16*, 227. DOI: 10.1007/BF00134616. URL http://dx.doi.org/10.1007/BF00134616

78. Chiang, C. J.; Lee, W. C.; Sheu, D. C.; Duan, K. J. Immobilization of β-fructofuranosidases from *Aspergillus* on Methacrylamide-based Polymeric Beads for Production of Fructooligosaccharides. *Biotechnol. Prog.* **1997,** *13*, 577–582.
79. Ghazi, I.; Fernández-Arrojo, L.; Garcia-Arellano, H.; Ferrer, M.; Ballesteros, A.; Plou, F. J. Purification and Kinetic Characterization of a Fructosyltransferase from *Aspergillus aculeatus. J. Biotechnol.* **2007,** *128*, 204–211.
80. Csanádi, Z.; Sisak, C. *Hung. J. Ind. Chem.* **2008,** *36* (1–2), 23.
81. *Markets & Markets*. Prebiotic Ingredients Market by Type (Oligosaccharides, Inulin, Polydextrose), Application (Food & Beverage, Dietary Supplements, Animal Feed), Source (Vegetables, Grains, Roots), Health Benefits (Heart, Bone, Immunity, Digestive, Skin, Weight Management), & by Region-Global Trends & Forecast to 2020; Report Code: FB 3567, 2015. Retrieved from: http://www.marketsandmarkets.com/Market-Reports/prebiotics-ingredients-market-219677001.html
82. *Transparency Market Research.* Prebiotic Ingredients Market (FOS, GOS, MOS, Inulin) for Food & Beverage, Dietary Supplements & Animal Feed-Global Industry Analysis, Market Size, Share, Trends, and Forecast 2012–2018, 2013. Retrieved from: http://www.transparencymarketresearch.com/prebiotics-market.html
83. *Grand View Research Inc.* Prebiotics Market Analysis by Ingredients (FOS, Inulin, GOS, MOS), by Application (Food & Beverages, Animal Feed, Dietary Supplements) and Segment Forecasts To 2020. ISBN Code: 978-1-68038-089-7, 2014. Retrieved from: http://www.grandviewresearch.com/industry-analysis/prebiotics-market
84. Bali, V.; Panesar, P. S.; Bera M. B.; Panesar R. Fructooligosaccharides: Production, Purification and Potential Applications. *Crit. Rev. Food Sci. Nutr.* **2013,** *55*, 1475–1490.
85. Nobre, C.; Teixeira, J. A.; Rodrigues, L. R. New Trends and Technological Challenges in the Industrial Production and Purification of Fructooligosaccharides. *Crit. Rev. Food Sci. Nutr.* **2013,** *55*, 1444–1455. DOI:10.1080/10408398.2012.697082.

CHAPTER 10

BIO-FUNCTIONAL PEPTIDES: BIOLOGICAL ACTIVITIES, PRODUCTION, AND APPLICATIONS

GLORIA ALICIA MARTÍNEZ-MEDINA[1], ARELY PRADO-BARRAGÁN[2], JOSÉ L. MARTÍNEZ[1], HÉCTOR RUIZ[1], ROSA M. RODRÍGUEZ -JASSO[1], JUAN C. CONTRERAS-ESQUIVEL[1], AND CRISTÓBAL N. AGUILAR[1*]

[1]*Food Research Department, Chemistry School, Coahuila Autonomous University, Saltillo Unit 25280, Coahuila, México*

[2]*Biotechnology Department, Biological and Health Sciences Division, Metropolitan Autonomous University, Iztapalapa Unit 09340, Ciudad de México, México*

[*]*Corresponding author. E-mail: cristobal.aguilar@uadec.edu.mx*

CONTENTS

ABSTRACT

Bio-functional peptides are molecules with biological properties, which represent an attractive, innovative, and original alternative, in the pharmaceutical and food industries, for use as nutraceutical additives with high added value. In the last decade, the most important challenges are standardization and implementation of production methodologies for peptide generation at industrial scale and further the purification of these compounds. The aims of this review are to have a close look at the production, biological properties characterization, and possible application described so far on bio-functional peptides.

10.1 INTRODUCTION

The relationship between food and health has been recognized since Hippocrates time,[25] secondary from this fact, arises concepts like "nutraceutical" or "functional food." In agreement with Institute of Food Technologist (IFT), functional food is a component that provides, besides the basic nutrition a health benefit. These components can afford essential elements in quantities that exceed the amount for individual maintenance, growth, and standard development; also can supply other biologically active components with desirable health effects.[26] Whereas that a nutraceutical, in accordance with Canadian Pharmacopeia is a compound purified or isolated from food matrix, generally sold in pharmaceutical form and has been demonstrating present positive physiological effects, preventive or protective function against a chronic disease.[16]

Proteins represent an integral food component, that provides essential amino acids that brings energy for healthy body's growth and maintenance, additionally, various proteins possess specific biological activities and making them potential functional food ingredients.[11] Recent researches on functional food has special interest on bioactive compounds including functional peptides.[30] This terminus includes short sequences from 2 to 40 amino acids units, that are inactive in precursor protein, however, at be released can exert a broad range of biological activities.[10]

Functional peptide can be liberated by hydrolysis or fermentative process starting from different protein sources, from animal origin among which are, bovine blood, gelatin, meat, and egg, some fishes like sardine and salmon, or vegetable protein like wheat, maize, soy, rice, mushroom, pumpkin, and sorghum.[38] Amino acid sequences in peptides confer different biological properties, besides some peptides can present multifunctionality.[19] Functional peptides can exhibit antioxidant, antihypertensive, immunomodulatory even antimicrobial, anti-fungal, or anticoagulant activities.[50]

After oral administration, peptides can act locally in gastrointestinal system or can be transported and overtake and impact in peripheral tissues through circulatory system, and exert directly their physiological properties in cardiovascular, digestive, immunologic, or nervous system,[40] and therefore can has therapeutic role and act like alternative to other pharmacological molecules in body systems; offer numerous advantages over conventional therapeutics due to their bio-activities, bio-specificity, broad spectrum, low toxicity, structural diversity, and low accumulation levels in body tissues.[2]

10.2 BIO-FUNCTIONALITIES

Functional peptides, derived from food protein, have been studied and show biological activities on digestive, nervous, and immune system, manifesting a positive effect on health.[11] Within biological activities most reported listed antimicrobial, antihypertensive, immunomodulatory, and antioxidant activities.

10.2.1 ANTIMICROBIAL ACTIVITY

Antimicrobial peptides relevance comes from the generated concerns for antibiotic excessive use, motivates new molecules searching with antimicrobial activity and effectiveness but, which involves less collateral effects. Antimicrobial peptides are characterized for being positive charged molecules and presence of hydrophobic amino acid residues, that can act against a microorganism broad spectrum, included Gram-positive and Gram-negative bacteria, viruses and fungi; this because an electrostatic attraction exists between peptides and microorganism cytoplasmic membrane which is negatively charged, and subsequently peptides oligomerize and form transmembrane pores, inducing a cellular content leak, even though can proceed through other mechanism like interruption of essential process for microorganism like DNA, protein an cell wall synthesis.[7]

10.2.2 ANTIHYPERTENSIVE ACTIVITY

Cardiovascular diseases (CVDs) are a set of conditions where heart and blood vessel can be affected; between we can found coronary heart disease, stroke, and heart failures further constitute world wide's leading death cause.[51] Chronic disease like hypertension is considered as the major risk in some CVDs development.[43]

Within mechanism that regulating blood pressure in organism can be found angiotensin I converting enzyme (ACE I) classified like carboxy-peptidase (EC3.4.15.1), that plays a substantial role, being that their action generates a potent vasoconstrictor peptide.[20] Hypertension treatment mainly has been used a set of synthetic drugs inhibiting ACE action and generally produce secondary effects like cough, cutaneous reaction, or taste perturbation;[56] actually an alternative is find molecules with ACE inhibition capacity from food origin sources.[15] ACE inhibitor peptides, act binding to the enzyme active site, coupling to an inhibitor site that promotes a change on conformational structure in protein, even joining to enzyme-substrate complex, avoiding enzyme functionality.[28] This molecule submit advantage like supply sources are cheap and they have the capacity to be incorporated in functional foods.

10.2.3 IMMUNOMODULATORY ACTIVITY

Peptides with immunomodulatory capacity exist, like in the case of some peptides derived from casein and whey milk proteins, that have capacity to improve and increase immune cells functions, promoting their proliferation, as well as antibodies and cytokines production.[34] Immunomodulatory activities fomented by peptides depends on the structure and amino acid type and charge present in this molecules.[3]

10.2.4 ANTIOXIDANT ACTIVITY

Free radical, normally, is generated in the organism during respiration, also can be produced by external stimuli like environmental pollution, tobacco components, or radiation, and they can act against infections, nevertheless this type of compounds excess may result on damage in biological molecules that integrate tissues like protein, lipids even DNA, and they translated in to developing diseases like atherosclerosis, arthritis, diabetes, or cancer.[53] Also, they can be added to food products in order to retard no desirable reaction like lipid-peroxidation that promote color, flavor, and aroma changes in substitution to synthetic antioxidant which are attributed toxicity and DNA damage.[8,11]

Antioxidant activity of this molecule is influenced by diverse structural factors like molecular size, amino acid composition, and sequence.[55] They can work through different mechanism, but most commons are electron donation, lipid radical neutralization, and pro-oxidant ion and metal chelation.[52]

10.2.5 OTHER FUNCTIONALITIES

Functional peptide also can exert broad diversity activities, like opiate activity, where that molecule present affinity for opiate receptors and act like hexogen modulators in hormone release, intestinal motility, and emotional behavior.[6]

On the other hand, anti-hypercholesterolemic peptides in particular obtained from soy hydrolysis have capacity to inhibit cholesterol absorption for their solubility repression.[40]

10.3 BIO-FUNCTIONAL PEPTIDES PRODUCTION

Due to the high therapeutic relevance attributed to emergent molecules like functional peptides have been developed a set of methodologies for their production, among which may be mentioned, chemical synthesis, microbial fermentation, or enzymatic hydrolysis even combinations of these techniques, also sought alternatives as pretreatments or implementation of new technologies for production like Table 10.1 demonstrate.

Critical parameters understanding for peptide with physiological activity generation and it is relevant; within these parameters can found: protein source and their characteristics, amino acid composition, process conditions, like temperature, pH, specificity, reaction time, and in the vegetal origin peptide, protein variation, may be compromised by environmental factors like temperature, moisture, and fertility of soils in which their growth,[21] nonetheless, adequate control of this points can generate multifunctional peptides or with a biological specific activity.[52]

10.3.1 MICROBIAL FERMENTATION

Fermentation is one of the most used processes for bio-functional peptide generation and lactic acid bacteria (LAB) constitute most commonly used strains like; *Lactobacillus helveticus*, *Lactobacillus delbrueckii* ssp. *bulgaricus*, *Lactobacillus lactis* ssp. *diacetylactis*, *Lactococcus lactis* ssp.

TABLE 10.1 Recent Works in Peptide Production.

Reference	Method	Microorganism/enzyme	Substrate	Peptide functionality
[39]	Submerged fermentation	*Aeribacillus pallidus* SAT4	Synthetic medium	Antimicrobial
[23]	Submerged fermentation	*Debaryomyces hansenii*	YPG medium supplemented with casein	ECA-I inhibitor
[63]	Solid state fermentation	*Bacillus subtilis*	Walnut protein meal	Antioxidant
[36]	Submerged fermentation	*Bacillus subtilis* A14h	Mineral medium supplemented with tomato pomace	Antioxidant and antibacterial
[12]	Enzymatic hydrolysis	Pepsin, trypsin, and alcalase.	Roe egg	Antioxidant
[54]	Enzymatic hydrolysis	Protease from hepatopancreas of Pacific white shrimp	Sea bass skin	Antioxidant
[4]	Enzymatic hydrolysis	Pepsin	Caseins and whey from goat milk	Antioxidant
[20]	Enzymatic hydrolysis	Subtilisin, trypsin, and combination	Casein and whey from goat milk	ECA-I Inhibitor
Huang et al. (2015)	Enzymatic hydrolysis	Papain and flavourzyme	Fish scales	Iron binding
Sabbione et al. (2015)	Enzymatic hydrolysis	Alcalase and trypsin	Amaranth protein isolate	Antithrombotic
[14]	Enzymatic hydrolysis	Trypsin, papain, neutrase, flavourzyme, and alcalase	Egg white	Antioxidant
[49]	MW assisted proteolysis	Bromelain	Gingko protein	Antioxidant
[61]	MW, US pretreatment and, enzymatic hydrolysis	Pepsin and trypsin	Milk protein	Antioxidant
[8]	SWH	–	Squid muscle	Antioxidant
[22]	HHP pretreatment and enzymatic hydrolysis	Alcalase, Savinase, Corolase 7089, and Protamex	Squid	Anti-inflammatory and antioxidant
[66]	HHP assisted proteolysis	Alcalase, Savinase, Corolase 7089, and Protamex	Lentil protein	ECA-I inhibitor and antioxidant

Cremoris o, *Streptococcus salivarius* ssp. *thermophilus*,[11] but also has been employed organisms like *Kluyveromyces marxianus* yeast,[24] fungus *Fusarium tricinctum*,[58] and also has been reported synthetic gene development for this kind of functional products,[65] using microorganisms like *S. thermophilus*[47] or *Escherichia coli*.[31]

Functional peptide can be founded naturally on dairy products due to are rich on precursor protein for this physiological active compounds type,[11] that derived from LAB proteolysis, conducted through; complex proteolytic system, that consists of three major components: (1) cell wall binding protease, that promote initial proteolysis, turning protein in oligopeptides, (2) specific transporters, that carry out peptides to cytoplasm, and (3) intracellular peptidases that transform low weight oligopeptides into free amino acid.[13] Likewise, has been presented a report set where peptide generation has been studied in traditional fermented food like Kefir[18] or Kapi (Thai traditional fermented shrimp pastes).[33]

10.3.2 ENZYMATIC HYDROLYSIS

Protein enzymatic hydrolysis is one of the most used methodologies for bio-functional peptide generation.[34] This process can be optimized, through certain physicochemical parameters control, like pH or temperature, providing ideal conditions for protease action.[32] Proteinases wide variety like chymotrypsin, alcalase, pepsin, thermolysin, or even enzymes from bacteria or fungal sources and also combination is employed in bio-functional peptides,[37] generally from animal or vegetable protein substrates, being milk proteins the most used;[57] but also bovine blood, meat, gelatin, egg, wheat, soy, rice, mushroom, pumpkin, and sorghum,[38] further, has been obtained peptides from marine substrates like algae, fishes, mollusks, crustaceans, or by-products like viscera, substandard muscles or skin.[41]

This peptide manufacture method has advantages in the generation of peptide defined profiles and also, highly concentrated alkalis or acid residues are not generated, that limits their use in products intended to human consumption, how takes place in chemical hydrolysis; as well as the fact that improve the generation of L-form amino acids that constitute molecules that promote biological activities, making them viable for their application in functional food formulation or nutraceuticals.[5]

Also exist methodologies that improve enzyme immobilization during the production process, in which highlights advantages like easy enzyme recuperation and high purity products.[45,62]

10.3.3 EMERGENT TECHNOLOGIES

Recent advances in functional peptides generation field are focus in search alternative protein sources and new production methodologies, like ultrasound (US) and microwave (MW), supercritical fluids hydrolysis (SPF), and high hydrostatic pressure (HHP) procedures.

10.3.3.1 ULTRASOUND

US is defined like acoustic wave, with above 20 kHz frequency, overpassing the human ear limits and that needs an external medium to propagate; this wave comes from a vibrational body, and when they propagated in surrounding medium, this start to oscillate, whereby oscillation too the particles transmit energy to each other.[35]

High intensity US technologies application in food matrix like proteins, implies the generation of physicochemical changes on this molecules,[42] due to cavitational, mechanical, and thermal effects generated, micro-jets formation, micro-turbulence, high velocity inter-particles collisions, and micropore perturbation, that induces the generation of elevated pressures and temperatures in medium, over 5000 K or 500 atm, also free radical generation, and result in a improve and increase in quality and velocity of extraction,[30] that can be exploited in food and pharmaceutical industry for functional peptide production, where US is implemented like a pretreatment.[29,67] On their simultaneous application during enzymatic hydrolysis,[64] due to the fact of this technology, have capacity to realize structural and conformational modifications in protein, affecting hydrogen bonds or hydrophilic interactions and therefore tertiary and quaternary structures, also, the physical and mechanical, caused by US can cause protein unfolding which can promote exposure of proteolysis susceptible sites and cause a major degreed of hydrolysis. But, between the adverse effects are conformation damage in the enzyme and therefore in the hydrolytic activity.[44]

10.3.3.2 MICROWAVE

On the other hand, technologies like MW, that basically, allows that molecules involves in reaction, fluctuate under magnetic field, where the energy is converted in heat and therefore it is possible to carry out chemical reactions,[9,46] which are accelerated, with increased yields and highly pure products,[27] has been reported be useful, like pretreatment or in assisted enzymatic or chemical hydrolysis for peptide production with functional potential from food proteins.

10.3.3.3 SUBCRITICAL WATER

Employing fluids like water under subcritical condition, that means use of water over their boiling point but under their critical point, keeping at liquid form with pressure,[48] this conditions modify natural characteristics of water regarding hydrogen bonds, ionic products, and dielectric constant, supplying a medium that facilitates some reactions like hydrolysis, this technology gained importance in extraction of products with functional properties like peptides, because it is an eco-friendly alternative.[8]

10.3.3.4 HIGH HYDROSTATIC PRESSURE

Non-thermal treatments like HHP, where materials are subjected to pressures from 100 to 1000 MPa in a vessel that promotes no covalent bonds rupture that improve set of structural modifications.[59] This is an alternative technology for food industry in interest compounds extraction from protein, using them like pretreatment, because this process has an effect where the protein structure is altered and exposing to enzyme action.[60]

10.4 RECOVERY PROCESS

Purification and separation process in functional peptide production allowing successful recovery, represent crucial in process, the purification method selection, depends largely of methodologies to generate, generally has been used methodologies widely applied on protein purification like solvent selective precipitation, ultrafiltration techniques, and chromatographic methods,[1] also properties like charges, variation in molecular weight, and affinity during purification methodologies are the most significant barriers, for this reason scientific propose methodologies that use complex instrumentation to solve this challenge set,[50] likewise has been applied combination on this type of methodologies, while other authors separate peptides with apparent anticancer properties through technologies named electrodialysis with an ultrafiltration membrane.[17] Usually, this technologies implementation are successful at laboratory level but when are treated to scale up, have repercussions in an enormous increase in bioprocess costs, being this stage demanding on new procedures and technologies that simplify and reduces cost on this phase.

10.5 CONCLUSIONS

Bioprocess implicated in functional peptide generation, implicate a set of complex steps, in which scientific research perform an effort to optimize the process and generate new production techniques, and also that in vitro functional activities expressed by compounds, remains after consumption, coupled to the intention of reducing their bitter taste; but especially the most outstanding challenge, like in other bioprocess, the raised cost implied in purification techniques of functional peptides intended to human consumption, whose propose presents a positive effect in health.

KEYWORDS

- **bio-functional peptides**
- **amino acids**
- **enzyme**
- **protein**
- **antihypertensive**

REFERENCES

1. Agyei, D., et al. Bioprocess Challenges to the Isolation and Purification of Bioactive Peptides. *Food Bioprod. Process.* **2016,** *98*, 244–256. Available at: http://dx.doi.org/10.1016/j.fbp.2016.02.003
2. Agyei, D.; Danquah, M. K. Industrial-scale Manufacturing of Pharmaceutical-grade Bioactive Peptides. *Biotechnol. Adv.* **2011,** *29* (3), 272–277. Available at: http://linkinghub.elsevier.com/retrieve/pii/S0734975011000024
3. Agyei, D.; Danquah, M. K. Rethinking Food-derived Bioactive Peptides for Antimicrobial and Immunomodulatory Activities. *Trends Food Sci. Technol.* **2012,** *23* (2), 62–69. Available at: http://linkinghub.elsevier.com/retrieve/pii/S0924224411001610

4. Ahmed, A. S., et al. Identification of Potent Antioxidant Bioactive Peptides from Goat Milk Proteins. *Food Res. Int.* **2015,** *74*, 80–88. Available at: http://www.sciencedirect.com/science/article/pii/S0963996915001957
5. Aluko, R. Bioactive Peptides. In *Functional Food and Nutraceuticals;* Springer: New York, NY, 2012; pp 37–61. Available at: http://link.springer.com/10.1007/978-1-4614-3480-1_3 (accessed Aug 25, 2016).
6. Alvarado Carrasco, C.; Guerra, M. Lactosuero Como Fuente de Péptidos Bioactivos. *An. Venez. Nutr.* **2010,** *23* (1), 45–50. Available at: http://www.scielo.org.ve/scielo.php?script=sci_arttext&amp\npid=S0798-07522010000100007
7. Aoki, W.; Kuroda, K.; Ueda, M. Next Generation of Antimicrobial Peptides as Molecular Targeted Medicines. *J. Biosci. Bioeng.* **2012,** *114* (4), 365–370. Available at: http://dx.doi.org/10.1016/j.jbiosc.2012.05.001
8. Asaduzzaman, A. K. M.; Chun, B. S. Recovery of Functional Materials with Thermally Stable Antioxidative Properties in Squid Muscle Hydrolyzates by Subcritical Water. *J. Food Sci. Technol.* **2013,** *52* (2), 793–802.
9. Budarin, V. L., et al. The Potential of Microwave Technology for the Recovery, Synthesis and Manufacturing of Chemicals from Bio-wastes. *Catal. Today.* **2015,** *239*, 80–89.
10. Carrasco-Castilla, J., et al. Use of Proteomics and Peptidomics Methods in Food Bioactive Peptide Science and Engineering. *Food Eng. Rev.* **2012,** *4* (4), 224–243. Available at: http://link.springer.com/10.1007/s12393-012-9058-8
11. de Castro, R. J. S.; Sato, H. H. Biologically Active Peptides: Processes for Their Generation, Purification and Identification and Applications as Natural Additives in the Food and Pharmaceutical Industries. *Food Res. Int.* **2015,** *74*, 185–198. Available at: http://linkinghub.elsevier.com/retrieve/pii/S0963996915300028
12. Chalamaiah, M., et al. Antioxidant Activity and Functional Properties of Enzymatic Protein Hydrolysates from Common Carp (*Cyprinus carpio*) Roe (Egg). *J. Food Sci. Technol.* **2015,** *52*, 8300–8307. Available at: http://www.scopus.com/inward/record.url?eid=2-s2.0-84921364741&partnerID=tZOtx3y1
13. Chaves-López, C., et al. Impact of Microbial Cultures on Proteolysis and Release of Bioactive Peptides in Fermented Milk. *Food Microbiol.* **2014,** *42*, 117–121. Available at: http://dx.doi.org/10.1016/j.fm.2014.03.005
14. Chen, C., et al. Purification and Identification of Antioxidant Peptides from Egg White Protein Hydrolysate. *Amino Acids.* **2012,** *43* (1), 457–466.
15. Chen, J., et al. Comparison of Analytical Methods to Assay Inhibitors of Angiotensin I-converting Enzyme. *Food Chem.* **2013,** *141* (4), 3329–3334. Available at: http://linkinghub.elsevier.com/retrieve/pii/S030881461300825X
16. Cicero, A. F. G.; Parini, A.; Rosticci, M. Nutraceuticals and Cholesterol-lowering Action. *IJC Metab. Endocr.* **2015,** *6*, 1–4. Available at: http://linkinghub.elsevier.com/retrieve/pii/S2214762414000516
17. Doyen, A., et al. Demonstration of In Vitro Anticancer Properties of Peptide Fractions from a Snow Crab By-products Hydrolysate after Separation by Electrodialysis with Ultrafiltration Membranes. *Sep. Purif. Technol.* **2011,** *78* (3), 321–329. Available at: http://dx.doi.org/10.1016/j.seppur.2011.01.037
18. Ebner, J., et al. Peptide Profiling of Bovine Kefir Reveals 236 Unique Peptides Released from Caseins During Its Production by Starter Culture or Kefir Grains. *J. Proteom.* **2015,** *117*, 41–57. Available at: http://www.sciencedirect.com/science/article/pii/S1874391915000135

19. Erdmann, K.; Cheung, B. W. Y.; Schröder, H. The Possible Roles of Food-derived Bioactive Peptides in Reducing the Risk of Cardiovascular Disease. *J. Nutr. Biochem.* **2008,** *19* (10), 643–654. Available at: http://linkinghub.elsevier.com/retrieve/pii/S0955286307002756
20. Espejo-Carpio, F. J., et al. Angiotensin I-converting Enzyme Inhibitory Activity of Enzymatic Hydrolysates of Goat Milk Protein Fractions. *Int. Dairy J.* **2013,** *32* (2), 175–183. Available at: http://linkinghub.elsevier.com/retrieve/pii/S0958694613001143
21. Fernandez Figares, I., et al. Amino-acid Composition and Protein and Carbohydrate Accumulation in the Grain of Triticale Grown Under Terminal Water Stress Simulated by a Senescing Agent. *J. Cereal Sci.* **2000,** *32* (3), 249–258. Available at: http://www.sciencedirect.com/science/article/pii/S0733521000903291
22. Garcia-Mora, P., et al. High-pressure Improves Enzymatic Proteolysis and the Release of Peptides with Angiotensin I Converting Enzyme Inhibitory and Antioxidant Activities from Lentil Proteins. *Food Chem.* **2015,** *171*, 224–232. Available at: http://dx.doi.org/10.1016/j.foodchem.2014.08.116
23. García-Tejedor, A., et al. Dairy Debaryomyces Hansenii Strains Produce the Antihypertensive Casein-derived Peptides LHLPLP and HLPLP. *LWT. Food Sci. Technol.* **2015,** *61* (2), 550–556.
24. Hamme, V., et al. Crude Goat Whey Fermentation by *Kluyveromyces marxianus* and *Lactobacillus rhamnosus:* Contribution to Proteolysis and ACE Inhibitory Activity. *J. Dairy Res.* **2009,** *76* (2), 152–157. Available at: http://www.ncbi.nlm.nih.gov/pubmed/19121243
25. Hasler, C. M. Functional Foods: Benefits, Concerns and Challenges—A Position Paper from the American Council on Science and Health. *J. Nutr.* **2002,** *132* (12), 3772–3781.
26. Hasler, C. M.; Brown, A. C. Position of the American Dietetic Association: Functional Foods. *J. Am. Diet Assoc.* **2009,** *109* (4), 735–746.
27. Hoz, A. de la.; Diaz-Ortiz, A.; Moreno, A. Microwaves in Organic Synthesis. Thermal and Non-thermal Microwave Effects. *Chem. Soc. Rev.* **2005,** *34*, 164–178. Available at: http://pubs.rsc.org/en/content/articlehtml/2005/cs/b411438h
28. Jao, C. L.; Huang, S. L.; Hsu, K. C. Angiotensin I-converting Enzyme Inhibitory Peptides: Inhibition Mode, Bioavailability, and Antihypertensive Effects. *BioMedicine.* **2012,** *2* (4), 130–136. Available at: http://www.sciencedirect.com/science/article/pii/S2211802012000526
29. Jia, J., et al. The Use of Ultrasound for Enzymatic Preparation of ACE-inhibitory Peptides from Wheat Germ Protein. *Food Chem.* **2010,** *119* (1), 336–342. Available at: http://dx.doi.org/10.1016/j.foodchem.2009.06.036
30. Kadam, S. U., et al. Ultrasound Applications for the Extraction, Identification and Delivery of Food Proteins and Bioactive Peptides. *Trends Food Sci. Technol.* **2015,** *46* (1), 60–67. Available at: http://linkinghub.elsevier.com/retrieve/pii/S092422441500179X
31. Kim, H. K., et al. Expression of the Cationic Antimicrobial Peptide Lactoferricin Fused with the Anionic Peptide in *Escherichia coli. Appl. Microbiol. Biotechnol.* **2006,** *72* (2), 330–338.
32. Kim, S. K.,Wijesekara, I. Development and Biological Activities of Marine-derived Bioactive Peptides: A Review. *J. Funct. Foods.* **2010,** *2* (1), 1–9.
33. Kleekayai, T., et al. Extraction of Antioxidant and ACE Inhibitory Peptides from Thai Traditional Fermented Shrimp Pastes. *Food Chem.* **2015,** *176*, 441–447. Available at: http://www.sciencedirect.com/science/article/pii/S0308814614019281

34. Korhonen, H.; Pihlanto, A. Bioactive Peptides: Production and Functionality. *Int. Dairy J.* **2006,** *16* (9), 945–960. Available at: http://linkinghub.elsevier.com/retrieve/pii/S0958694605002426
35. Mason, T. J.; Lorimer, J. P. *Applied Sonochemistry;* FRG: Wiley-VCH Verlag GmbH & Co. KGaA: Weinheim, Germeny, 2002. Available at: http://doi.wiley.com/10.1002/352760054X (accessed Aug 25, 2016).
36. Moayedi, A.; Hashemi, M.; Safari, M. Valorization of Tomato Waste Proteins Through Production of Antioxidant and Antibacterial Hydrolysates by Proteolytic *Bacillus subtilis:* Optimization of Fermentation Conditions. *J. Food Sci. Technol.* **2016,** *53* (1), 391–400.
37. Mohanty, D. P., et al. Milk Derived Bioactive Peptides and Their Impact on Human Health—A Review. *Saudi J. Biol. Sci.* **2015,** *23*, 577–583. Available at: http://linkinghub.elsevier.com/retrieve/pii/S1319562X15001382
38. Möller, N. P., et al. Bioactive Peptides and Proteins from Foods: Indication for Health Effects. *Eur. J. Nutr.* **2008,** *47* (4), 171–182. Available at: http://link.springer.com/10.1007/s00394-008-0710-2
39. Muhammad, S. A.; Ahmed, S. Production and Characterization of a New Antibacterial Peptide Obtained from *Aeribacillus pallidus* SAT4. *Biotechnol. Rep.* **2015,** *8*, 72–80.
40. Mulero Cánovas, J., et al. Péptidos Bioactivos. *Clin. Investig. Arterioscler.* **2011,** *23* (5), 219–227.
41. Ngo, D. H., et al. Biological Activities and Potential Health Benefits of Bioactive Peptides Derived from Marine Organisms. *Int. J. Biol. Macromol.* **2012,** *51* (4), 378–383. Available at: http://dx.doi.org/10.1016/j.ijbiomac.2012.06.001
42. O'Donnell, C. P., et al. Effect of Ultrasonic Processing on Food Enzymes of Industrial Importance. *Trends Food Sci. Technol.* **2010,** *21* (7), 358–367.
43. Ogedegbe, G.; Pickering, T. G. Epidemiology of Hypertension, Chapter 68. In *Hurst's The Heart*; Fuster, V., Walsh, R. A., Harrington, R. A., Eds.; 13th ed.; McGraw Hill Education: USA, 2011.
44. Ozuna, C., et al. Innovative Applications of High-intensity Ultrasound in the Development of Functional Food Ingredients: Production of Protein Hydrolysates and Bioactive Peptides. *Food Res. Int.* **2015,** *77*, 685–696. Available at: http://dx.doi.org/10.1016/j.foodres.2015.10.015
45. Pedroche, J., et al. Obtaining of *Brassica carinata* Protein Hydrolysates Enriched in Bioactive Peptides Using Immobilized Digestive Proteases. *Food Res. Int.* **2007,** *40* (7), 931–938.
46. Reddy, P. M., et al. Evaluating the Potential Nonthermal Microwave Effects of Microwave-assisted Proteolytic Reactions. *J. Proteom.* **2013,** *80*, 160–170. Available at: http://dx.doi.org/10.1016/j.jprot.2013.01.005
47. Renye, J. A., Somkuti, G. A.Cloning of Milk-derived Bioactive Peptides in *Streptococcus thermophilus. Biotechnol. Lett.* **2008,** *30* (4), 723–730.
48. Rogalinski, T.; Herrmann, S.; Brunner, G. Production of Amino Acids from Bovine Serum Albumin by Continuous Sub-critical Water Hydrolysis. *J. Supercrit. Fluids.* **2005,** *36* (1), 49–58.
49. Ruan, G., et al. The Study on Microwave Assisted Enzymatic Digestion of Ginkgo Protein. *J. Mol. Catal B Enzym.* **2013,** *94*, 23–28. Available at: http://dx.doi.org/10.1016/j.molcatb.2013.04.010

50. Saadi, S., et al. Recent Advances in Food Biopeptides: Production, Biological Functionalities and Therapeutic Applications. *Biotechnol. Adv.* **2014,** *33* (1), 80–116. Available at: http://www.sciencedirect.com/science/article/pii/S0734975014001906
51. Salehi-Abargouei, A., et al. Effects of Dietary Approaches to Stop Hypertension (DASH)-style Diet on Fatal or Nonfatal Cardiovascular Diseases-incidence: A Systematic Review and Meta-analysis on Observational Prospective Studies. *Nutrition.* **2013,** *29* (4), 611–618. Available at: http://dx.doi.org/10.1016/j.nut.2012.12.018
52. Samaranayaka, A. G. P.; Li-Chan, E. C. Y. Food-derived Peptidic Antioxidants: A Review of Their Production, Assessment, and Potential Applications. *J. Funct. Food.* **2011,** *3* (4), 229–254.
53. Sarmadi, B. H.; Ismail, A. Antioxidative Peptides from Food Proteins: A Review. *Peptides.* **2010,** *31* (10), 1949–1956. Available at: http://linkinghub.elsevier.com/retrieve/pii/S0196978110002640
54. Senphan, T.; Benjakul, S. Antioxidative Activities of Hydrolysates from Seabass Skin Prepared Using Protease from Hepatopancreas of Pacific White Shrimp. *J. Funct. Food.* **2014,** *6,* 147–156. Available at: http://linkinghub.elsevier.com/retrieve/pii/S1756464613002284
55. Sila, A.; Bougatef, A. Antioxidant Peptides from Marine By-products: Isolation, Identification and Application in Food Systems. A Review. *J. Funct. Food.* **2016,** *21*, 10–26. Available at: http://dx.doi.org/10.1016/j.jff.2015.11.007
56. Spiller, H. A. Angiotensin Converting Enzyme (ACE) Inhibitors. In *Encyclopedia of Toxicology;* 2014; Elsevier. Vol 4, Set. 1-4, pp 14–16.
57. Srinivas, S.; Prakash, V. Bioactive Peptides from Bovine Milk α-Casein: Isolation, Characterization and Multifunctional Properties. *Int. J. Pept. Res. Ther.* **2010,** *16* (1), 7–15. Available at: http://link.springer.com/10.1007/s10989-009-9196-x
58. Tejesvi, M. V., et al. An Antimicrobial Peptide from Endophytic *Fusarium tricinctum* of *Rhododendron tomentosum* Harmaja. *Fungal Divers.* **2013,** *60* (1), 153–159.
59. Tian, Y., et al. Effect of High Hydrostatic Pressure (HHP) on Slowly Digestible Properties of Rice Starches. *Food Chem.* **2014,** *152*, 225–229. Available at: http://dx.doi.org/10.1016/j.foodchem.2013.11.162
60. Toldrà, M., et al. Hemoglobin Hydrolysates from Porcine Blood Obtained through Enzymatic Hydrolysis Assisted by High Hydrostatic Pressure Processing. *Innov. Food Sci. Emerg. Technol.* **2011,** *12* (4), 435–442. Available at: http://dx.doi.org/10.1016/j.ifset.2011.05.002
61. Uluko, H., et al. Effects of Thermal, Microwave, and Ultrasound Pretreatments on Antioxidative Capacity of Enzymatic Milk Protein Concentrate Hydrolysates. *J. Funct. Food.* **2015,** *18* (2), 1138–1146. Available at: http://dx.doi.org/10.1016/j.jff.2014.11.024
62. Wang, Y., et al. Preparation of Active Corn Peptides from Zein Through Double Enzymes Immobilized with Calcium Alginate-chitosan Beads. *Process Biochem.* **2014,** *49* (10), 1682–1690.
63. Wu, W., et al. Optimization of Production Conditions for Antioxidant Peptides from Walnut Protein Meal Using Solid-state Fermentation. *Food Sci. Biotechnol.* **2014,** *23* (6), 1941–1949.
64. Yang, B., et al. Amino Acid Composition, Molecular Weight Distribution and Antioxidant Activity of Protein Hydrolysates of Soy Sauce Lees. *Food Chem.* **2011,** *124* (2), 551–555. Available at: http://dx.doi.org/10.1016/j.foodchem.2010.06.069

65. Zambrowicz, A., et al. Manufacturing of Peptides Exhibiting Biological Activity. *Amino Acids.* **2013,** *44* (2), 315–320.
66. Zhang, Y., et al. In Vitro Anti-inflammatory and Antioxidant Activities and Protein Quality of High Hydrostatic Pressure Treated Squids (*Todarodes pacificus*). *Food Chem.* **2016,** *203*, 258–266. Available at: http://dx.doi.org/10.1016/j.foodchem.2016.02.072
67. Zou, Y., et al. Enzymolysis Kinetics, Thermodynamics and Model of Porcine Cerebral Protein with Single-frequency Countercurrent and Pulsed Ultrasound-assisted Processing. *Ultrason. Sonochem.* **2016,** *28*, 294–301. Available at: http://dx.doi.org/10.1016/j.ultsonch.2015.08.006.

CHAPTER 11

GUAR GUM AS A PROMISING HYDROCOLLOID: PROPERTIES AND INDUSTRY OVERVIEW

CECILIA CASTRO-LÓPEZ[1], JUAN C. CONTRERAS-ESQUIVEL[1], GUILLERMO C. G. MARTINEZ-AVILA[2], ROMEO ROJAS[2], DANIEL BOONE-VILLA[3], CRISTÓBAL N. AGUILAR[1], and JANETH M. VENTURA-SOBREVILLA[1*]

[1]*Departamento de Investigación en Alimentos, Facultad de Ciencias Químicas, Universidad Autónoma de Coahuila, Venustiano Carranza e Ing, José Cárdenas s/n, Col. República, Saltillo 25280, Coahuila, México*

[2]*Facultad de Agronomía, Universidad Autónoma de Nuevo León, Francisco Villa s/n, Col. Ex-Hacienda El Canadá, Escobedo 66050, Nuevo León, México*

[3]*School of Health Sciences, University of the Valley of Mexico Campus Saltillo, Tezcatlipoca 2301, Saltillo 25204, Coahuila, Mexico*

[*]*Corresponding author. E-mail: janethventura@uadec.edu.mx*

CONTENTS

ABSTRACT

Guar gum (GG) is a galactomannan, obtained from *Cyamopsis tetragonolobus* (Leguminosae) which has been cultivated in India and Pakistan for centuries. Chemically, GG is a hydrocolloid polysaccharide composed by mannose and galactose in molecular ratio of 2:1. It has wide applications in pharmaceutical formulations, cosmetics, foods, paper, mining, and many other industries, and is used as a natural thickener, emulsifier, stabilizer, bonding and gelling agent, soil stabilizer, natural fiber, flocculant, and fracturing agent. GG is white to yellowish white powder, nearly odorless with bland taste that is manufactured by mechanical extraction of endosperm from the guar seed. These seeds are separated from the plant and dried. Refined guar splits are first obtained by roasting, dehusking, and polishing. These splits are then pulverized and tailor made in various mesh sizes for usage in the different industries. It is practically insoluble in organic solvents. In cold or hot water, it disperses and swells almost immediately to form some highly viscous thixotropic solutions. Through references reported in the literature about the GG, the aim of this article was to review the occurrence of this gum, its production, physicochemical properties, identification, and industrial applications.

11.1 INTRODUCTION

The term hydrocolloid is derived from the Greek "hydro" (water) and "kolla" (glue). Hydrocolloids can be defined as molecules of high molecular weight that usually have colloidal properties, capable of producing gels when are combined with the appropriate solvent. The presence of a large number of hydroxyl groups in their structure significantly increases their affinity for water molecules, making them hydrophilic compounds.[37] This term is applied to a variety of substances with gummy characteristics. However, it is more common to use the term to refer to polysaccharides or their derivatives, obtained from plants or microbiological processing and which may or may not be chemically modified to change or improve their technological capabilities.[77,78,80]

The heterogeneous group consists of polysaccharides and proteins and the researchers have mainly studied them due to their range of use in industry and the functionalities that impart to whatever system or product into which they are incorporated.[4,59,81] The commercially important hydrocolloids and their origins are given in Figure 11.1. In recent years, hydrocolloids from some seeds like quince gum (*Cydonia oblonga*), vinal gum (*Prosopis ruscifolia*), and espina corona gum (*Gleditsia amorphoides*),[6,56,61] especially guar gum (GG) *(Cyamopsis tetragonolobus)* have evoked tremendous interest due to its several industry applications.[60,64] GG has a wide range of functional properties, these include thickening, gelling, emulsifying, stabilization, and controlling the crystal growth, among others, but the functionality in these capacities can be attributed to the chemical characteristics of the monomers and the polymer's molecular structure (including the chain length and branching pattern where pertinent). These characteristics also determine how the GG yield is affected by factors such as temperature, pH, presence of certain ions, etc.[45,46,54] Consequently, there is a large body of knowledge of how to exploit GG properties in order to improve future applications. In this chapter, information on GG, origin, manufacturing, structure, properties, and applications are described in subsequent sections to understand better its importance and potential.

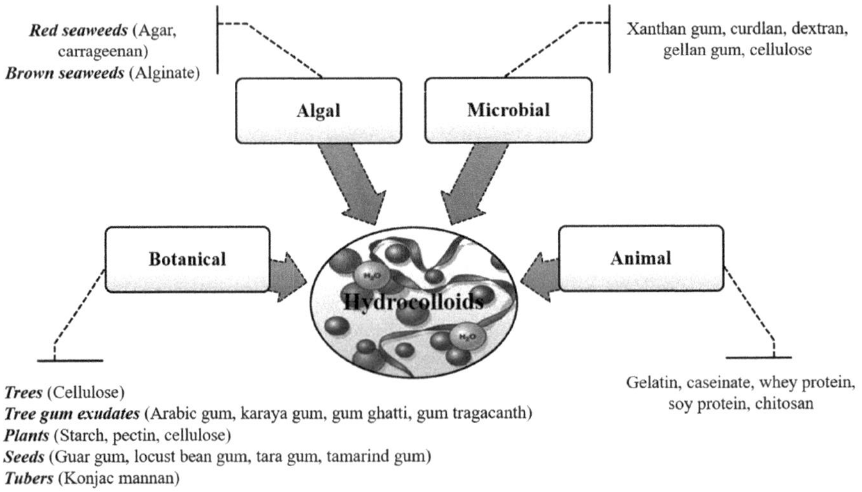

FIGURE 11.1 Sources of important hydrocolloids.

11.2 GUAR GUM

11.2.1 CULTIVATION

The guar plant, *C. tetragonolobus* (L.) Taub. (family Leguminosae), commonly known as guaran or cluster bean, is an important leguminous annual crop. It grows upright, reaching a height of 2–3 m. It has a main single stem with either basal branching or fine branching along the stem. It is a robust, bushy, semi-upright type of plant. Guar has well-developed tap root system. Stems and branches are angular, grooved, forked hairs, and sometimes glaucous. Guar has branched and un-branched growth habit. It has pointed saw-toothed, alternate, and trifoliate leaves with small purple and white flowers borne along the axis of spikelet. It bears hairy pods in clusters of 4–12 cm length, each pod with 7–8 seeds. Seed is hard, flinty, flattened, ovoid, and about 5 mm long, also are white, grey or black in color (Fig. 11.2). Guar plant grows in specific climatic condition, which ensures a soil temperature around 25°C for proper germination, long photoperiod, with humid air during its growth period and finally short photo-period with cool dry air at flowering and pod formation.[30] This crop prefers a well-drained sandy loam soil. It can tolerate saline and moderately alkaline soils with pH ranging between 7.5 and 8.0. Heavy clay soils, poor in nodulation, and bacterial activities are not suitable. Finally, the plants are sown after the first rains in July and harvested in October–November, being a short-cycle crop that is harvested within 3–4 months of its plantation.[43,66]

FIGURE 11.2 Guar (a) pods, (b) seeds, (c) splits, and (d) powder.

Guar plant is native to the Indian subcontinent; however, this crop is also grown in other parts of the world, like, Pakistan, Australia, Brazil, USA, Malawi, and South Africa.[83] India is the world leader with 80% production of guar with its cultivation in semiarid, Northwestern parts of country encompassing states of Rajasthan, Gujarat, Haryana, and Punjab. Pakistan with 15% of world production is next to India. Remaining 5% guar is produced in rest of the world with a total production of 15,000 Tons per annum. Efforts have been made to promote cultivation of guar in Australia by the Department of Agriculture and Rural Industrial Development Agency where the researchers found that guar could be grown in the northern region relatively successfully. Similarly, it is reported that countries like China and Thailand are also trying to grow guar. Therefore, in the future, economy developed by cultivation of guar may not remain monopolized by India and Pakistan.[3,40]

11.2.2 EXTRACTION

In general, the guar seeds consist of three parts: the hull (14–17%), germ (43–47%), and endosperm (35–42%). GG, the principal marketable processed product of the plant, comes from the endosperm.[62] Several methods have been used for the manufacture of different grades of GG, but due to its complex nature, the thermo-mechanical process is generally used for the manufacture of edible grade and industrial grade GG.[28,48]

This process is normally undertaken by using unit operations of roasting, differential attrition, sieving, and polishing (Fig. 11.3).[28] It is very important to select guar split in this process. The split is screened to clean it and then soaked to pre-hydrate it in a double cone mixer. The prehydrating stage is very important because it determines the rate of hydration of the final product. Soaked splits, which have reasonably high moisture content, are passed through a flaker. Flaked guar splits are ground and then dried. Obtained powder is screened through rotary screens to deliver required particle size. Oversized particles are either recycled to main ultra fine or reground in a separate regrind plant, according to the viscosity requirement. This stage helps to reduce load at the grind step. Soaked splits are difficult to grind. Direct grinding of those generates undesirable overheating in the grinder, which reduces hydration of the final product. Through the heating, grinding, and polishing process, husk is separated from the endosperm halves and refined GG splits are obtained. After grinding process, refined guar splits are then treated and converted into powder. The split manufacturing process yields husk and germ called "guar meal," widely sold in the international market as cattle feed. Manufacturers define different quality grades of GG by its particle size, the viscosity generated with a given concentration, and the rate at which that viscosity is developed.[12,27,52,53,67]

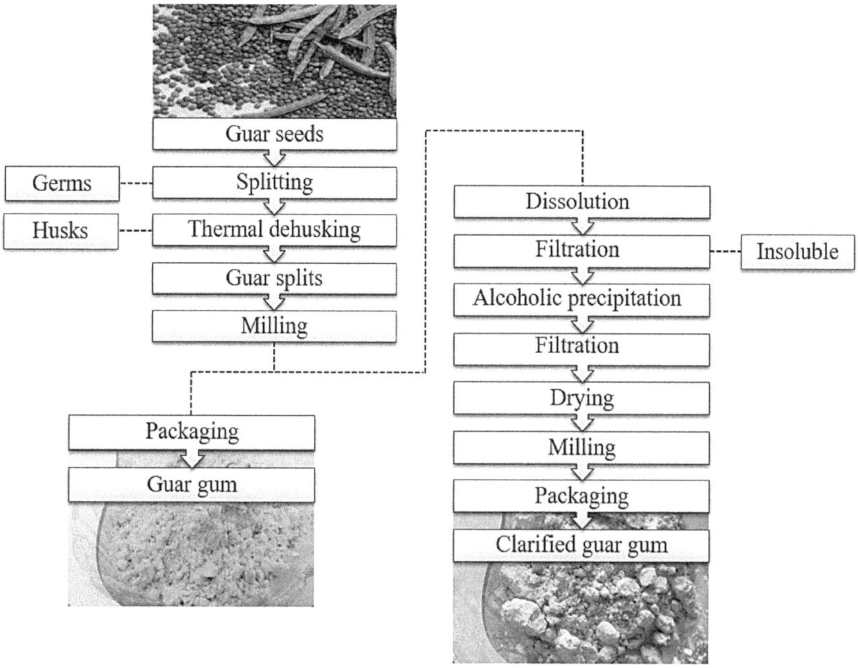

FIGURE 11.3 Guar gum flowchart.

11.2.3 DERIVATIVES

Modification of GG is done by two ways either substitution process that involves replacement of free hydroxyl group of the carbohydrate backbone by different cationic, anionic, amphoteric, and nonionic groups, or hydrolyzation process that means enzymatic, acidic, and basic hydrolysis of guar to yield low molecular weight gum. These modifications enhance properties and applications of guar in a broad spectrum of industries.[34,58,74]

Many derivatives of GG have been prepared and reported in the literature, some of these are carboxymethyl guar gum,[51] hydroxymethyl guar gum,[31] hydroxypropyl guar gum,[32] O-carboxymethyl-O-hydroxypropyl guar gum (CMHPG),[69] O-2-hydroxy-3-(trimethylammonium propyl) guar gum (HTPG), O-carboxymethyl-O-2-hydroxy-3-(trimethylammonium propyl) guar gum (CMHTPG),[33] acryloyloxy guar gum,[68] methacryloyl guar gum,[82] sulfated guar gum,[36] and GG esters.[19] The derivatives of GG and some applications are presented in Table 11.1.

TABLE 11.1 Currently Manufactured Guar Gum Derivatives.

Guar gum derivatives				
Type of derivative	**Substituent group**	**Ionic charge**	**Application**	**Reference**
Carboxymethyl guar gum (CMG)	$-CH_2-COO^-Na$	Anionic	Nanoparticles for drug delivery	[18]
			Treatment of wastewater	[49]
Hydroxypropyl guar gum (HPG)	$-CH_2-CH(OH)CH_3$	Nonionic	Lubricant drops for eye treatment	[58]
Carboxymethyl-hydroxypropyl guar gum (CMHPG)	$-CH_2-COO^-Na$ $-CH_2-CH(OH)CH_3$	Anionic	Fracturing fluid in mining industry and oil recovery	[43]

11.3 CHEMISTRY OF GUAR GUM

GG (CAS number 9000-30-0) is a galactomannan that contains 34.6% D-galactopyranosyl units and 64.4% D-mannopyranosyl units. The chemistry of the galactomannan unit confirmed that GG molecule is a linear carbohydrate polymer with a molecular weight range of 50,000–8,000,000 Da.[28]

It has a straight chain of D-mannose units linked together by β (1–4) glycoside bond and D-galactose units are joined at each alternate position by a (1–6) glycosidic linkage[23] (Fig. 11.4). Therefore, GG forms a rod-like polymeric structure with a mannose backbone linked to galactose side chains, which are randomly placed on backbone with an average ratio of galactose to mannose of 1:2. The polymeric structure of GG contains numerous hydroxyl

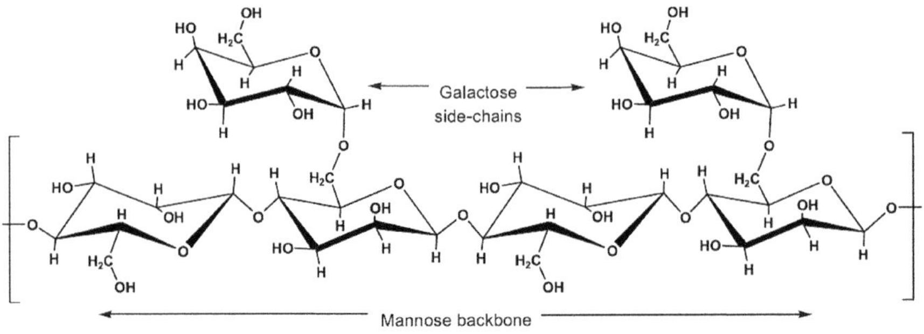

FIGURE 11.4 Chemical structure of guar gum.

groups, which are treated for manufacturing different derivatives important in several industries. Their properties mainly depend upon their chemical features like chain length, abundance of cis-OH group, steric hindrance, degree of polymerization, and additional substitution.[12,15,50,75]

11.4 PROPERTIES OF GUAR GUM

GG is a white to yellowish powder and is nearly odorless. The most important property of GG is its ability to hydrate rapidly in water to attain uniform and very high viscosity at relatively low concentrations. Another advantage associated with GG is that it is soluble in hot and cold water and provides full viscosity. Apart from being the most cost-effective stabilizer and emulsifier, it provides texture improvement, and water bonding, enhances mouthfeel, and controls crystal formation. There is a wide range of physical and organoleptic properties of GG so the principal characteristics of this gum are summarized in Table 11.2 and are widely discussed in subsequent sections.

TABLE 11.2 Characteristics of Guar Gum.

	Guar gum
Origin	Extract of endosperm of seed (Leguminosae)
Chemical composition	• Medium-galactose galactomannan; • Mannose + galactose (ratio M: G = 1.6: 1)
Nutritional value (in 100 g)	292 kJ (70 kcal); slow resorption
Fiber content	Approx. 80% (contains 10% protein)
Toxicology	No health concerns, no ADI value defined
Solubility at low temperature (H_2O)	Highly soluble
Appearance of an aqueous solution	Opaque, grey, and cloudy
Viscosity of solution in water	High in cold water and low in hot water
Impact of heat on viscosity in water (pH 7)	Viscosity decrease
Shear stability	Pseudoplastic > 0.5% concentration, shear thinning
Thickening effect	High
pH stability	High (pH 2–10)
Film formation	Low
Emulsion stabilization	Medium

TABLE 11.2 *(Continued)*

	Guar gum
Gelation	No (only with borate ions)
Crystallization control	High
Synergistic effects with other hydrocolloids	+ Starch/xanthan/CMC: viscosity increase; + gelling polysaccharides (e.g., agar): increased gel strength and elasticity
Negative interactions	Viscosity reduction with polyols
Dosage level in foods	Low to medium (0.05–2%, mostly 0.2–0.5%)

11.4.1 RHEOLOGY

GG is the most efficient aqueous thickener known. Solutions of GG and its derivatives are non-Newtonian, classified as pseudoplastic. They become fluid reversibly when heat is applied, but irreversibly degrade when prolonged high temperature is applied at prolonged times. Some of the hydroxyalkylated derivatives, recently developed, resist this degradation to a much greater degree. Resist solutions well shear degradation, compared to other water-soluble polymers, but degrade with time under high shear.[11]

11.4.2 VISCOSITY

The most important characteristic of GG is its ability to be dispersed in water and hydrate or swell rapidly and almost completely in cold water to form viscous colloidal dispersions. The viscosity attained is dependent on time, temperature, concentration, pH, rate of agitation, degree of purification, and practical size of the powdered gum used (Fig. 11.5). Temperature keeps a proportional relationship with viscosity of GG dispersion: the lower the temperature, lower the rate at which viscosity increases and lower the final viscosity. Heating the gum at temperatures above 60°C tends to provide a high initial viscosity but leads to an inferior stability (in terms of time-dependent changes in viscosity). The most convenient temperature depends on the source. For example, the optimal conditions to disperse GG involve heating at 25–40°C for 2 h.[72]

The recommended use level of GG in aqueous systems is generally much less than 1%, since at higher concentrations the viscosity becomes excessive for most applications. For a typical solution, if the concentration (e.g., 1–2%) bends a tenfold increase in viscosity is obtained (4100–44,000 cps, respectively)

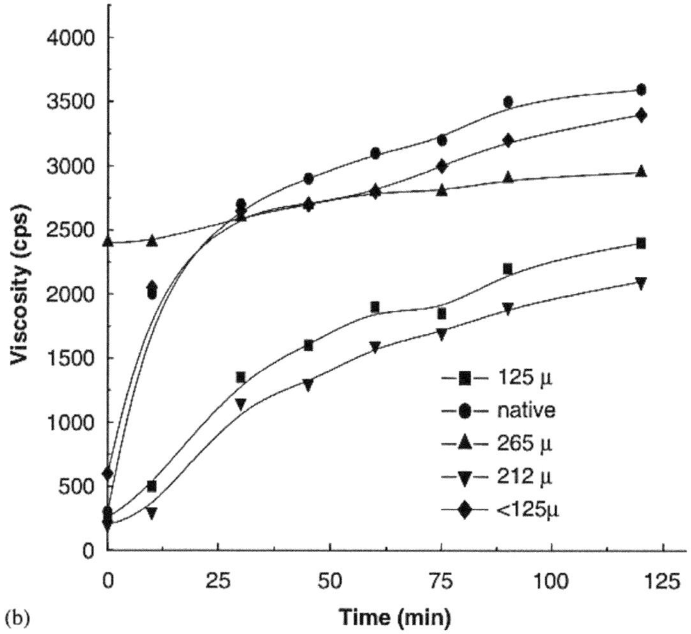

FIGURE 11.5 Changes in viscosity of guar gum as a function of time.

High viscosity products with a concentration of 3% are thick solutions that seem gels. There are guar derivatives with low viscosities for special applications, for example, when a high solids content is favored when desired electrically charged molecules thickening power and controlled behavior, or at least pseudoplasticity desired, or more Newtonian flow.[14,49]

11.4.3 HYDRATION RATE

Rate of hydration of GG varies. A number of factors are known to influence the hydration or dissolution process including the molecular weight and concentration of the galactomannan in the guar powder and also the environmental conditions such as temperature and pH and the presence of co-solutes such as sucrose and salts. The major determinant of hydration kinetics is particle size, which reflects the change in surface area exposed to water. The rate and degree of hydration of GG are critical variables in influencing its biological activity.[42] A hydration time of about 2 h is required in practical applications in order to reach maximum viscosity. For some applications in which there is a need for a quick initial viscosity, very fine mesh GGs are available. However, a considerable period of time is still required for maximum hydration and viscosity to be achieved.[48]

11.4.4 HYDROGEN BONDING ACTIVITY

The hydrogen bonding activity is generally attributed to the presence and behavior of the hydroxyl group in GG molecule. The straight chain structure of GG, along with the regularity of the single membered galactose branches, produces a molecule that exhibits unusual effects on hydrated colloidal systems due to the formation of hydrogen bond. GG shows hydrogen bonding with cellulosic material and hydrated minerals. With a slight addition of GG, there are marked alterations in electrokinetic properties of any system.[20]

11.4.5 REFRACTIVE INDEX

Studies on the trends of some specific physical properties as refractive index of GG aqueous systems at different total gum concentration, polymer ratio, and temperatures are not yet reported. The relationships among refractive index and gum concentration can be employed to determine the real gum concentration quickly and with acceptable accuracy. Recently, Moreira et al.[39] studied the values of GG aqueous dispersions at different concentrations (0.05, 0.10, 0.20, 0.40, 0.60, and 0.80% w/w) and temperatures (15, 20, 25, 30, 35, and 40°C). They found that in all cases, the values decrease with increasing temperature at each gum concentration with refractive indices for solutions ranging between 1.3310 and 1.3352.

11.4.6 EFFECT OF PH

GG is stable in solution over a wide pH range. Guar solutions have an almost constant viscosity over pH range about 1.0–10.5. This stability is believed due to non-ionic, uncharged nature of the molecule. While the pH does not affect the final viscosity, maximum hydration takes place at pH 8.0–9.0. Slowest hydration is present at high (above 10.0) and low (below 4.0) pH values. The preferred method of preparing guar solutions is dissolving at fastest hydration rate pH and then adjusts the pH to desired value (Fig. 11.6). Maximum viscosities achieved at both acid and alkaline pH are the same despite the difference in hydration rates.[76]

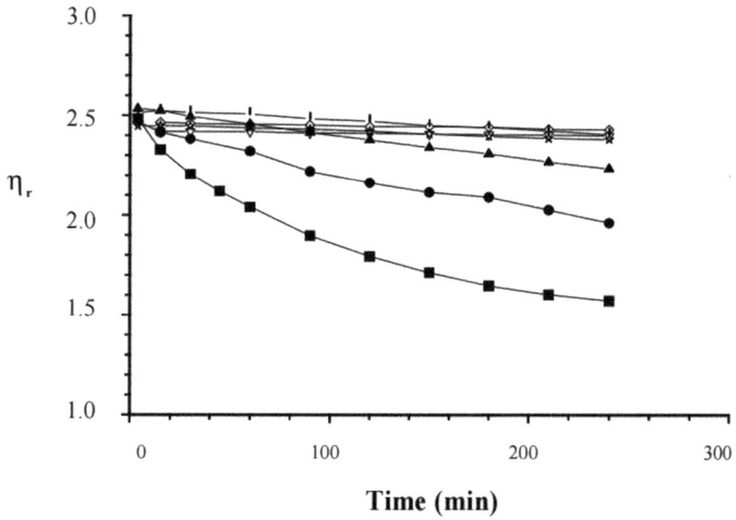

FIGURE 11.6 The relative viscosity (η_r) changes of GG solutions (0.07% w/ w) at different pH levels, temperature = 50°C: (■) pH 1.5; (●) 2.0; (▲) 2.2; (¤) 2.5; (Δ) 3.0; (◊) 3.5; (□) 4.0.

11.4.7 REACTIONS WITH SALTS

Since salt and sugar are probably the two most widely used ingredients in food other than water, their effect on GG has been extensively investigated.[8] The behavior of GG in brine is essentially the same as in water. The hydration rate is not affected, although the final viscosity is somewhat increased by the presence of sodium chloride. This property has made it a very valuable component of oil-well drilling muds, where the capacity of maintaining high viscosity in the presence of brine encountered during drilling operation is absolutely essential.[21]

11.4.8 REACTIONS WITH SUGARS

In the presence of sugar, GG has a competition with it for the available water, triggering a delaying action on the hydration of the gum as sugar concentration grows.

Additionally, viscosity of GG and sugar preparations increases directly to the sugar concentration.[8] Viscosities of guar solutions containing sugar continue to increase for several days and this delay of hydration rate may be due to a reduction in the mobility of the water, proportional to the sugar concentration. In such systems, the full value of GG as a thickening and stabilizing agent is developed after about a week of storage. Sugar is effective in protecting GG against hydrolysis and loss of viscosity when heated or autoclaved. The presence of 5–10% sugar in the liquid offers maximum protection with maximum viscosity. Sugar also offers protection from the hydrolyzing effect at low pH values (down to pH 3.0) under cooking conditions.[8]

11.4.9 GEL FORMATION

Borate ion acts as a cross-linking agent with hydrated GG to form cohesive and structured gels. Formation and strength of these gels depend on the pH, temperature, and concentrations of reactants (Fig. 11.7). The optimum pH range for gel formation is 7.5–10.5. The solution-gel transformation is reversible; the gel can be liquefied by decreasing the pH below 7.0 or by

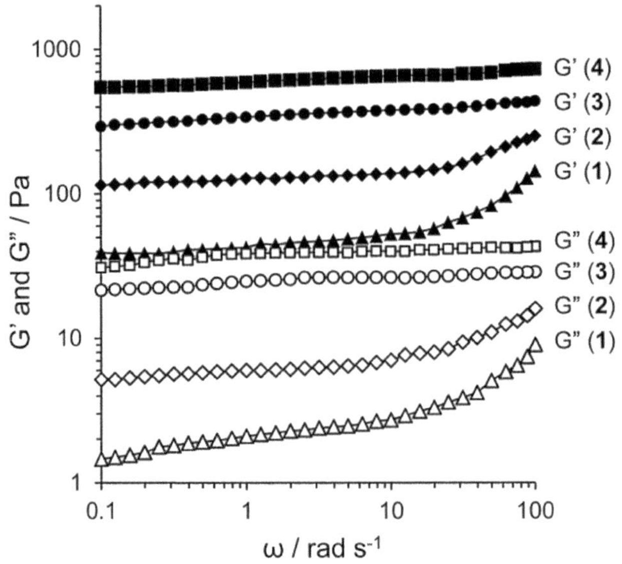

FIGURE 11.7 Viscoelastic behavior of guar gum gels (samples 1–4100 mg) swollen with PBS buffer, pH 7.4 (2 mL) at 298°K.

heating.[21] Borated gels can also be liquefied by the addition of glycerol or mannitol, capable of reaction with the borate ion. Borate ion will inhibit the hydration of GG if it is present at the time the powdered gum is added to water. The minimum concentrations necessary to inhibit hydration are dependent on pH. For example, with 1.0% of GG, 0.25–0.5% (based on guar weight) of borax (sodium tetraborate) is needed at pH 10.0–10.5, while at pH 7.5–8.0, 1.5–2.0% of borax is required. The complexing reaction is reversible and lowering the pH below 7.0 permits the gum to hydrate normally. This technique is often used to provide better mixing and easier dispersion.[76]

11.4.10 SYNERGISTIC EFFECT

Synergisms of GG with other materials, including others gums like xanthan gum, agar, carrageenan, or starches, are well studied.[13] The degree of the synergism is believed to be related to mannose/galactose (M/G) ratio and the galactosyl distribution on the mannan backbone which basically is determined by source and method employed for the extraction of this kind of gum.[10,17,35] Ali Razavi et al.[1] examined the synergistic effect between sage seed gum (SSG) and GG solutions using various steady and dynamic rheological parameters. They found that increase SSG fraction, the extent of viscosity reduction in the range of 0.01–316 s^{-1} increased from 58.68 for GG to 832.73 times for SSG which was not the same at different ranges of shear rate. Steady and dynamic shear tests suggested interaction with longer timescale in SSG chains in comparison with that in GG. The synergist effect of all viscoelastic parameters from frequency sweep test was observed for 3–1 SSG-GG blend, concluding that this mixture can be attractive commercially because they offer the potential to create new textures and to manipulate the rheology of products in the industry.[44]

11.5 TOXICITY

The only available data for GG that relates to absorption, distribution, metabolism, excretion, and toxicity were found in dietary studies.[79] In this way, an acute oral toxicity study on partially hydrolyzed GG (PHGG) was performed using groups of 16 (8 males AND 8 females per group) 4-week-old Jcl: ICR mice and Jcl: SD rats. PHGG was administered by gavage at a concentration of 30% in distilled water (dose = 6000 mg kg^{-1} body weight; dose volume = 20 mL kg^{-1}) to one group per species. The control group was dosed orally with distilled water. Dosing was followed by a 14-day observation period, after which all animals were killed and examined macroscopically. Soft feces were reported for male and female mice, but no abnormal signs were reported for rats. There were no test substance-related effects on body weight (rats and mice), food consumption (rats), or necropsy findings (mice and rats). None of the animals died, and the LD50 was >6000 mg kg^{-1} in both species.[29] On the other hand, in a 28-day oral feeding study, 2 groups of 10 rats (5 males and 5 females per group) were fed with PHGG in the diet (500 and 2500 mg/kg doses, respectively), daily. Body weights and food consumption were measured, and gross and microscopic pathology were evaluated. No adverse effects were observed at either administered dose.[73] The available information reveals that there are no adverse short-term toxicological consequences in animals of consuming GG in amounts exceeding those currently consumed in the normal diet. However, it should be noted that no long-term animal feeding studies of GG have been reported. It may be advisable in due course to conduct adequate feeding studies in several species, including pregnant animals, at dosage levels that approximate and exceed the current estimated maximum daily human intakes.[24]

11.6 INDUSTRIAL APPLICATIONS: OVERVIEW

11.6.1 FOOD INDUSTRY

GG stands as one of the cheapest hydrocolloids in food industry and about 40% of the total is presently used as food additive. The importance of this gum in food application is due to its various unique functional properties like water retention capacity, reduction in evaporation rate, alteration in freezing rate, modification in ice crystal formation, regulation of rheological properties, and involvement in chemical transformation.[62] United States Food and Drug Administration (FDA) regulates the use of gums and classify them as either food additives or generally recognized as safe substances (GRAS number 2537).

The use of GG at a concentration not exceeding 2% is allowed in food application.[24] Tomato ketchup serum loss and flow values decrease when this gum is added, which makes it a novel thickener for this product.[26] It has also been incorporated into pizzas, biscuits, and pastries.[25] GG prevents staleness and crumb formation in baked foods, provides unparallel moisture preservation to the dough, retards fat penetration, increases the dough volume, provides greater resiliency, and improves texture and shelf life. In wheat bread dough, addition of this gum results in significant increase in loaf volume on baking.[9] The depolymerized GG is used in the preparation of low-calorie food. It enhances the creaming stability and control rheology of emulsion prepared by egg yolk.[22] In beverages, is often used as a thickening or viscosity control agent in levels of 0.25–0.75%. GG is useful due to its resistance to break down at low pH conditions. In addition, since GG is soluble in cold water, it is easy to use in most beverages processing plants.[38] In dairy products, it thickens milk, yogurt, and liquid cheese products, and helps to maintain the homogeneity and texture of ice cream and sherbets.[5]

11.6.2 PHARMACEUTICAL INDUSTRY

GG or its derivatives are used in pharmaceutical industries as gelling, viscosifying, thickening, suspension, stabilization, emulsification, preservation, water retention/water phase control, binding, clouding, process aid, pour control for suspensions, antacid formulations, tablet binding, disintegration agent, controlled drug delivery systems, slimming aids, nutritional foods, etc. Its hydrophilic properties and the ability to form gel make it useful in gastric ulcer treatment.

Studies revealed that a diet supplemented with this gum decreased the appetite, hunger, and desire for eating.[7] In tablets, the gum is used as binder and increases the mechanical strength of tablets during pressing, and in jellies and ointments it is used as thickener. The depolymerized GG has been good bulking agent for dietetic food and as a food fiber; it has been used in sugar and lipid metabolic control particularly in diabetic and heart patients.[63] GG is widely used in capsules as dietary fiber that decreases hypercholesterolemia, hyperglycemia, and obesity.[16] Sufficient gum intake as dietary fiber helps in bowel regularization, total and LDL cholesterol reduction, diabetes control, enhancement of mineral absorption, and prevention of digestive problems like constipation and enhances bowel movement.[84] Some pharmaceutical companies are using it for making bandage paste and in dentistry formulations. The formulations like inhalable, injectable, beads, microparticles, nanoparticles, solid monolithic matrix films, and implants also facilitate the use of the gum.[70] The high swelling characteristics of this gum sometimes hinder its use as a drug delivery carrier but it can be improved by derivatization, grafting and network formation and can be satisfactorily used for targeted drug delivery by forming coating matrix systems, hydrogels and nano/microparticles.[57]

11.6.3 METALLURGICAL AND MINING INDUSTRY

GG is used in froth flotation of potash as an auxiliary reagent, depressing the gangue minerals, which might be clay, talc, or shale. GG is also used as a flocculant or settling agent to concentrate ores or tailings in the mining industry.[65]

Also is approved by many public organizations for use in potable water treatment as a coagulant aid in conjunction with such coagulants as alum (potassium aluminum sulfate), iron (III) sulfate, and lime (calcium oxide). In industrial waters, GG flocculates clays, carbonates, hydroxides, and silica when used alone or in conjunction with inorganic coagulants.[47]

11.6.4 PAPER INDUSTRY

GG is used as a size for paper and textiles. The major use of this gum in papermaking is in the wet end of the process. The gum is added to the pulp suspension just before the sheet is formed. GG replaces or supplements the natural hemicelluloses in paper bonding. Advantages gained by addition of guar to pulp include improved sheet formation with a more regular distribution of pulp fibers (less fiber bundles); increased mullen bursting strength; increased fold strength; increased tensile strength; increased pick; easier pulp hydration; improved finish; decreased porosity; increased flat crush of corrugating medium; increased machine speed with maintenance of test results; and increased retention of fines.[2]

11.6.5 COSMETIC INDUSTRY

The unique cosmetic properties of GG include cold solubility, viscosity enhancing, solvent resistance film forming, protective colloid, wide pH range resistance, stability, non-toxic nature, safe, etc. So it is a choice of thickener, suspending agent, binder and emulsifier agent in various hair/skin care cosmetic products like creams, shampoos, premium quality soaps, lotions, conditioners, and moisturizer.[65]

In the manufacturing of toothpaste, this gum binds the aqueous phase of the paste and is used in sizeable scale to impart flowing nature so that the paste can be extruded from the collapsible tubes with the application of a little force. In saving cream preparation, it does the same work besides providing stabilizing the system; import slip during shaving and improve skin aftershave.[12]

In emulsion systems like cream and lotions, GG prevents phase separation, sudden release of moisture, increase emulsion stability, prevent water loss, and is used as protective colloid. It stabilizes the emulsion during freeze-thaw cycle, where the water phase condenses out of the system. In lotion, it provides additional spreadability and an agreeable feel.[71] Cationic GG is used to thicken various cosmetics and toiletries products, especially to impart thickening, conditioning, foam stability, softening, and lubricity. In aerosol dispensing aqueous liquid preparation as spray or mist, it reduces fog migration. The inert and compatible nature of guar with the detergents makes it suitable for use in shampoo and cleansing preparation. Hair colorants contain guar as thickener. It is also available in self-emulsifying grades and can be used to prepare dry face mask preparation.[3]

11.7 CONCLUSIONS

Guar is an agricultural produce spreading over into Indo-Pakistan subcontinent many generations ago. GG is a high molecular weight natural hydrocolloid composed of galactose and mannan units that is obtained from the endosperm of the guar plant (*C. tetragonoloba*). Usually, it is produced as a whitish powder similar to flour and functions as thickener, stabilizer, and emulsifier. Due to its low cost, non-toxicity, biodegradability, biocompatibility, high viscosity, and high water solubility, GG has a peculiar interest for researchers due to its availability in terms of volumes and opportunities in many industrial applications such as the food, textile, printing, explosives, pharmaceutical, and even hydraulic fracturing fields. This review states that GG has multiple physical and chemical properties that make it a strong candidate for its application in food science and technology research.

ACKNOWLEDGMENTS

The authors would like to thanks to the Mexican Council for Science and Technology (CONACyT) for providing financial support in the INFR 2015: 254178 project. This work is part of the activities in the Master degree in food science and technology of the Faculty of Chemistry Sciences.

KEYWORDS

- *Cyamopsis tetragonolobus*
- **guar gum**
- **properties**
- **industry**
- **potential**

REFERENCES

1. Ali-Razavi, S. M.; Alghooneh, A.; Behrouzian, F.; Cui, S. W. Investigation of the Interaction between Sage Seed Gum and Guar Gum: Steady and Dynamic Shear Rheology. *Food Hydrocol.* **2016,** *60,* 67–76.
2. Anderson, K. R.; Larson, B.; Thoresson, H. O. (EKAAB POT Int. Applied WO 8500, 100 (CL D2H3I20) SE Appl. 84/306207, 1986, pp 35. (Cf. Guar Res Ann 5:44).
3. APEDA. APEDA Annual Export Report. [Online], 2011, Available at: http://agriexchange.apeda.gov.in/index/product_description_32head.aspx?gcode=0502 (accessed Feb 28, 2016).
4. BeMiller, J. Gums and Related Polysaccharides. In *Glycoscience;* Fraser-Reid, B., Tatsuta, K., Thiem, J., Eds.; Springer Berlin Heidelberg: New York, NY, 2008.
5. Brennan, C. S.; Tudorica, C. M. Carbohydrate Based Fat Replacers in the Modification of the Rheological, Textural and Sensory Quality of Yoghurt: Comparative Study of the Utilization of Barley Beta-glucan, Guar Gum and Inulin. *Int. J. Food Sci. Technol.* **2008,** *43,* 824–833.
6. Busch, V. M.; Kolender, A. A.; Santagapita, P. R.; Buera, M. P. Vinal Gum, a Galactomannan from *Prosopis ruscifolia* Seeds: Physicochemical Characterization. *Food Hydrocol.* **2015,** *51,* 495–502.
7. Butt, M. S.; Shahzadi, N.; Sharif, M. K.; Nasir, M. Guar Gum: A Miracle Therapy for Hypercholesterolemia, Hyperglycemia and Obesity. *Crit. Rev. Food Sci. Nutr.* **2007,** *47,* 389–396.
8. Carlson, W. A.; Zeigenfuss, E. M. The Effect of Sugar on Guar Gum as a Thickening Agent. *J. Food Technol.* **1965,** *19,* 954–958.
9. Cawley, R. W. The Role of Wheat Flour Pentosans in Baking. II. Effect of Added Flour Pentosans and Other Gums on Gluten Starch Loaves. *J. Sci. Food Agric.* **1964,** *15,* 834–838.
10. Cerqueira, M. A.; Pinheiro, A. C.; Souza, B. W. S.; Lima, A. M. P.; Teixeira, J. A.; Moreira, R. A. Extraction, Purification and Characterization of Galactomannans from Non-traditional Sources. *Carbohydr. Polym.* **2009,** *75,* 408–414.
11. Chenlo, F.; Moreira, R.; Silva, C. Rheological Behavior of Aqueous Systems of Tragacanth and Guar Gums with Storage Time. *J. Food Eng.* **2010,** *96,* 107–113.
12. Chudzikowski, R. J. Guar Gum and Its Applications. *J. Soc. Cosmet. Chem.* **1971,** *22,* 43–60.

13. Cui, S. W.; Eskin, M. A. N.; Wu, Y.; Ding, S. Synergisms between Yellow Mustard Mucilage and Galactomannans and Applications in Food Products–A Mini Review. *Adv. Colloid Interface Sci.* **2006,** *128–130,* 249–256.
14. Cunha, P. L.; Castro, R. R.; Rocha, F. A.; de Paula, R. C.; Feitosa, J. P. Low Viscosity Hydrogel of Guar Gum: Preparation and Physicochemical Characterization. *Int. J. Biol. Macromol.* **2005,** *37* (1–2), 99–104.
15. Daas, P. J. H.; Schols, H. A.; De-Jongh, H. H. J. On the Galactosyl Distribution of Commercial Galactomannans. *Carbohydr. Res.* **2000,** *329,* 609–619.
16. Dall'alba, V.; Silva, F. M.; Antonio, J. P. Improvement of the Metabolic Syndrome Profile by Soluble Fiber-guar Gum-in Patients with Type 2 Diabetes a Randomised Clinical Trial. *Br. J. Nutr.* **2013,** *110,* 1601–1610.
17. Dea, I. C. M.; Morris, E. R.; Rees, D. A.; Welsh, E. J.; Barnes, H. A.; Price, J. Associations of Like and Unlike Polysaccharides: Mechanism and Specificity in Galactomannans, Interacting Bacterial Polysaccharides, and Related Systems. *Carbohydr. Res.* **1977,** *57,* 249–272.
18. Dodi, G.; Pala, A.; Barbu, E.; Peptanariu, D.; Hritcu, D.; Popa, M. I.; Tamba, B. I. Carboxymethyl Guar Gum Nanoparticles for Drug Delivery Applications: Preparation and Preliminary In Vitro Investigations. *Mater. Sci. Eng.* **2016,** *63,* 628–636.
19. Dong, C.; Tian, B. Studies on Preparation and Emulsifying Properties of Guar Galactomannan Ester of Palmitic Acid. *J. Appl. Polym. Sci.* **1999,** *72* (5), 639–645.
20. Doyle, J. P.; Giannouli, P.; Martin, E. J.; Brooks, M.; Morris, E. R. Effect of Sugars, Galactose Content and Chain Length on Freeze-thaw Gelation of Galactomannans. *Carbohydr. Polym.* **2006,** *64,* 391–401.
21. El-awad, G. *A Study of Guar Seed and Guar Gum Properties* (*Cyamopsis tetragonolobus*) [Online]; 1998. Available at: http://www.iaea.org/inis/collection/NCLCollectionStore/_Public/31/037/31037745.pdf (accessed Feb 10, 2016).
22. Ercelebi, E. A.; Ibanoglu, E. Stability and Rheological Properties of Egg Yolk Granule Stabilized Emulsions with Pectin and Guar Gum. *Int. J. Food Prop.* **2010,** *13,* 618–630.
23. FAO. [Online]; 2016. Available at: http://www.fao.org/ag/agn/jecfa-additives/specs/monograph3/additive-218.pdf (accessed Feb 28, 2016).
24. FDA *GRAS Substances (SCOGS) Database* [Online]; 2016. Available at: http://www.fda.gov/Food/IngredientsPackagingLabeling/GRAS/SCOGS/ucm2006852.htm (accessed Feb 10, 2016).
25. Griffith, A. J.; Kennedy, J. F. Biotechnology of Polysaccharide. In *Chemistry Carbohydrate;* Kennedy, J. I., Ed.; Publication, Oxford Science: Oxford, England, 1988; pp 597–635.
26. Gujral, H. S.; Sharma, A.; Singh, N. Effects of Hydrocolloids, Storage Temperature and Duration on the Consistency of Tomato Ketchup. *Int. J. Food Prop.* **2002,** *5,* 179–191.
27. Gunjal, B. B.; Kadam, S. S. *CRC Hand Book of World Food Legumes;* CRC Press: USA, 1991; Vol. 1, pp 289–299.
28. Kawamura, Y. *Guar Gum Chemical and Technical Assessment,* 2008, Prepared for the 69th Joint FAO/WHO Expert Committee on Food Additives.
29. Koujitani, T.; Oishi, H.; Kubo, Y.; Maeda, T.; Sekiya, K.; Yasuba, M.; Matsuoka, N.; Nishimura, K. Absence of Detectable Toxicity in Rats Fed Partially Hydrolyzed Guar Gum (K-13) for 13 Weeks. *Int. J. Toxicol.* **1997,** *16* (6), 611–623.
30. Kumar, V. *Perspective Production Technologies of Arid Legumes. Perspective Research Activities of Arid Legumes in India* Kumar, D., Henary, A., Eds.; India Arid Legumes Society: Hisar, India, 2009; pp 119–155.

31. Lapasin, R.; Pricl, S.; Tracanelli, P. Rheology of Hydroxyethyl Guar Gum Derivatives. *Carbohydr. Polym.* **1991,** *14* (4), 411–427.
32. Lapasin, R.; Pricl, S.; Lorenzi, L. D.; Torriano, G. Flow Properties of Hydroxypropyl Guar Gum and Its Long-chain Hydrophobic Derivatives. *Carbohydr. Polym.* **1995,** *28* (3), 195–202.
33. Li-Ming, Z.; Jian-Fang, Z.; Peter, S. H. A Comparative Study on Viscosity Behavior of Water-soluble Chemically Modified Guar Gum Derivatives with Different Functional Lateral Groups. *J. Sci. Food Agric.* **2005,** *85* (15), 2638–2644.
34. Mccleary, B. V.; Neukom, H. Effect of Enzymic Modification on the Solution and Interaction Properties of Galactomannans. *Prog. Food Nutr. Sci.* **1982,** *6,* 109–118.
35. McCleary, B. V.; Clark, A. H.; Dea, I. C. M.; Rees, D. A. The Fine Structures of Carob and Guar Galactomannans. *Carbohydr. Res.* **1985,** *139,* 237–260.
36. Mestechkina, N. M.; Egorov, A. V.; Shcherbukhin, V. D. Synthesis of Galactomannan Sulfates. *J. Appl. Biochem. Microbiol.* **2010,** *42* (3), 326–327.
37. Milani, J.; Gisoo, M. *Hydrocolloids in Food Industry, Food Industrial Processes. Methods and Equipment;* InTech. ISBN: 978-953-307-905-9. [Online]; 2012. Available from: http://www.intechopen.com/books/food-industrial-processes-methods-and equipment/hydrocolloids-in-foodindustry (accessed Feb 10, 2016).
38. Miyazawa, T. Hydrocolloid Structures, Which Allow More Water Interactions Through Hydrogen Bonding. *Carbohydr. Res.* **2006,** *341,* 870–877.
39. Moreira, R.; Chenlo, F.; Silva, C.; Torres, M. D.; Díaz-Varela, D.; Hilliou, L.; Argence, H. Surface Tension and Refractive Index of Guar and Tragacanth Gums Aqueous Dispersions at Different Polymer Concentrations, Polymer Ratios and Temperatures. *Food Hydrocol.* **2012,** *28,* 284–290.
40. Mudgil, D.; Barak, S. Guar Gum: Processing, Properties and Food Applications-a Review. *J. Food Sci. Technol.* **2014,** *51* (3), 409–418.
41. Mukherjee, I.; Sarkar, D.; Moulik, S. P. Interaction of Gums (Guar, Carboxymethylhydroxypropyl Guar, Diutan, and Xanthan) with Surfactants (DTAB, CTAB and TX-100) in Aqueous Medium. *Langmuir.* **2010,** *26,* 17906–17912.
42. Nandhini-Venugopal, K.; Abhilash, M. Study of Hydration Kinetics and Rheological Behaviour of Guar Gum. *Int. J. Pharma. Sci. Res.* **2010,** *1* (1), 28–39.
43. Nemade, S. N.; Sawarkar, S. B. Recovery and Synthesis of Guar Gum and Its Derivatives. *Int. J. Adv. Res. Chem. Sci.* **2015,** *2* (5), 33–40.
44. Nor-Hayati, I.; Ching, C. W.; Rozaini, M. Z. H. Flow Properties of o/w Emulsions as Affected by Xanthan Gum, Guar Gum and Carboxymethyl Cellulose Interactions Studied by a Mixture Regression Modeling. *Food Hydrocol.* **2016,** *53,* 199–208.
45. Nussinovitch, A. *Hydrocolloid Applications: Gum Technology in the Food and Other Industries;* Chapman and Hall: London, 1997.
46. Nussinovitch, A. Water Soluble Polymer Applications in Foods. Blackwell Science: Oxford, England, 2003.
47. Pala, S.; Ghoraia, S.; Dasha, M. K. Flocculation Properties of Polyacrylamide Grafted Carboxymethyl Guar Gum (CMG-g-PAM) Synthesized by Conventional and Microwave Assisted Method. *J. Hazard. Mater.* **2011,** *192,* 1580–1588.
48. Panda, H. *The Complete Book on Gums and Stabilizers for Food Industry*; Asia Pacific Business Press Inc.: Kamla Nagar, Delhi, 2010, pp 107–109. ISBN:9788178331317.
49. Parija, S.; Misra, M.; Mohanty, A. K. Studies of Natural Gum Adhesive Extracts: An Overview. *Polym. Rev.* **2001,** *41,* 175–197.

50. Pasha, M.; Swamy, N. G. N. Derivatization of Guar to Sodium Carboxymethyl Hydroxyl Propyl Derivative, Characterization and Evaluation. *Pak. J. Pharm. Sci.* **2008,** *21* (1), 40–44.
51. Patel, J. J.; Karve, M.; Patel, N. K. A Novel Approach to Synthesize Carboxymethyl Guar Gum via Friedel Craft Acylation Method. **2014,** *10* (1),18–22.
52. Patel, M. B.; McGinnis, J. The Effect of Autoclaving and Enzyme Supplementation of Guar Meal on the Performance of Chicks and Laying Hens. *Poult. Sci.* **1985,** *64,* 1148–1156.
53. Pathak, R. Clusterbean: Physiology, Genetics and Cultivation. Springer International Publishing AG: Basel, Switzerland, 2015. DOI 10.1007/978-981-287-907-3_3
54. Peleg, M. On Fundamental Issues in Texture Evaluation and Texturization. *Food Hydrocol.* **2006,** *20,* 405–414.
55. Petricek, I.; Berta, A.; Higazy, M. T.; Németh, J.; Prost, M. E. Hydroxypropyl-guar Gellable Lubricant Eye Drops for Dry Eye Treatment. *Exp. Opin. Pharm.* **2008,** *9* (8), 1431–1436.
56. Perduca, M. J.; Spotti, M. J.; Santiago, L. G.; Judis, M. A.; Rubiolo, A. C.; Carrara, C. R. Rheological Characterization of the Hydrocolloid from *Gleditsia amorphoides* Seeds. *LWT-Food Sci. Technol.* **2013,** *51,* 143–147.
57. Prabhaharan, M. Prospective of Guar Gum and Its Derivatives as Controlled Drug Delivery System. *Int. J. Biol. Macromol.* **2011,** *49* (2), 117–124.
58. Prabhanjan, H.; Gharia, M. M.; Srivastava, H. C. Guar Gum Derivatives. Part II: Preparation and Properties. *Carbohydr. Polym.* **1990,** *12,* 1–7.
59. Rana, V.; Rai, P.; Tiwary, A. K.; Singh, R. S.; Kennedy, J. F.; Knells, C. J. Modified Gums: Approaches and Applications in Drug Delivery. *Carbohydr. Polym.* **2011,** *83,* 1031–1047.
60. Reddy, K.; Krishna-Mohan, G.; Satla, S.; Gaikwad, S. Natural Polysaccharides: Versatile Excipients for Controlled Drug Delivery Systems. *Asian J. Pharm. Sci.* **2011,** *6* (6), 275–286.
61. Ritzoulis, C.; Marini, E.; Aslanidou, A.; Georgiadis, N.; Karayannakidis, P. D.; Koukiotis, C.; Filotheou, A.; Lousinian, S.; Tzimpilis, E. Hydrocolloids from Quince Seed: Extraction, Characterization, and Study of their Emulsifying/stabilizing Capacity. *Food Hydrocol.* **2014,** *42,* 178–186.
62. Rodge, A. B.; Sonkamble, S. M.; Salve, R. V. Effect of Hydrocolloid (Guar Gum) Incorporation on the Quality Characteristics of Bread. *J. Food Process Technol.* **2012,** *3* (2), 1–7.
63. Saeed, S.; Mosa-Al-Reza, H.; Fatemeh, A. N. Antihyperglycemic and Antihyperlipidemic Effects of Guar Gum on Streptozotocin-induced Diabetes in Male Rats. *Pharmacogn. Mag.* **2012,** *8,* 65–72.
64. Scherz, H. Hydrocolloids: Stabilizers, Thickening and Gelling Agents in Food Products. Food Chemistry and Food Quality. *Food Chemical Society GDCh;* Behr's Verlag GmbH: Hamburg, Germany, 1996; Vol. 2,
65. Sharma, B. R.; Chechani, V.; Dhuldhoya, N. C.; Merchant, U. C. *Guar Gum* [Online]; 2007. Available at: http://www.lucidcolloids.com/pdf/814_guar-gum.pdf (accessed Feb 10, 2016).
66. Sharma, P. *Guar Industry Vision 2020: Single Vision Strategies;* CCS National Institute of Agricultural Marketing: Jaipur, Rajasthan, 2010.
67. Sharma, P.; Gummagolmath, K. C. Reforming Guar Industry in India: Issues and Strategies. *Agric. Econ. Res. Rev.* **2012,** *25* (1), 37–48.

68. Shenoy, M. A.; D'Melo, D. J. Synthesis and Characterization of Acryloyloxy Guar Gum. *J. Appl. Polym. Sci.* **2010,** *117* (1), 148–154.
69. Shi, H. Y.; Li, M. Z. New Grafted Polysaccharide Based on O-carboxymethyl-O-hydroxypropyl Guar Gum and Nisopropylacrylamide: Synthesis and Phase Transition Behavior in Aqueous Media. *Carbohydr. Polym.* **2007,** *67* (3), 337–342.
70. Soumya, R. S.; Ghosh, S.; Abraham, E. I. Preparation and Characterization of Guar Gum Nanoparticles. *Int. J. Biol. Macromol.* **2010,** *46* (2), 267–269.
71. Srichamroen, A. Influence of Temperature and Salt on the Viscosity Property of Guar Gum. *Naresuan Univ. J.* **2007,** *15* (2), 55–62.
72. Srivastava, M.; Kapoor, V. P. Seed Galactomannans: An Overview. *Chem. Biodivers.* **2005,** *2* (3), 295–317. DOI: 10.1002/cbdv.200590013
73. Takahashi, H.; Yang, S.; Fujiki, M.; Kim, M.; Yamamoto, T.; Greenberg. N. A. Toxicity Studies of Partially Hydrolyzed Guar Gum. *J. Am. Coll. Toxicol.* **1994,** *13* (4), 273–278.
74. Thomas, T. A.; Dabas, B. S.; Chopra, D. D. Guar Gum Has Many Uses. *Indian Farm.* **1980,** *32,* 7–10.
75. Tripathy, S.; Das, M. K. Guar Gum: Present Status and Applications. *J. Pharm. Sci. Innov.* **2013,** *2* (4), 24–28.
76. Whistler, R. L.; Hymowitiz, T. *Guar Agronomy, Production, Industrial Use and Nutrition*; Purdue University Press: West Lafayette, India, 1979; pp 1–96.
77. Whistler, R. L.; Daniel, J. R. Carbohydrates. In *Food Chemistry,* 2nd ed.; Marcel Dekker: New York, NY, 1985.
78. Whistler, R. L. Factors Influencing Gum Costs and Applications. In *Industrial Gums,* 2nd ed.; Whistler, R. L., BeMiller, J. N., Eds.; Academic Press: San Diego, CA, 1973.
79. WHO. *Toxicological Evaluation of Certain Food Additives and Contaminants. WHO Food Additives Series 21* [Online]; Cambridge University Press, 2016. Available at: http://www.inchem.org/documents/jecfa/jecmono/v21je01.htm (accessed Feb 10, 2016).
80. Williams, P. A.; Phillips, G. O. Introduction to Food Hydrocolloids. In *Handbook of hydrocolloids;* Phillips, G. O., Williams, P. A., Eds.; CRC Press: New York, NY, 2000; pp 1–19.
81. Wüstenberg, T. *Cellulose and Cellulose Derivatives in the Food Industry: Fundamentals and Applications,* 1st ed.; Wiley-VCH Verlag GmbH & Co: Weinheim, Germany, 2015; pp 1–2.
82. Xiao, W.; Dong, L. In *Novel Excellent Property Film Prepared from Methacryloyl Chloride-Graft-Guar Gum Matrixes.* Xian Ning, China, 16–18 April 2011; 1442–1445, 2011.
83. Yadav, H.; Shalendra, N. *An Analysis of Guar Crop in India: United States Department of Agriculture (USDA);* GAIN Report IN4035; New Delhi, India. 2014.
84. Yoon, S. J.; Chu, D. C.; Juneja, L. R. Chemical and Physical Properties. Safety and Application of Partially Hydrolyzed Guar Gum as Dietary Fiber. *J. Clin. Biochem. Nutr.* **2008,** *42,* 1–7.

PART III
Special Topics

CHAPTER 12

WHEY PROTEIN-BASED EDIBLE FILMS: PROGRESS AND PROSPECTS

OLGA B. ALVAREZ-PÉREZ[1], RAÚL RODRÍGUEZ-HERRERA[1], ROSA M. RODRÍGUEZ-JASSO[1], ROMEO ROJAS[2], MIGUEL A. AGUILAR-GONZÁLEZ[3], and CRISTÓBAL N. AGUILAR[1*]

[1]*Department of Food Research, School of Chemistry, Universidad Autónoma de Coahuila, Saltillo 25280, Coahuila, Mexico*

[2]*School of Agronomy, Research Center and Development for Food Industry, Universidad Autónoma de Nuevo León, General Escobedo 66050, Nuevo León, Mexico*

[3]*CINVESTAV–Center for Research and Advanced Studies, IPN Unit Ramos Arizpe, Coahuila, Mexico*

[*]*Corresponding author. E-mail: cristobal.aguilar@uadec.edu.mx*

CONTENTS

ABSTRACT

The dairy industry generates large amounts of serum that are not used in generating high amounts of pollutants being disposed of improperly causing environmental damage. In Mexico, whey is a by-product of the dairy industry that has no added value and produced in large quantities. This product has high potential to be reused as a raw material for other processes, the use of this product currently discarded, solves a clear problem of pollution, and is a potential residue for integrating products into production chains, so source that promotes good sustainable management of this resource, with the use of products that are currently discarded and that cause pollution and could be used as feedstock for other processes, avoiding ecological imbalance and reinstating materials to the production chain in the dairy sector. Manufacturing of edible films from whey protein products might represent an effective means of utilization of excess whey. The formation of the heat-induced gel structure involves a complex series of chemical reactions involving dissociation, denaturation, and exposure of hydrophobic amino acid residues. These reactions are influenced by experimental conditions such as protein concentration, pH, heating temperature, and ionic strength. The aim of this study was to determine the progress in the use of whey and interactions that are generated when mixed in a food matrix.

12.1 INTRODUCTION

Most foods are highly perishable and subject to alterations and modifications caused by various factors (chemical, physical, and biological) that are primarily responsible for its deterioration. It has been said that a food is rot when it loses its normal characteristics. To avoid or delay senescence, a lot of methods have been developed for preserving and processing foods taking as a fundamental principle to prevent the alteration or decomposition, mainly in their organoleptic properties (taste, odor, color, texture, etc.) or avoiding the process of putrefaction, typical of the breakdown of protein foods of animal origin.

In the last 50 years, new technologies have been developed. These techniques enable a great food distribution worldwide. Some methods have been used over the years such as heat, cooling, added sugar, acidification, fermentation, drying or dehydration, modified atmospheres, etc.[1] One of the conservation methods used recently are modified atmospheres, using inert gases that reduce the maturation process using specialized materials for containing a product to generate a micro atmosphere.[2] This arises due to the trend toward production and/or consumption of minimally processed food or organic food that is forcing to the food industry, research centers and regulatory agencies, and so on to transform their technologies and find different ways to processing foods.

A viable alternative is the formulation of edible films that help in controlling the deterioration occurred in foods, without neglecting that the requirements to support the control shall be based on the nature of food in which will be applied. Some of desirable characteristics in coatings and films are controlling the water vapor permeability to gases and volatile compounds, among others. Currently there are a variety of films or edible coatings based on various polymers as pectin,[3] Arabic gum,[4] galactomannans,[5] chitosan,[6] starch,[7,8] alginate,[9,10] xanthan gum,[11,12] zein,[13,14] pullulan,[15,16] hydroxypropyl methylcellulose,[17,18] locus bean gum,[19] etc. However, protein-based films are characterized by its important functional properties among which is the delay or decrease the mass transfer through it, because it possess a complex structure. Also, serve as an alternative to synthetic materials used as packings.[20]

In Mexico, the whey is a by-product of the dairy industry which has no added value although is produced in large quantities and in spite of the diversity of applications and products being developed, wastage has become the main source of pollution in this industry, making it a serious environmental problem, and this is mainly due to its high biochemical oxygen demand (BOD). For example, when the serum is pouring in a water body, the microorganisms need a large amount of oxygen to degrade it and consequently reduce the concentration of dissolved oxygen killing the fauna that exist in this ecosystems.[21]

The application or use of biodegradable resources coming from renewable sources is a strategy that reduces environmental problems and adds value. In recent years, the recovery of proteins from renewable, agricultural, or industrial waste, as is the case of some effluents of the dairy industry (mainly whey), are an important part of the most severe contaminants that exist, that despite of its many uses, its discharged into the soil, drains, and water bodies, becoming a serious problem for the environment. Its use as a source of conservation and recycling becomes an excellent choice for innovation to develop new biodegradables products.[22] Applications include the alcoholic fermentation, demineralization, hydrolyzed, forming edible films, etc. This latest packaging technology enables the development of products with specific barrier, mechanical, and thermal characteristics in certain packaging as films.

Natural biopolymers come from four main sources: animal origin (e.g., collagen and gelatin), seafood (e.g., chitón and chitosan), agricultural origin (e.g., lipids such as triglycerides and hydrocolloids such as protein and polysaccharides), and microbial origin (such as polylactic acid, PLA and polyhydroxyalkanoates, PHA).[22,23] Proteins provide the opportunity to be used as raw material for the production of films. The films are formed by different types of amino acids that allow the development of intermolecular interactions (as ionic interactions), that be combined with other compounds and carried with different temperatures, allows the obtention of coatings with different chemical and physical properties. Optimizing interactions between amino acids, the formation of polymers with improved stability, barrier, mechanical, and solubility properties is favored.[20,24]

The coatings prepared from whey protein have advantages over coatings made from other biopolymers due to the excellent nutritional value, imperceptible taste, and carrying capacity of food additives.[25] The objective of this review is to have knowledge in progress of the use of whey proteins in the manufacture of packaging materials and/or edible coatings as well as the interactions that occur in the matrix formed from various materials as lipids.

12.2 THE MILK WHEY, A BY-PRODUCT OF THE DAIRY INDUSTRY

Each sector of the food industry generates waste in different amounts depending on the type of product that is produced. In Mexico, the food industry has become one of the productive sectors of higher socioeconomic and environmental impacts from their production chains to their degree of pollution. In the dairy industry, it is necessary to subjected the raw material to various processes to obtain the desired product with prolonged periods of storage and good preservation, which generates a large volume of air pollutants, solid waste, hazardous toxic waste, and liquid effluents.[26–28]

Only in 2013, Mexico generated about 10,926,771 L of milk leaving as residue an estimated of 4,964,099 L of serum. From January to March 2014, milk production generate an estimated of 2,595,134 L that will generate 848,427 L of serum.[29] Within the dairy industry, cheese is a primary product, which uses about 25% of total world production in its preparation and undoubtedly the main product of the dairy industry is the serum used in the preparation of cheese, which retains about 55% of the components of milk.

Whey is defined as one liquid translucent green substance obtained by removal of clot milk in the elaboration of cheese after precipitation of the protein.[30–32] There are several types depending of whey casein removal.

The first is sweet whey, which is based on coagulation of renin at pH 6.5. The second is the acid whey that results from the fermentation process in which organic or mineral acids to coagulate casein are added. The nutritional composition of both types may vary slightly, as the content of lactose and protein that are prevalent in sweet whey.[31,32] (Table 12.1). It is estimated that for every kilogram of cheese are produced 9 kg of whey, which represents about 85–90% of the volume of milk.[28,33–35]

TABLE 12.1 Composition of Whey.

Component	Sweet whey (g/L)	Whey (g/L)
Total solids	63.0–70.0	63.0–70.0
Lactose	46.0–52.0	44.0–46.0
Protein	6.0–10.0	6.0–8.0
Calcium	0.4–0.6	1.2–1.6
Phosphates	1.0–3.0	2.0–4.5
Lactate	2.0	6.4
Chlorides	1.1	1.1

The chemical composition of whey contains: lactose (4.5–5% w/v), soluble protein (0.6–0.8% w/v), lipids (0.4–0.5% w/v), and minerals (8–10% dry matter) as potassium, calcium, phosphorus, sodium, and magnesium. It also has B vitamins (thiamin, pantothenic acid, riboflavin, pyridoxine, nicotinic acid, and cobalamin) and ascorbic acid.[32,34,36,37] The high nutrient content generates BOD and chemical oxygen demand (COD), 3.5 and 6.8 kg per 100 kg of liquid whey, being the lactose the main component contributing to the high BOD and COD.[32,34,38–41]

Although proteins are not the most abundant component in whey, it is the most important economically and nutritionally[32,42] It has about 20% of the proteins in bovine milk, being the main component the β-lactoglobulin (β-Lg) (10%) and α-lactalbumin (α-La) (4%) of whole milk protein, also contains other proteins such as lactoferrin, lactoperoxidase, immunoglobulins, and glycomacropeptide. Its nutritional property and high biological value is attributed to the amount of essential amino acids (leucine, lysine, tryptophan, threonine, cysteine, methionine, histidine, valine, isoleucine, and phenylalanine) present in about 26%.[32,43,44]

However, despite their composition, they are discarded into rivers, sewage, and industrial water recollection centers without given or provide them any added value. In Mexico, the penalties provided in the General Law of Ecological Equilibrium and Environmental Protection, National

Water Law and other applicable breach of the Official Mexican Standard NOM-PA-CCA-009/93 that establishes the maximum permissible limits of entities of pollutants in wastewater discharges into receiving bodies of the processing industry of milk and its derivatives, leading to increased demand for innovation in technological processes for treating effluents, that due to the constant increase in dairies industry, will not cover. Unlike developed countries (e.g., USA, Germany, China, etc.), serum is dehydrated for use in making beverages, dairy products, and meat extenders.[45–47]

In order to reduce the environmental problem caused by the deposition of whey, some alternatives have been proposed to transform this problem of waste generation in a potential economic resource that could give to this waste source a high value added. Traditionally, whey is used to feed the population of cattle and pigs to provide them energy, protein and minerals, however, there are some techniques use the serum.

12.2.1 ALTERNATIVES TO THE USE OF SERUM

12.2.1.1 ETHANOL PRODUCTION

Ethanol production from serum has been widely studied and some industrial processes have been implemented in developed countries in milk production. Industrial plants are operating in Ireland, USA, New Zealand, etc.[48,49] Generally, deproteinized serum is used before the ultrafiltration. The first studies were conducted in the 1930s, using yeast capable of fermenting lactose.[50] The most used species can ferment this disaccharide are *Kluyveromyces marxianus* (*Kluyveromyces fragilis* before), *Kluyveromyces lactis*, and *Candida kefyr* (formerly *Candida pseudotropicalis*). The main limitation of this process is the low concentration of ethanol obtained by intolerance of some strains and the low lactose level that generates between 2% and 3% as maximum of ethanol at the end of fermentation.[51]

12.2.1.2 DEMINERALIZATION

Serum has a high content of salt and other minerals, calculating the dry weight; this is approximately 8–12%, what makes its application, as a food additive is limited. By the serum demineralization it is possible to open new avenues for use as partially demineralized whey (25–30%) or very demineralized whey (90–95%).[52] The partially demineralized whey concentrate can be used, for example, in the manufacture of ice cream and bakery products, while very demineralized whey concentrate or powder may be used in infant formulas and a large degree of products. Demineralization involves removing certain organic salts with organic ions reduction of citrates and lactates using techniques as nano- and/or ultrafiltration.[53] Ion exchange resins or electrodialysis can perform this procedure.[54]

12.2.1.3 PROTEIN CONCENTRATES WHEY

These are prepared by ultrafiltration consisting of a semipermeable membrane, which selectively allows passage of materials of low molecular weight as water, ions, and lactose, while retaining high molecular weight materials such as protein. The retentate is concentrated by evaporation and lyophilized[32,34,55] These concentrates are prepared as substitutes of skim milk and are used in the production of yogurt, processed cheese, in several applications of drinks, sauces, noodles, cookies, ice cream, cakes, dairy, bakery, acrne, beverages, and products of infant formulas due to their excellent functional properties of the proteins and their nutritional benefits.[30,32,34]

12.2.1.4 HYDROLYSATES

The preparations of enzymatic hydrolysates rich in oligopeptides represent a way to improve protein utilization. These preparations have been used in countries such as dietary supplements or physiological needs, to senior age, premature babies, athletes who control the weight through dieting, and children with diarrhea. This is because the amino acids provided by protein hydrolysates are completely absorbed in the digestive system compared with the intact protein without solubilizing.[32,56,57]

12.2.1.5 ISOLATED

Isolated whey protein is one that contains no fat, sugar, and lactose which includes major bovine proteins such as beta-lactalbumin, lactoglobulin, and lactoferrin. Protein supplements based protein isolated from whey provides a pure source of high quality protein with minimal amounts of fat, carbohydrates, and lactose. This protein also has the characteristic of being fast digestion and increase levels of amino acids available in the tissues needed to build muscle,[58–60] has a high antioxidant capacity.[61,62] In this way, several technologies have been used. Thus, concentration of whey may be realized by heating and drying (evaporation, spray-drying, and freeze-drying) or by reverse osmosis. Membrane separation technologies have been used for obtain proteins ingredients from whey.[63]

12.2.1.6 INFANT FORMULAS

The interest in improving the biological and nutritional milk yield that is modified to resemble human and thus be used in infant formulas has increased in recent years. For this, the ingredients are isolated from bovine milk and are adapted using the human milk as suitable reference. The composition of the bovine milk, despite differs in many aspects relating to human milk, such as to the content of casein, lactose, mineral salts, and specifically in the proportion of proteins found (β-Lg is not in the breast milk), is still the main source of nutrition for infant formulas.[32]

12.2.1.7 EDIBLE FILMS

Environmental pollution caused by packaging made of polyethylene has been a major reason for the development of research to increase the interest in the use of biodegradable polymers forming food packaging. One of the most studied polymers was the whey proteins due to the excellent nutritional value, taste, and soft capacity to serve as a means to add color, flavor, and functional food ingredients.[25]

12.3 FEATURES OF PROTEINS AS A STARTING MATERIAL

Proteins are biomolecules consisting of carbon, nitrogen, hydrogen, and oxygen, which may also contain sulfur, phosphorus, iron, magnesium, copper, and so on. They consist of 20–500 amino acids, in addition to the alpha amino and carboxyl alpha groups involved in the peptide bonds have a lateral chain with different functional groups, which give a distinctive. These are able to form three-dimensional amorphous networks mainly stabilized by non-covalent interactions and the functional properties of these materials are very dependent on the structural heterogeneity (due to amino acids which form the primary structure), thermal sensitivity, and the hydrophilic behavior of proteins.[64]

Depending on their own-primary sequence of each polypeptide chain structure, assumes different organizations on its axis—secondary structure—stabilized by hydrogen bonds and the tertiary structure—the three-dimensional organization reflects the polypeptide chain, based on the hydrogen bonding, Van der Waals forces, electrostatic and hydrophobic interactions, and disulfide bonds, to form globular protein or random fibrous structures. Finally, quaternary structure occurs as a consequence of the association of different polypeptide, equal or not to each other, which interact through non-covalent bonds resulting in single molecules. The functionality of proteins is known as the expression of their physicochemical properties, which affect systems or food matrices.[65] These properties can be classified based on the ability of proteins to interact with water molecules to establish interactions with other proteins and with its molecular surface characteristics.[66] From materials, technology are important to those properties related to protein–protein interactions, which give rise to the formation of matrices with specific characteristics with outstanding properties such as gel-forming ability and the ability of forming materials such as films, coatings, fibers, etc.[64, 67]

The inherent properties of these biopolymers make them excellent starting materials for films and/or biodegradable coatings. Within amino acids, polar and nonpolar, the load distribution along the chain creates a chemical potential. In β-Lg, the domains of the polar and nonpolar areas can generate a matrix in a protein-based film through the interactive forces. These arrays or systems can be stabilized from electrostatic interactions, hydrogen bonding, Van der Waals force, covalent, and disulfide bonds.[68,69]

The functional properties of composite films, which take advantage of each component and decrease its disadvantages, depend on their composition and formation process. Knowledge of how each component interacts each other (physically or chemically) facilitates the design of the films or coatings with structural features and properties specific barrier.[70] Among the side benefits for using proteins to form films and coatings is the existence of multiple sites for its chemical interactions as function of its various functional groups of amino acid.

12.3.1 WHEY PROTEIN

The Food and Drug Administration (FDA), mentions that milk proteins must have all the proteins found naturally in milk, and must be in the same relations, while that isolates and concentrates of whey protein, are those obtained by removing nonprotein components and are free of casein. These milk proteins have properties that can provide desirable texture and other attributes to the final product. They have multiple applications in traditional foodstuffs. Various types of milk protein, like whey protein concentrates (WPC), whey protein isolates (WPI), casein, caseinates, etc. can be obtained from the waste generated by industrial complexes. To concentrate and separate the protein, various techniques have been used as ultrafiltration or ion exchange technology. Then subjected to a drying process to obtain the WPC and WPI, which are highly soluble, probably because in water, acquire colloidal dimensions, are amphoteric and the complete hydrolysis produces a mixture of amino acids. Depending on the influence of different pH and ionizable groups to which they are, they can develop loads, either positive or negative, or reach a net charge of zero to reach the isoelectric point.[65]

The whey proteins differ from the caseins in their net negative charge that is uniformly distributed along the chain. Hydrophobic, polar, and charged amino acids are likewise uniformly distributed. Consequently proteins fold and so most hydrophobic groups are enclosed within the protein molecule. Protein interactions that occur between chains determine the formation of the network of the films and their properties.[71]

Whey is a significant source of functional proteins like β-Lg and α-La mainly obtained as a by-product of the cheese industry and casein.[72] These functional proteins confer to whey some distinctive characteristics such as solubility, stability of systems interphase, and protein thermal stability.[73] The whey proteins possess an excellent nutritional value, variable solubility in water, and aptitude as emulsifier agent.[74] There is research on the use of milk proteins as edible films.[75,76] For their high protein content, both the concentrate (WPC, 80% protein approx.) and isolated (WPI, > 90% protein) of whey proteins, are ideal for the formation of edible films.[77] However, the prior denaturation of β-Lg and α-La, is necessary to expose the hidden sulfhydryl (—SH) and disulfide (SS) groups in the hydrophobic center of the native globular tertiary structure of these proteins. The subsequent formation of intermolecular disulfide bonds, mainly in the nanomeric units of β-Lg,[78] promotes the generation of a stable three-dimensional network. It is therefore necessary to incorporate a plasticizer (e.g., glycerol or sorbitol) to decrease the density and reversibility of intermolecular interactions and increase chain mobility and thus flexibility of the film.[79] Films whey based plasticized with glycerol are excellent barriers to O_2, CO_2 and C_2H_4,[75,80] unfortunately its highly hydrophilic characteristic make it turns into not good barriers to water vapor.[77] Using this type of film in conjunction with hydrophobic membranes makes this type of coatings the ideal ones for the study of postharvest fruits and vegetables, especially in those highly perishable fruit.

12.4 INTERFACIAL AND ELECTROSTATIC INTERACTIONS

Homogenization is a technique widely used in the food industry that obtains consistent uniformly mixing, in order to obtain a dispersion of oil in these products, that is, a system having a dispersed phase and a continuous phase. In the process, aggregates are formed that are absorbed into the surface with the oil droplets newly formed. Proteins being the amphipathic molecules, have polar and nonpolar parts that are oriented in the interface so that a portion of nonpolar amino acids are in contact with the oil phase and the polar groups with the aqueous phase.[81–83]

The bilayer film formation is formed by the homogenization of lipid in a concentrated protein solution to form an emulsion, which will allow dehydration.[84] The result is a continuous protein matrix covered by lipid droplets embedded on. Although edible films are not effective as barriers bilayers films, these are superior to those of a single layer in the mechanical properties. The presence of the emulsion droplets in the film increases the distance traversed by water molecules, which diffuse through the film, therefore the water vapor permeability decreases. This is called tortuosity effect; part of the protein in the film is partially immobilized at the interface of the immobilized lipid droplets. This interfacial protein can adopt few configurations than the protein in the film mass. Biopolymer segments are less mobile, so the diffusion of water through the interfacial protein is reduced. Therefore, the water vapor permeability of the interfacial protein is lower than that of the mass of protein. This effect has been termed interfacial interaction.[75] Tortuosity has been thought to depend only on the volume fraction of the lipid film, while the interfacial interaction is a function of the surface area of emulsified lipid, and this depends on the volume fraction and particle size.[75] Many researchers have attempted to obtain the lowest possible particle size in the emulsions in order to maximize stability and interfacial interactions. The phases of dispersed solids should provide a good effect of interfacial interaction because the absorbed protein cannot move in the plane of the interface. However, production of very small droplets of dispersed solid takes energy.[85]

A microstructural study of mixtures of isolated whey protein and mesquite gum stressed the importance of taking into account the electrostatic interactions of the components.[70] The electrostatic interactions of the polymer are determined by the physicochemical characteristics of each polymer (charge density, molecular weight, etc.), their concentrations and solution conditions (pH, ionic strength, ion type, etc.).[86] For this study, the mixture of an anionic polysaccharide and a protein, electrostatic interaction has to be based on the net charge of each molecule that is pH dependent. In the study conditions, isolate whey protein is negatively charged, like mesquite gum, favoring the electrostatic repulsion between the two molecules, and consequently, the formation of aggregates.

Furthermore, the mixture of two different polymers often results in a phase separation of two domains in only a biopolymer. This happens due to the tendency of these molecules to be associated with other similar structures.[87] As a rule, the mixtures of biopolymers tend to segregate[88] and thermodynamically, protein and polysaccharides can be compatible or incompatible in aqueous solution. Incompatibility occurs when the repulsion between the biopolymers (as an example, when both are negatively charged) are in solution. In this case, the solvent/biopolymer interactions are favored as opposed to those biopolymer/biopolymer and solvent/solvent interactions. Finally becomes the system in two phases, with each phase becoming rich in one of the biopolymers.[89]

In a pH acid range from 1 to 3, protein molecules and k-carrageenan have opposite charges, thereby leaving a strong attraction there between. Under these conditions, it is difficult for the molecules of whey protein to form a three-dimensional protein matrix. The pH 6 provides the optimum conditions to obtain gels of mixture. A pH value of 6 is relatively close to the isoelectric point of the whey protein (Ip 5.2) where these have a low net charge and therefore tend to aggregate.[90]

The pH plays an important role in the formation of protein networks. The behavior of whey protein is highly related to the isoelectric point (PI), which is 5.2 to WPI. Below this, the total charge of WPI is mainly positive, while if the PI is greater, the overall charge is negative. WPI gels formed are transparent after heating, indicating protein denaturation in a pattern of fine wire. This behavior can be related to electrostatic repulsion between the protein molecules. In this case, proteins are negatively charged, when cations are added the bridges are enabled and the gelation is allowed.[91]

Complex formation of beta casein and guar rubber causes a disruption of a small proportion of the secondary structure of the protein between galactomannans of *Mimosa scabrella* and milk proteins. Doublier [92] suggested that the mechanism of interaction and compatibility involve creating weak electrostatic complexes soluble in water, which can be destroyed in increasing ionic strength. Their rheological studies on dynamic and static systems showed that the fraction of whey protein (α-La and β-Lg) do not have interaction with the galactomannan of *M. scabrella*, shown by the absence of variation in the values of viscosity.[93]

12.4.1 HEATING

The application of heating in protein solutions improves its properties as a matrix, which is due to the connections between the protein chains. One of modifications frequently used are disulfide bonds or unions, which occur with heat treatment, followed by polymerization of protein chains.[94] This allows expose chains, sulfhydryl, and hydrophobic groups. Hydrogen bonds and nonpolar hydrophobic groups in the molecule of the protein are disrupted, which expose amino groups solvent a more open structure.[95,96] The presence of thiol groups (β-Lg) is of importance to changes occurring in the solution during heating because engage in reactions with other proteins.[97] Polymerization occurs through the exchange of intermolecular disulfide bonds groups.[69,98] After heating, there may be non-covalent type interactions for exposure of new groups in the whey proteins, these interactions may be ionic, hydrophobic, and van der Waals force.[99]

12.5 PROTEIN–LIPID INTERACTIONS

A study of diversity and versatility of lipid–protein interactions in biological systems demonstrated associations between these two models may be only in the surface of the hydrophobic membrane (located in biological systems) or involve penetration in a hydrophobic membrane. The regulation occurs through conformational changes resulting from phosphorylation, ligand binding, etc. Proteins utilize amphipathic helices to test the lipid composition of the membrane, particularly the overall content of anionic lipid in a specific anionic lipid.[100]

Biological studies of lipids in the membrane protein structures show that there are three general principles of lipid binding proteins that are distinguished by their lipid interactions with proteins. First, there is an annular cover of lipids bound to the surface of the protein, which resembles the bilayer structure. Second, the lipid molecules are embedded in cavities and crevices of the surface of the protein, frequently by multisubunit complexes and multimeric ensembles. These surface lipids not override are typically present in oligomeric subunits or interfaces. Finally, few examples represent lipids, which reside in the membrane protein or protein complex membrane and these are in unusual positions.[101]

12.6 PROTEIN–POLYSACCHARIDE INTERACTIONS

Proteins may present some modifications to their functional properties after they have been conjugated with some polysaccharides. These complexes require covalent bond formation, usually after a heat treatment and conditions of low water activity.[102,103]

Over time various techniques have been used for better understanding of the possible interactions between proteins and polysaccharides present. Depending on aqueous environmental conditions such as pH, ionic strength, etc., it is possible to describe four different behaviors for interactions of milk proteins and anionic polysaccharides in the aqueous phase: (1) At neutral pH and low ionic strength, both proteins and polysaccharides are negatively charged and although electrostatic attractive interactions might exist between protein parts of positive and negative charges of the polysaccharide, these biopolymers are co-soluble in low concentrations, (2) pH values near to the isoelectric point or relatively low values, electrostatic protein–polysaccharide complexes are formed, (3) in a high reduction of aqueous phase, pH allows the aggregation of soluble complexes and complex coacervation, and finally, and (4) to less than pH 2.5, electrostatic complexes of biopolymers are generally removed due to the protonation of acidic functional groups of polysaccharide.[104,105]

Interactions exist thanks to the presence of attractive electrostatic and covalent bonding forces, which occur in the positively charged proteins and anionic polysaccharides with low energy that may result in an insoluble precipitate of both polymers.[103] The polysaccharide interactions with β-Lg, show that when β-Lg adsorbed onto the air–water interface in the presence of polysaccharides three phenomena can occur. First is that the polysaccharides are adsorbed on the interface competing with the protein for the interface (competitive adsorption). Second, the complexes of polysaccharides with adsorbed protein, are linked mainly by electrostatic interactions or hydrogen bonds[106,107] and finally the existence of limited thermodynamic compatibility between protein and polysaccharide. The polysaccharide adsorbed protein concentrate.

The performance of mixtures of polysaccharides is determined by the pH and the concentration and also by the ionic strength of the system under study. The pH generally strongly affects the net charge of the protein and plays an important role in interactions between protein and anionic polysaccharides.[86] Protein at the interface is partially deployed forming intermolecular hydrophobic associations or disulfide bonds.[108] Murray[109] mention that some properties of multilayer films are enhanced by the mixture of proteins and polysaccharides. Protein–polysaccharide interactions usually are formed by the binding of the protein to the amino groups to reduce the carboxyl end groups of polysaccharides via a controlled Maillard reaction.[110,111] Frequently protein complexes—polysaccharide arise from noncovalent associations mainly by electrostatic attractive interactions. In a study of interfacial rheology measurements using the expansion method drop, an elastic behavior for the whey protein was found. Showed that at low concentrations of protein, where the first biopolymer molecules reach the interface at optimum conditions can completely absorb any locus and occupy a high interface area because the interface is still free of molecules. However, at high concentrations, abundant protein–protein interactions, side to side, at the interface, induce a more compact conformation of the protein.[112]

12.7 CONCLUDING REMARKS

Numerous studies have been conducted to elucidate the interactions of proteins with other polysaccharides, as well as for changes and/or modifications that occur in the protein structure when subjected to different processes to produce a final product. These studies have generated different models that allow predictions of what happens within a matrix made of proteins, and even further, theoretically elucidate what interactions with other components are distorted when the denaturalization is reached. It is important to note that all data shown were carried out under defined conditions of work; any changes or modifications can directly influence the behavior of interactions. Studies remain to be done to learn more about this topic and thus know if there are any changes or if there are new functional properties that can take advantage in proteins.

12.8 LOOKING AHEAD

Industrial and environmental practices require alternatives to the reutilization of serum for constructive purposes, since the degree of pollution caused by pouring the drain is high. The dairy industry it is constantly growing each year and it will so far, so it is necessary to further research and technological development to take new technological achievement of this waste, and thereby extend the diversification of products to reach the international market with innovative products and be able to handle large-scale production benefiting both the environmental and economic sector. The dairy sector has suggested the implementation of a plant receiving serum from various companies to conduct dewatering processes thereof, so that represents a solution to the problem of disposal, which in turn would generate jobs and increased revenue.

KEYWORDS

- **dairy industry**
- **isolates**
- **reutilization**
- **edible films**
- **interactions**

REFERENCES

1. FAO. Manual de Capacitación en Nutrición y Alimentación de Peses Y Camarones Cultivados. 2004. http://www.fao.org/docrep/field/003/ab492s/AB492S01.htm. 15/11/2014.
2. Ruiz-Martínez, J. Cubiertas Comestibles óleo Proteicas Para Prolongar La Vida De Anaquel Del Tomate. M.S. Thesis, Universidad Autónoma de Coahuila, 2013.
3. Maftoonazad, N.; Ramaswamy, H. S.; Marcotte, M. valuation of Factors Affecting Barrier, Mechanical and Optical Properties of Pectin-based Films Using Response Surface Methodology. *J. Food Proc. Eng.* **2007,** *30* (5), 539–563.
4. Ali, A.; Maqbool, M.; Ramachandran, S.; Alderson, P. G. Gum Arabic as a Novel Edible Coating for Enhancing Shelf Life and Improving Postharvest Quality of Tomato (*Solanum lycopersicum* L.) Fruit. *Postharvest Biol. Technol.* **2010,** *58*, 42–47.
5. Cerqueira, M. A.; Lima, A. M.; Texeira, J. A.; Moreira, R. A.; Vicente, A. A. Suitability of Novel Galactomannans as Edible Coatings for Tropical Fruits. *J. Food Eng.* **2009,** *94*, 372–378.
6. Bourbon, A. I.; Pinheiro, A. C.; Cerqueira, M. A.; Rocha, C.; Avides, M. C.; Quintas, M.; Vicente, A. A. Physico Chemical Characterization of Chitosan Based Edible Films Incorporating Bioactive Compounds of Different Molecular Weight. *J. Food Eng.* **2011,** *106*, 111–118.
7. De Aquino, A. B.; Blank, A. F.; De Aquino Santana, L. C. L. Impact of Edible Chitosan–Cassava Starch Coatings Enriched with Lippia Gracilis Schauer Genotype Mixtures on

the Shelf Life of Guavas (*Psidium guajava* L.) During Storage at Room Temperature. *Food Chem.* **2015,** *171*, 108–116.
8. Slavutsky, A. M.; Bertuzzi, M. A. Formulation and Characterization of Nanolaminated Starch Based Film. *LWT. Food Sci. Technol.* **2015,** *61* (2), 407–413.
9. Guerreiro, A. C.; Gago, C. M. L.; Faleiro, M. L.; Miguel, M. G. C.; Antunes, M. D. C. The Effect of Alginate-based Edible Coatings Enriched with Essential Oils Constituents on Arbutus Unedo L. Fresh Fruit Storage. *Postharvest Biol. Technol.* **2015,** *100*, 226–233.
10. Narsaiah, K.; Wilson, R. A.; Gokul, K.; Mandge, H. M.; Jha, S. N.; Bhadwal, S.; Anurag, R. K.; Malik, R. K.; Vij, S. Effect of Bacteriocin-incorporated Alginate Coating on Shelf-life of Minimally Processed Papaya (Carica papaya L.). *Postharvest Biol. Technol.* **2015,** *100*, 212–218.
11. Arismendi, C.; Chillo, S.; Conte, A.; Del Nobile, M. A.; Flores, S.; Gerschenson, L. N. Optimization of Physical Properties of Xanthan Gum/tapioca Starch Edible Matrices Containing Potassium Sorbate and Evaluation of Its Antimicrobial Effectiveness. *LWT. Food Sci. Technol.* **2013**, *53,* (1), 290–296.
12. Zambrano-Zaragoza, M. L.; Mercado-Silva, E.; Del Real, L. A.; Gutiérrez-Cortez, E.; Cornejo-Villegas, M. A.; Quintanar-Guerrero, D. The Effect of Nano-coatings with α-tocopherol and Xanthan Gum on Shelf-life and Browning Index of Fresh-cut "Red Delicious" Apples. *Innov. Food Sci. Emerg. Technol.* **2014**, *22*, 188–196.
13. Ünalan, I. U.; Arcan, I.; Korel, F.; Yemenicioğlu, A. Application of Active Zein-based Films with Controlled Release Properties to Control Listeria Monocytogenes Growth and Lipid Oxidation in Fresh Kashar Cheese. *Innov. Food Sci. Emerg. Technol.* **2013,** *20*, 208–214.
14. Yin, Y. C.; Yin, S. W.; Yang, X. Q.; Tang, C. H.; Wen, S. H.; Chen, Z.; Xiao, B, J.; Wu, L. Y. Surface Modification of Sodium Caseinate Films by Zein Coatings. *Food Hydrocoll.* **2014,** *36*, 1–8.
15. Khanzadi, M.; Jafari, S. M.; Mirzaei, H.; Chegin, F. K.; Maghsoudlou, Y.; Dehnad, D. Physical and Mechanical Properties in Biodegradable Films of Whey Protein Concentrate-pullulan by Application of Beeswax. *Carbohydr. Polym.* **2015,** *118*, 24–29.
16. Synowiec, A.; Gniewosz, M.; Kraśniewska, K.; Przybył, J. L.; Bączek, K.; Węglarz, Z. Antimicrobial and Antioxidant Properties of Pullulan Film Containing Sweet Basil Extract and an Evaluation of Coating Effectiveness in the Prolongation of the Shelf Life of Apples Stored in Refrigeration Conditions. *Innov. Food Sci. Emerg. Technol.* **2014,** *23*, 171–181.
17. Ding, C.; Zhang, M.; Li, G. Preparation and Characterization of Collagen/Hydroxypropyl Methylcellulose (HPMC) Blend Film. *Carbohydr. Polym.* **2015,** *119*, 194–201.
18. Fagundes, C.; Palou, L.; Monteiro, A. R.; Pérez-Gago, M. B. Effect of Antifungal Hydroxypropyl Methylcellulose-beeswax Edible Coatings on Gray Mold Development and Quality Attributes of Cold-stored Cherry Tomato Fruit. *Postharvest Biol. Technol.* **2014,** *92*, 1–8.
19. Martins, J. T.; Cerqueira, M. A.; Bourbon, A. I.; Pinheiro, A. C.; Souza, B. W. S.; Vicente, A. A. Synergistic Effects Between κ-carrageenan and Locust Bean Gum on Physicochemical Properties of Edible Films Made Thereof. *Food Hydrocoll.* **2012,** *29* (2), 280–289.
20. Montalvo, C.; López, A.; Palou, E. Películas Comestibles De Proteína: Características, Propiedades y Aplicaciones. *Temas Selectos de Ingeniería de Alimentos* **2012,** *6–2*, 32–46.

21. Carrillo Aguado, J. L. Tratamiento y Reutilización Del Suero De Leche. *Mundo Lácteo y Cámico.* **2006,** *6,* 27–30.
22. Villada, H. S.; Acosta, H. A.; Velasco, R. J. Biopolymers Naturals Used in Biodegradable Packaging. *Temas Agrarios.* **2007,** *12* (2), 5–13.
23. Tharanathan, R. N. Biodegradable Films and Composite Coatings: Past, Present and Future. *Trends Food Sci. Technol.* **2003,** *14,* 71–78.
24. Vroman, I.; Tighzert, L. Biodegradable Polymers. *Materials.* **2009,** *2* (2), 307–344.
25. Li, C.; Chen, H. Biodegradation of Whey Protein-based Edible Films. Department of Nutrition and Food Sciences. University of Vermont, Buirlington. *J. Polym. Environ.* **2000,** *8,* 135–143.
26. Comisión Nacional del Medio Ambiente Región Metropolitana (CONAMA/RM); Guía Para el Control y Prevención de la Contaminación Industrial, Santiago, Chile, 1998; pp 58.
27. Restrepo, M. Producción Más Limpia En La Industria Alimentaria. *Rev Prod. Limpia.* **2006,** *1* (1), 88–101.
28. González Cáceres, M. J. Aspectos Medio Ambientales Asociados a Los Procesos De La Industria Láctea. *Mundo Pecuario.* **2012,** *8* (1), 16–32.
29. SIAP Resumen Nacional de Producción Agrícola. Servicio de Información Agroalimentaria y pesquera (SIAP)-SAGARPA, Gobierno de Mexico. Delegacion Hidalgo, CP 11800, Ciudad de México.
30. Foegeding, S.; Luck, P. Whey Protein Products. In *Encyclopedia of Dairy Sciences*; Academic Press: Cambridge, MA, 2002; pp 1957–1960.
31. Jelen, P. Whey Processing, In *Encyclopedia of Dairy Sciences*; Roginski, H., Fuquay, J. F., Fox, P. F., Eds.; Academic Press–An Imprint of Elsevier: Amsterdam, Boston, London, 2003; Vol. 4, pp 2740.
32. Parra, R. Importancia en la Industria de Alimentos. *Rev. Fac. Nac. Agron. Medellin.* **2009,** *62* (1), 4967–4982.
33. Berruga, M. I. Desarrollos de Procedimientos Para el Tratamiento de Efluentes de Quesería. Ph.D. Thesis. Universidad Complutense de Madrid. Facultad de Veterinaria, 1999, p 337.
34. Muñl, A.; Paez, G.; Faria, J.; Ferrer, J.; Ramones, E. Eficiencia de Un Sistema de Ultrafiltración/nanofiltración Tangencial en Serie Para el Fraccionamiento y Concentración Del Lactosuero. *Rev. Científica FCV-LUZ XV.* **2012,** *4,* 361–367.
35. Liu, X.; Powers, J. R.; Swanson, B. G.; Hill, H. H.; Clark, S. High Hydrostatic Pressure Affects Flavor-binding Properties of Whey Protein Concentrate. *J. Food Sci.* **2005,** *70,* C581–C585.
36. Londoño, M. Aprovechamiento Del Suero Acido de Queso Doble Crema Para La Elaboración de Quesillo Utilizando Tres Métodos de Complementación de Acidez Con Tres Acidos Orgánicos. In *Revista Perspectivas en Nutrición Humana-Escuela de Nutrición y Dietética;* Universidad de Antioquia: Medellín, Antioquia, Colombia, 2006; *16,* pp. 11–20.
37. Panesar, P.; Kennedy, J.; Gandhi, D.; Bunko, K. Bioutilisation of Whey for Lactic Acid Production. *Food Chem.* **2007,** *105,* 1–14.
38. Ghaly, A.; Kamal, M. Submerged yeast Fermentation of Acid Cheese Whey for Protein Production and Pollution Potential Reduction. *Water Res.* **2004,** *38* (3), 631–644.
39. Mukhopadhyay, R.; Chatterjee, S.; Chatterjee, B. P.; Banerjee, P.; Guha, A. Production of Gluconic Acid from Whey by Free and Immobilized *Aspergillus niger. Int. Dairy J.* **2005,** *15* (3), 299–303.

40. Koutinas, A.; Papapostolou, A.; Dimitrellou, D.; Kopsahelis, N.; Katechaki, E.; Bekatorou, A.; Bosnea, L. Whey Valorisation: A Complete and Novel Technology Development for Dairy Industry Starter Culture Production. *Bioresour. Technol.* **2009,** *100* (15), 3734–3739.
41. Almeida, K. E.; Tamime, A. Y.; Oliveira, M. N. Influence of Total Solids Contents of Milk Whey on the Acidifying Profile and Viability of Various Lactic Acid Bacteria. *LWT. Food Sci. Technol.* **2009,** *42* (2), 672–678.
42. Linden, G.; Lorient, D. *Bioquímica Agroindustrial: Revalorización Alimentaria de la Producción Agrícola;* Editorial Acribia: Zaragoza, España, 1996; p 454.
43. Baro, L.; Jiménez, J.; Martínez, A.; Bouza, J. Péptidos y Proteínas de la Leche Con Propiedades Funcionales. *J. Ars. Pharm.* **2001,** *42* (3–4), 135–145.
44. Hinrichs, R.; Gotz, J.; Noll, M.; Wolfschoon, A.; Eibel, H.; Weisser, H. Characterization of Different Treated Whey Protein Concentrates by Means of Low-resolution Nuclear Magnetic Resonance. *Int. Dairy J.* **2004,** *14* (9), 817–827.
45. Andrade, L. Efecto Del Flujo de Alimentación Sobre la Ultrafiltración Del Suero Pasteurizado de Queso. Engineer Thesis, Escuela Agrícola Panamericana Zamorano Honduras, 1999, pp 1–24.
46. Johnson, B. Los Concentrados de Proteína de Suero y Sus Aplicaciones en Productos Bajos en Grasa. Alfa Editores Técnicos México: Mexico City, Mexico, 2004; Vol. 19, pp 1–3.
47. Teniza García, O. Estudio del Suero de Queso de Leche de Vaca y Propuesta Para el Reusó del Mismo. M.S. Thesis, Instituto Politécnico Nacional, Centro de Investigación en Biotecnología Aplicada Unidad Tlaxcala México, 2008, pp 36–40.
48. Marwaha, S. S.; Kennedy, J. F.; Sehgal, V. K. Simulation of Process Conditions of Continuous Ethanol Fermentation of Whey Permeate Using Alginate Entrapped *Kluyveromyces marxianus* NCYC 179 Cells in a Packed-bed Reactor System. *Proc. Biochem.* **1988,** *23* (2), 17–22.
49. Castillo, F. J. Lactose Metabolism Yeasts. In *Yeast: Biotechnology and Biocatalysis*; Verachtert, H., De Mot, R., Eds.; Marcel Dekker Inc.: New York, NY, 1990; pp 297–320.
50. Marth, E. H. Fermentation Products from Whey. In *Byproducts of Milk;* Webb, B. H., Whittier, E. O., Eds.; Avi Publishing Company Inc.: Westport, CT, 1970; pp 43–82.
51. García, M.; Quintero, R.; López, A. *Biotecnología Alimentaria;* Lamusa, S. A., Ed.; México, 2004; 186–207.
52. Gosta, B. *Manual de Industrias Lácteas;* Ediciones Mundi-prensa: Madrid, Spain, 2003.
53. Suárez, E.; Lobo, A.; Alvarez, S.; Riera, F. A.; Álvarez, R. Demineralization of Whey and Milk Ultrafiltration Permeate by Means of Nanofiltration. *Desalination* **2009,** *241* (1–3), 272–280.
54. Brandelli, A.; Daroit, D. J.; Corrêa, A. P. E. Whey as a Source of Peptides with Remarkable Biological Activities. *Food Res. Int.* **2015,** *73,* 149–116.
55. Zadow, J. Protein Concentrates and Fractions. 6152–6156. In *Encyclopedia of Food Science and Technology;* Francis, F., Ed.; Wiley: New York, NY, 2003.
56. Santana, M.; Rolim, E.; Carreiras, R.; Oliveira, W.; Medeiros, V.; Pinto, M. Obtaining Oligopeptides from Whey: Use of Subtilisin and Pancreatin. *Am. J. Food Technol.* **2008,** *3* (5), 315–324.
57. Spellman, D.; O'Cuinn, G.; FitzGerald, R. Bitterness in Bacillus Proteinase Hydrolysates of Whey Proteins. *Food Chem.* **2009,** *114* (2), 440–446.
58. Paul, G. L. The Rationale for Consuming Protein Blends in Sports Nutrition. *J. Am. Coll. Nutr.* **2009,** *28* (4), 464S–472S.

59. Tang, J.; Moore, D.; Kujbida, G.; Tarnopolsky, M.; Phillips, S. Ingestion of Whey Hydrolysate, Casein, or Soy Protein Isolate: Effects on Mixed Muscle Protein Synthesis at Rest and Following Resistance Exercise in Young Men. *Appl. Physiol.* **2009,** *107*, 987–992.
60. Phillips, S.; Hartman, J.; Wilkinson, S. Dietary Protein to Support Anabolism with Resistance Exercise in Young Men. *J. Am. Coll. Nutr.* **2005,** *24* (2), 134S–139S.
61. Bayram, T.; Pekmez, M.; Arda, N.; Yalçın, A. S. Antioxidant Activity of Whey Protein Fractions Isolated by Gel Exclusion Chromatography and Protease Treatment. *Talanta.* **2008,** *75* (3), 705–709.
62. Kong, B.; Peng, X.; Xiong, Y. L.; Zhao, X. Protection of Lung Fibroblast MRC-5 Cells against Hydrogen Peroxide-induced Oxidative Damage By 0.1–2.8 kDa Antioxidative Peptides Isolated from Whey Protein Hydrolysate. *Food Chem.* **2012,** *135* (2), 540–547.
63. Brans, G.; Schroën, C. G. P. H.; van der Sman, R. G. M.; Boom, R. M. Membrane Fractionation of Milk: State of the Art and Challenges. *J. Membrane Sci.* **2004,** *243* (1–2), 263–272.
64. Mauri, A. N.; Añon, M. C. Effect of Solution pH on Solubility and Some Structural Properties of Soybean Protein Isolate Films. *J. Sci. Food Agric.* **2006,** *86* (7), 1064–1072.
65. Badui, S. *Química de Los Alimentos*; México, D. F., Ed.; Alhambra Mexicana: México, 1999; 648.
66. Cheftel, J. C.; Cuq, J. L.; Lorient, D. Amino Acids, Peptides and Proteins. In *Food Chemistry*; Fennema, O. R., Ed.; Marcel Dekker, Inc.: New York, NY, 1985; pp 246.
67. Petruccelli, S. Modificaciones Estructurales de Aislados Proteicos de Soja Producidas Por Tratamientos Reductores y Térmicos y su Relación Con Propiedades Funcionales. Ph.D. Thesis, Universidad Nacional De La Plata, 1993.
68. Krochta, J. M.; Baldwin, E. A.; Nisperos-Carriedo, M. *Edible Coatings and Films to Improve Food Quality;* Technomic Publishing Co.: Estados, Unidos, 1994.
69. Tomasula, P. M.; Qi, P.; Dangaran, K. L. Structure and Function of Protein-based Edible Films and Coatings. In *Edible Films and Coatings for Food Applications;* Embuscado, M. E., Huber, K. C., Eds.; Springer Science Media: New York, NY, 2009; Chapter 2, pp 25–26.
70. Osés, J.; Fabregat Vázquez, M.; Pedroza Islas, R.; Tomas, S. A.; Cruz Orea, A.; Maté. J. I. Development and Characterization of Composite Edible Films Based on Whey Protein Isolate and Mesquite Gum. *J. Food Eng.* **2009,** *92*, 56–62.
71. Swaisgood, H. E. Characteristics of Milk. In *Food Chemistry;* Fennema, O. R., Ed.; Marcel Dekker: New York, NY, 1996; pp 841–878.
72. Fox, P. F.; McSweeney, P. L. H. *Dairy Chemistry and Biochemistry;* Springer Science and Business Media: Berlin, Germany, 1998; p 478.
73. Capitani, C. D.; Pacheco, M. T. B.; Gumerato, H. F.; Vitali, A.; Schmidt, F. L. Recuperação de Proteínas do Soro de Leite Por Meio de Coacervação Com Polissacarídeo (Milk Whey Protein Recuperation by Coacervation with Polysaccharide). *Braz. J. Agric. Res.* **2005,** *40*, 1123–1128.
74. Walstra, P.; Geurts, T. J.; Noomen, A.; Jellema, A. *Dairy Technology: Principles of Milk Properties and Processes;* Marcel Dekker: New York, NY, 1999; p 727.
75. McHugh, T.; Krochta, J. Water Vapor Permeability Properties of Edible Whey Protein-lipid Emulsion Films. *J. Am. Oil Chem. Soc.* **1994,** *71*, 307–312.
76. Krochta, J. M.; de Milder-Johnson, C. Edible Films and Biodegradable Polymer Films: Challenges and Opportunities. *Food Technol.* **1997,** *51* (2), 61–74.

77. Gallieta, G. Formación y Caracterización de Películas Comestibles en Base a Suero de Leche. M.S. Thesis, Universidad de la República Oriental del Uruguay, Montevideo. Uruguay, 2001, p 135.
78. Monahan, F.; McClements, J. D. J.; Kinsella, J. E. Polymerization of Whey Protein in Whey Protein Stabilized Emulsion. *J. Agric. Food Chem.* **1993,** *41*, 1826.
79. Banker, G. S. Film Coating Theory and Practice. *J. Pharm. Sci.* **1996,** *55*, 81–89.
80. Gallieta, G.; Vanaya, F.; Ferrari, N.; Diano, W. Barrier Properties of Whey Protein Isolate Films to Carbón Dioxide and Ethylene at Carious Water Activities. In *Biopolymer Science: Food and Non Food Applications*; Colonna, P., Guilbert, S., Eds.; Les Colloques N.91, INRA Editions: Montpellier, France, 1998; pp 327–335.
81. Singh, H.; Ye, A. Interactions and Functionality of Milk Proteins in Food Emulsions. In *Milk Proteins: From Expression to Food;* Academic Press: Cambridge, MA, 2009; 321–340.
82. Dickinson, E. Structure and Composition of Adsorbed Protein Layers and the Relationship to Emulsion stability. *J. Chem. Soc. Faraday Tra.* **1992,** *88*, 2973–2983.
83. Dickinson, E. Stability and Rheological Implications of Electrostatic Milk Protein Polysaccharide Interactions. *Trends Food Sci. Technol.* **1998,** *9*, 347–354.
84. Krochta, J. M. Food Emulsion and Foams; Theory and Practice. Wan, P. J., Cavallo, J. L., Saleeb, F. Z., McCarthy, M. J., Eds.; American Institute of Chemical Engineers: New York, NY, 1993; pp 57–61.
85. Fairley, P.; Krochta, J. M.; German, J. B. Interfacial Interactions in Edible Films from Whey Protein Isolate. *Food Hydrocoll.* **1997,** *11* (3), 245–252.
86. Schmitt, C.; Sanchez, C.; Desobry-Banon, S.; Hardy, J. Structure and Technofunctional Properties of Protein–Polysaccharide Complexes: A Review. *Crit. Rev. Food Sci. Nutr.* **1998,** *38* (8), 689–753.
87. Norton, I. T.; Frith, W. J. Microstructure Design in Mixed Biopolymer Composites. *Food Hydrocoll.* **2001,** *15*, 543–553.
88. Kruif, C, G.; Tuinier, R. Polysaccharide Protein Interactions. *Food Hydrocoll.* **2001,** *15*, 555–563.
89. Nayebzadeh, K.; Chen, J.; Mousavi, M. S. M. Interaction of WPI and Xanthan in and Rheological Properties of Protein Gels and O/W Emulsions. *Int. J. Food Eng.* **2007,** *3* (4), 1–17.
90. Mleko, S.; Li-Chan, E. C. Y.; Pikus, S. Interactions of k-carrageenan with Whey Proteins in Gels Formed at Different pH. *Food Res. Int.* **1997,** *30* (6), 427–433.
91. Turgeon, S. L.; Beaulieu, M. Improvement and Modification of Whey Protein Gel Texture Using Polysaccharides. *Food Hydrocoll.* **2001,** *15*, 583–591.
92. Doublier, J. L.; Garnier, C.; Renard, D.; Sanchez, C. Protein–polysaccharide Interactions. *Curr. Opin. Colloid Interface Sci.* **2000,** *5*, 202–14.
93. Perissutti, G. E.; Bresolin, T. M. B.; Ganter, J. L. M. S. Interaction between the Galactomannan from *Mimosa Scabrella* and Milk Proteins. *Food Hydrocoll.* **2002,** *16*, 403–417.
94. Pérez-Gago, M. B.; Nadaud, P.; Krochta, J. M. Water Vapor Permeability Solubility, and Tensile Properties of Heat Denatured versus Native Whey Protein Films. *J. Food Sci.* **1999,** *64*, 1034–1037.
95. Damodaran, S. Amino Acids, Peptides and Proteins. In *Fennema's Food Chemistry*, 4th ed.; Damodaran, S., Parkin, K. L., Fennema, O. R., Eds.; CRC Press, Taylor and Francis Group: Boca Raton, FL, 2008; pp 217–330.
96. Wihodo, M.; Moraru, C. I. Physical and Chemical Methods Used to Enhance the Structure and Mechanical Properties of Protein Films: A Review. *J. Food Eng.* **2013,** *114*, 292–302.

97. Chae, S. I.; Heo, T. R. Production and Properties of Edible Film Using Whey Protein. *Biotechnol. Bioprocess Eng.* **1997,** *2*, 122–125.
98. Galani, D.; Apenten, R. K. O. Heat Induced Denaturation and Aggregation of Beta Lactoglobulin: Kinetics of Formation of Hydrophobic and Disulphide Linked Aggregates. *Int. J. Food Sci. Technol.* **1999,** *34*, 467–476.
99. Kinsella, J. E. Milk Proteins: Physicochemical and Functional Properties. *Crt. Rev. Food Sci. Nat.* **1984,** *21*, 197–262.
100. Dowhan, W.; Mileykovskaya, E.; Bogdanov, M. Diversity and Versatility of Lipid Protein Interactions Revealed by Molecular Genetic Approaches. *Biochim. Biophys. Acta.* **2004,** *1666*, 19–39.
101. Palsdottir, H.; Hunte, C. Lipids in Membrane Protein Structures. *Biochim. Biophys. Acta.* **2004,** *1666*, 2–18.
102. Alfaro Arismendi, M. J. Desarrollo de Metodologías de Encapsulación Utilizando Aislado Proteico de Suero Lácteo Modificado Con Azúcares Para La Protección de Ingredientes Activos en Alimentos. Ph.D. Thesis, Universidad Central de Venezuela, Facultad de Ciencias en Ciencia y Tecnología de Alimentos, 2012.
103. Muñoz, J.; del, M.; Alfaro, C.; Zapata, I. Avances En La Formulación de Emulsions. *Grasas y Aceites.* **2007,** *58* (1), 64–73.
104. Weinbreck, F.; de Vries, R.; Schroogen, P.; de Kruif, C. G. Complex Coacervation of Whey Proteins and Gum Arabic. *Biomacromolecules.* **2003,** *4* (2), 293–303.
105. Perez, A. A.; Carrara, C. R.; Carrera, C.; Rodriguez Patino, J. M. Interactions between Milk Whey Protein and Polysaccharide in Solution. *Food Chem.* **2009,** *116*, 104–113.
106. Dickinson, E. Hydrocolloids at Interfaces and the Influence on the Properties of Dispersed Systems. *Food Hydrocoll.* **2003,** *17* (1), 25–40.
107. Baeza, R.; Carrera Sánchez, C.; Pilosof, A. M. R.; Rodríguez Patino, J. M. Interactions of Polysaccharides with b-lactoglobulin Adsorbed Films at the Air–water Interface. *Food Hydrocoll.* **2005,** *19*, 239–248.
108. Funtenberger, S.; Dumay, E.; Cheftel, J. C. Pressure Induced Aggregation of Beta Lactoglobulin in pH 7.0 Buffers. *Food Sci. Technol.* **1995,** *28* (4), 410–418.
109. Murray, B. S. Rheological Properties of Protein Films. *Curr. Opin. Colloid Interface Sci.* **2011,** *16*, 27–35.
110. Kato, A.; Sasaki, Y.; Furuta, R.; Kobayashi, K. Functional Protein/polysaccharide Conjugate Prepared by Controlled Dry Heating of Ovalbumin/Dextran Mixtures. *Agric. Biol. Chem.* **1990,** *54*, 107–112.
111. Bouyer, E.; Mekhloufi, G.; Rosilio, V.; Grossiord, J. L.; Agnely, F. Proteins, Polysaccharides, and Their Complexes Used as Stabilizers for Emulsions: Alternatives to Synthetic Surfactants in the Pharmaceutical Field. *Int. J. Pharm.* **2012,** *436*, 359–378.
112. Wüstnek, R.; Moser, B.; Muschiolik, G. Interfacial Dilational Behaviour of Adsorbed Beta Lactoglobulin Layers at the Different Fluid Interfaces. *Colloids Surf. B. Biointerfaces.* **1999,** *15*, 263–273.

CHAPTER 13

GRAFTED CINNAMIC ACID: A NOVEL MATERIAL FOR SUGARCANE JUICE CLARIFICATION

PRITI RANI, PINKI PAL*, SUMIT MISHRA, JAY PRAKASH PANDEY, and GAUTAM SEN

Chemistry Department, Birla Institute of Technology, Mesra, Ranchi 835215, Jharkhand, India

**Corresponding author. E-mail: pinkipalhbti@gmail.com*

CONTENTS

ABSTRACT

In the processing of sugarcane juice for the manufacture of white sugar, clarification is an essential step to eliminate unwanted solids from the essential sugar solution. Applicability of graft copolymer of sodium alginate (SAG) was studied for this clarification step. Cinnamic acid grafted sodium alginate (SAG-g-P (CA)) was prepared by microwave-assisted method, that is, microwave radiation in conjunction with chemical free radical initiator used to initiate grafting reaction. The clarification efficacy in sugar cane juice was evaluated by standard jar test method. Polysaccharide graft copolymer as clarifying agent in sugar industry (unlike conventional process of sulfitation, phosphatation, carbonation, etc.) is green and competitive substitute as manifested by biodegradability, physiological inertness, and immense flocculation competencies. The optimal dosage of SAG-g-P (CA) as flocculant in juice was at 20 ppm.

13.1 INTRODUCTION

The major challenges in sugar processing involve removal of colloidal as well as soluble macromolecular substances imparting haziness to the raw sugarcane juice. The clarification mechanism necessitates separation of suspended and colloidal particles in sugar cane juice before processing to produce white sugar. Since the system consists of a complicated and dynamic colloidal solution, the clarification by simple decantation/filtration process is not highly effective. The most common practices employed in sugar manufacturing units are liming—sulphitation-boiling process,[1] phosphatation-carbonation process,[2] etc. The juice obtained by such process is dark brown in color. Consequently, highly stained product is obtained which cannot correlate with white sugar. Besides this, sludge/solid waste formed, poses a disposal difficulty, and deliberate environmental issues.

Therefore, some backup techniques like flocculation,[2] coagulation,[3] adsorption,[4] ion exchange,[5] electrodialysis,[6] including membrane filtration[7] are also involved to figure out these affairs. The most commonly used coagulant (aluminum sulfate) in water treatment is also used in clarification of cane juice[8] but chronic exposure to Al^{+3} ions has been associated with retrogressive diseases like Alzheimer's disease.[9] Some other complications have discouraged the use of these methods specially those concerning food regulations,[10] cost and energy inputs, operational feasibility, and the planet's environment.

Among the various operations discussed, membrane filtration process is most effective one, but it also suffers from economical as well as technical obstacles like high sucrose loss,[11] membrane fouling,[12,13] low permeate flux, etc.[14] Therefore the production of juice having persistent high purity and low color is a quite challenging and tedious job.

To overcome all these concerns, we propose a safer and competent flocculation process aided by novel graft copolymer of sodium alginate (SAG) to remove the impurities of cane juice. Flocculation is a surface phenomenon which involves the accumulation of individual dispersed particles leading to splitting of the solid–liquid interface.[15–17] By this process, the sedimentation or filterability of small and fine particulates improves due to the formation of larger flocs.

Cinnamic acid grafted sodium alginate (SAG-g-P (CA)) was prepared by microwave-assisted process in which microwave radiation was applied in association with chemical free radical initiator (ceric ammonium nitrate (CAN)) to inaugurate grafting reaction.[15–17] The intended grafting has been confirmed in terms of intrinsic viscosity measurement, FTIR spectroscopy, analysis of constituent elements, surface morphological, and thermal analysis. The prepared graft copolymer has been investigated in clarification of sugarcane juice through standard jar test method, for intended application as flocculant which results in enhanced control on wastages of sucrose and turbidity of clarified cane juice.

13.2 EXPERIMENTAL

13.2.1 MATERIALS

SAG was purchased from CDH, New Delhi, India. Cinnamic acid and CAN was supplied by E. Merck, Mumbai, India. Methanol was procured from Merck, Germany. Sugarcane juice was brought from local market of BIT, Mesra, Ranchi, India (zeta potential: −12.9 mV).

13.2.2 SYNTHESIS OF SAG-G-P(CA) BY MICROWAVE-ASSISTED PROCESS

A series of graft copolymer of SAG with different weight percentages of cinnamic acid and catalytic amount of CAN were prepared by microwave-assisted process. Monomer was added to polysaccharide solution in increasing amount. After homogenizing, CAN was added to it. The reaction mixture was irradiated during continuous stirring using microwave reactor (Catalyst™ system CATA 4 R, 700 watts) until gelling sets in. The graft copolymer was then precipitated in an excess of methanol.[18,19] Any occluded homopolymer was removed by soaking the gel in methanol for about 24 h followed by washing with methanol thrice. The graft copolymer of SAG-g-P (CA) was dried in an oven at 50°C until a constant weight was obtained. For each condition, experiments were conducted with three independent replicates and reported as a mean of three readings. The hypothesis of synthesis mechanism has been interpreted in Scheme 13.1 and details have been given in Table 13.1.

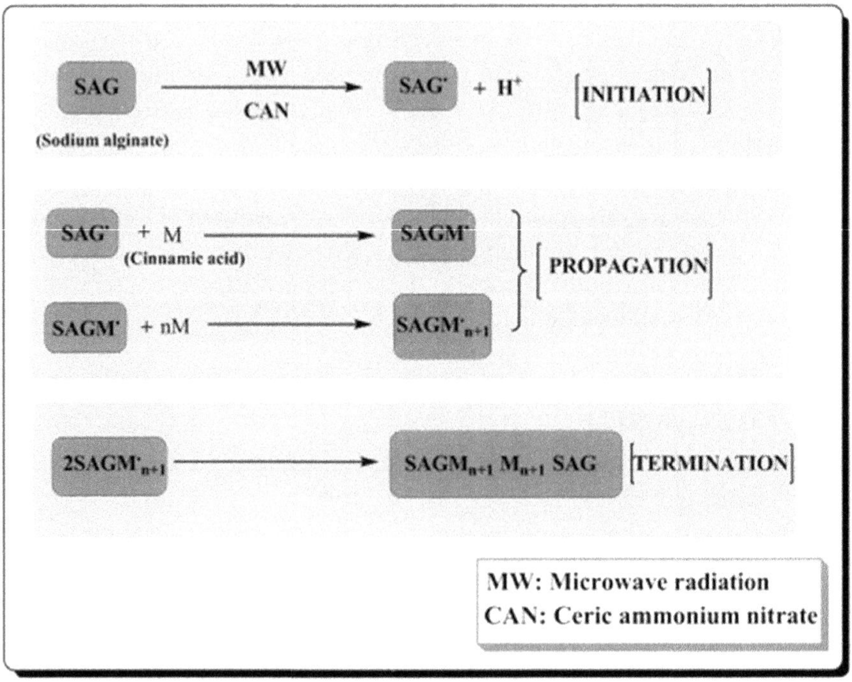

SCHEME 13.1 The hypothesis of synthesis mechanism of SAG-g-P (CA).

13.2.3 CHARACTERIZATION

13.2.3.1 GRAVIMETRIC ANALYSIS

The graft copolymers were characterized according to Fanta's definition.[20,21]

TABLE 13.1 Synthesis Details of Various Grades of SAG-g-P (CA). The Results Are Mean ($N = 3$) Including Standard Deviation.

Polymer grades	Wt. of sodium alginate (g)	Wt. of cinnamic acid (g)	Wt. of CAN (g)	%G	%GE	Intrinsic viscosity (dL/g)	Number average molecular weight (KDa)
SAG-g-P (CA)1	1	5	0.2	19.66 ± 1.527	3.90 ± 0.655	2.9	450
SAG-g-P (CA)2	1	5	0.3	42.33 ± 0.577	8.46 ± 0.115	3	600
SAG-g-P (CA)3	1	5	0.4	70.33 ± 1.527	14.06 ± 0.305	3.8	875
SAG-g-P (CA) 4	1	7.5	0.4	177.66 ± 1.154	23.66 ± 0.115	5	6140
SAG-g-P (CA) 5	1	10	0.4	121.66 ± 1.154	12.16 ± 0.115	4.2	3100
SAG-g-P (CA)6	1	5	0.5	28.33 ± 0.577	5.66 ± 0.114	3.08	500
Sodium alginate (SAG)	–	–	–	–	–	2.79	100

Note: The bold values represent the best grade of graft copolymer.

$$\%G = \frac{\text{Wt. of Graft copolymer} - \text{Wt. of polysaccharide}}{\text{Wt. of polysaccharide}} \times 100 \quad (13.1)$$

$$\%GE = \frac{\text{Wt. of Graft copolymer} - \text{Wt. of polysaccharide}}{\text{Wt. of monomer}} \times 100 \quad (13.2)$$

where $\%G$ stands for percentage grafting and $\%\ GE$ for grafting efficiency, respectively.

13.2.3.2 EVALUATION OF INTRINSIC VISCOSITY

Viscosity of the aqueous polymer solutions (pH = 7) was measured with an Ubbelohde viscometer at 25°C. The efflux time for solutions was determined at four different concentrations (0.1%, 0.05%, 0.025%, and 0.0125%). Relative viscosity ($\eta_{rel} = t/t_0$) was computed from the efflux time of polymer solutions (t) and that of the solvent (t_0, for distilled water). Specific viscosity was measured with the equation $\eta_{sp} = \eta_{rel} - 1$. Then, the reduced viscosity (η_{sp}/C) and the inherent viscosity ($\ln \eta_{rel}/C$) were measured, where "C" stands for concentration of solution in g/dL. The intrinsic viscosity was then evaluated graphically from the point of intersection after extrapolation of two plots, that is, $\ln \eta_{rel}/C$ versus C and η_{sp}/C versus C, to zero concentration of solution.[22] The intrinsic viscosity of different grades of graft copolymers has been reported in Table 13.1. The interaction between percentage grafting and intrinsic viscosity has been graphically illustrated in Figure 13.1.

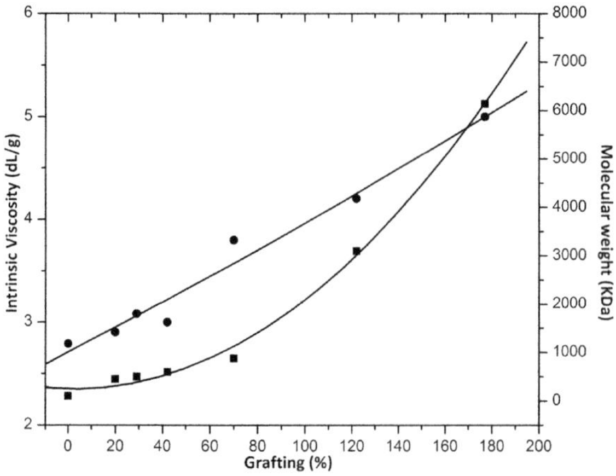

FIGURE 13.1 The interaction between percentage grafting, intrinsic viscosity, and molecular weight.

13.2.3.3 SOLUBILITY OF GRAFT COPOLYMER IN POLAR AND NONPOLAR SOLVENT

Solubility of parent polysaccharide (SAG) and various grades of SAG-g-P(CA) was determined in distilled water and in n-hexane by gravimetric method Figure 13.2.

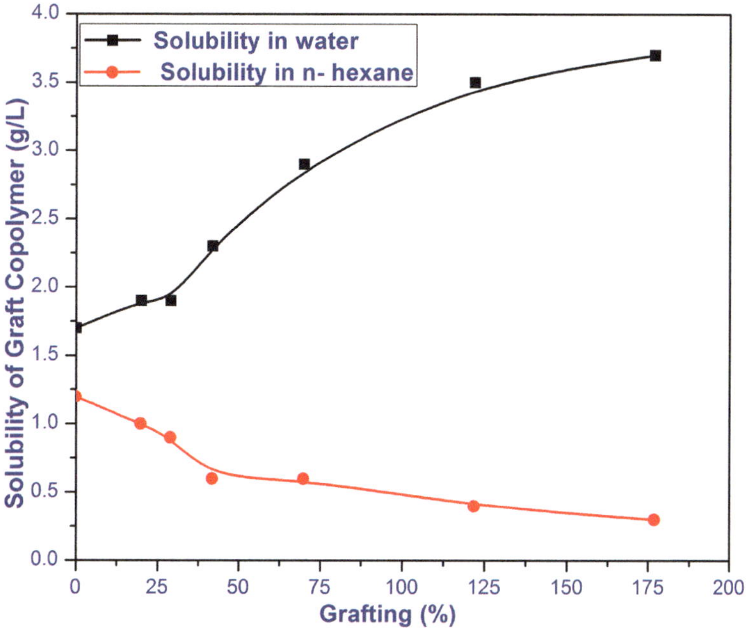

FIGURE 13.2 The correlation between percentage grafting and solubility of synthesized graft copolymers.

13.2.3.4 INSTRUMENTAL ANALYSIS

The grafting of SAG with acrylamide was examined by numerous physicochemical methods. The number average molecular weight of SAG and various grades of SAG-g-P (CA) were examined by osmometry (Model: 3320, Osmometer, A + Adv. Instruments, Inc.) and was computed by the given equation.[23]

$$\frac{\pi}{C_{\text{dry}}} = \frac{\Phi \times n}{M_n} \times 10^3 \tag{13.3}$$

where M_n is the number average molecular weight, n is the number of components into which a molecule dissociates, π is the osmosis per kilogram solvent, C_{dry} is the concentration of the dry sample in the solution, Φ is the osmotic coefficient, which accounts for the nonideal behavior of the solution (assuming ideal behavior). The results have been reported in Table 13.1. The interaction between number average molecular weight, percentage grafting, and intrinsic viscosity has been given in Figure 13.1.

Elemental composition of SAG and the best grade of graft copolymer were analyzed by elemental analyzer (Model–Vario EL III; Make–M/s Elementar, Germany). Sodium percentage was evaluated by flame photometry (Model no: CL378 ELICO). The data have been shown in Table 13.2.

TABLE 13.2 Elemental Analysis of Sodium Alginate, Cinnamic Acid, and SAG-g-P (CA) 4 (Best Grade).

Polymer grade	%C	%H	%N	%O	%Na
Sodium alginate	32.34	5.304	0.00	56.85	5.225
Cinnamic acid	74.24	6.435	0.00	20.22	0.00
SAG-g-P (CA) 4	51.16	5.701	0.00	41.70	1.062

The FTIR study of specimen was performed with FTIR spectrophotometer (Model IR-Prestige 21, Shimadzu Corporation, Japan) by KBr dispersion process. The spectra were recorded in solid state and have been illustrated in Figure 13.3a,b.

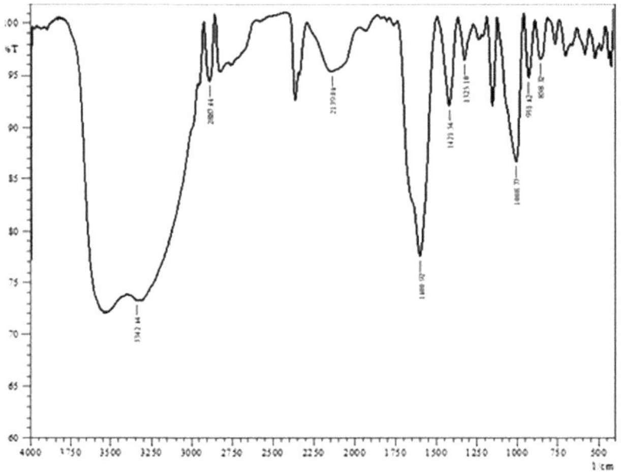

FIGURE 13.3a FTIR of sodium alginate.

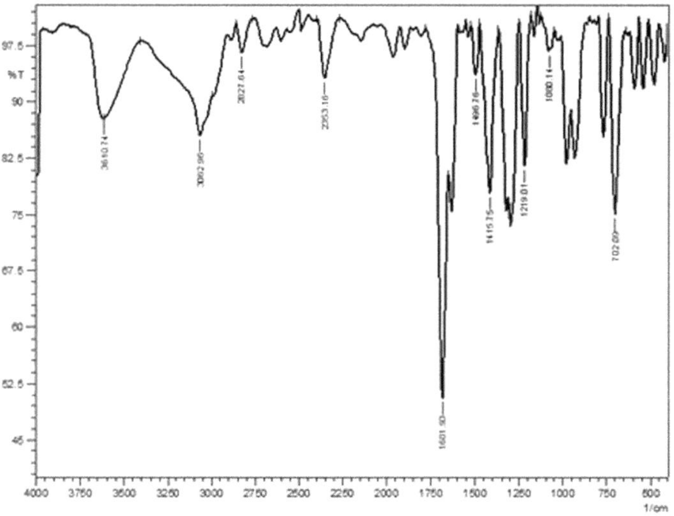

FIGURE 13.3b FTIR of SAG-g-P (CA) 4.

Surface morphology of the synthesized polymers was examined through scanning electron microscope (SEM) (Model: JSM-6390LV, Jeol, Japan). Images have been shown in Figure 13.4a,b.

FIGURE 13.4a SEM of sodium alginate.

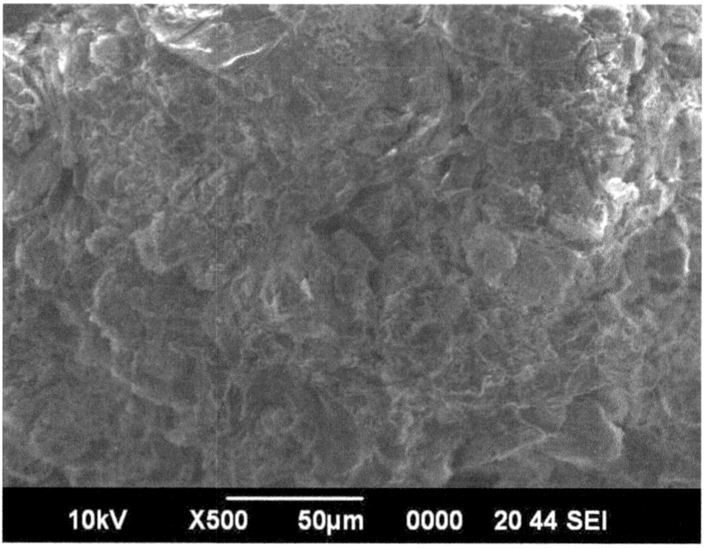

FIGURE 13.4b SEM of SAG-g-P (CA) 4.

Thermogravimetric analysis (TGA) was performed using TGA instrument (Model: DTG-60; Shimadzu, Japan). A heating rate of 5°C/min was maintained from 25°C to 600°C in an inert atmosphere of nitrogen. The thermogram has been depicted in Figure 13.5a,b.

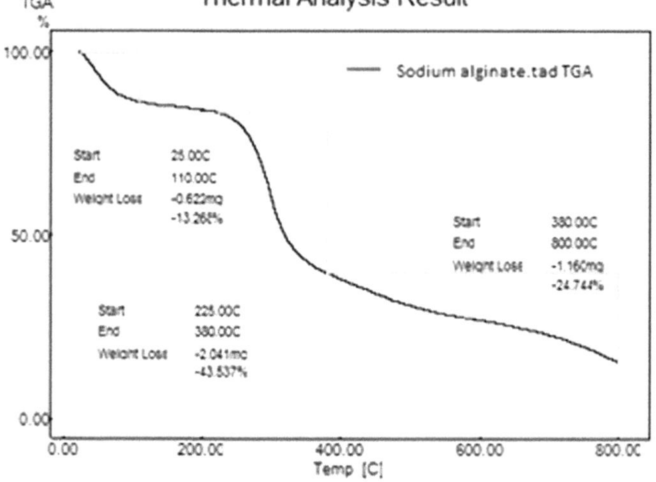

FIGURE 13.5a TGA of sodium alginate.

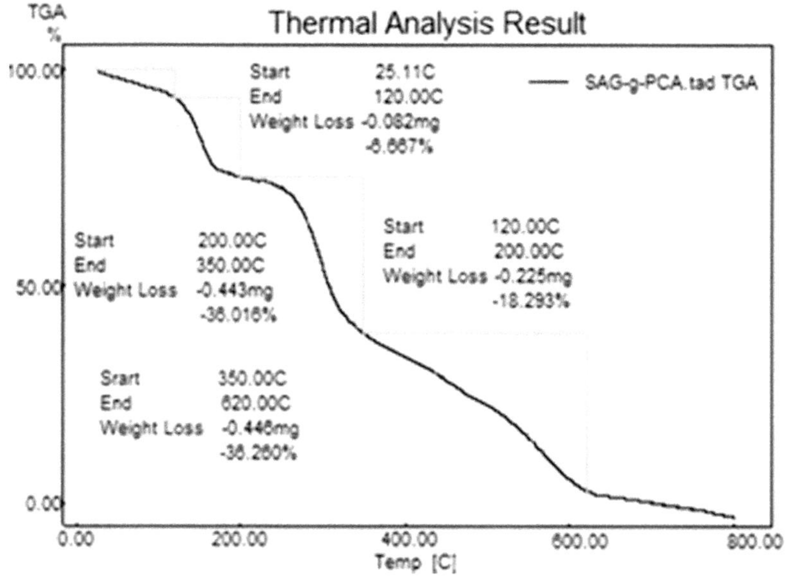

FIGURE 13.5b TGA of SAG-g-P (CA) 4.

13.3 CLARIFICATION OF CANE JUICE BY FLOCCULATION USING GRAFT COPOLYMER

Cane juice clarification was done by applying micro jar test apparatus (Make: Simeco, Kolkata, India).The protocol includes having a uniform amount (100 mL) of the sugarcane juice in each of six identical 100 mL beakers. Required quantity of the flocculant (SAG or various grades of SAG-g-P(CA)) was mixed in concentrated solution form (except in case of blank) to bring about the desired dosage (from 0 to 25 ppm).The contents of these beakers were stirred similarly at 150 rpm for 30 s, 60 rpm for 5 min, followed by 25 min of settling time. Consequently, supernatant liquid was withdrawn and turbidity was analyzed in a calibrated nephelo-turbidity meter (Model: Digital Nephelo-Turbidity Meter 132, Systronics, India). Simultaneously, optical density of the same supernatant liquid was also analyzed in a calibrated spectrophotometer (DR/2400, Hach®) at λ_{max} 520 nm. For each condition, experiments were conducted with three independent replicates and reported as a mean of three readings. The clarification competency thus evaluated for SAG and various grades of graft copolymer have been graphically plotted in Figure 13.6a,b. The optimized dose in each condition is revealed by the minima of the curve.

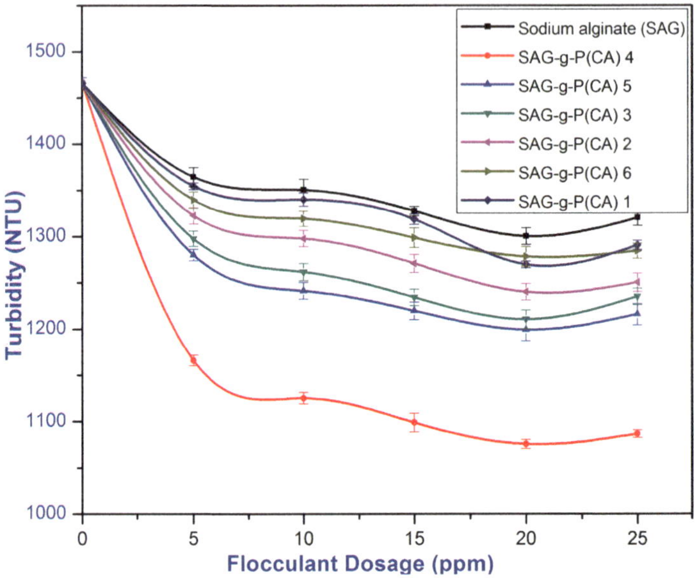

FIGURE 13.6a Flocculation characteristics of SAG and synthesized grades of SAG-g-P (CA) by determination of turbidity.

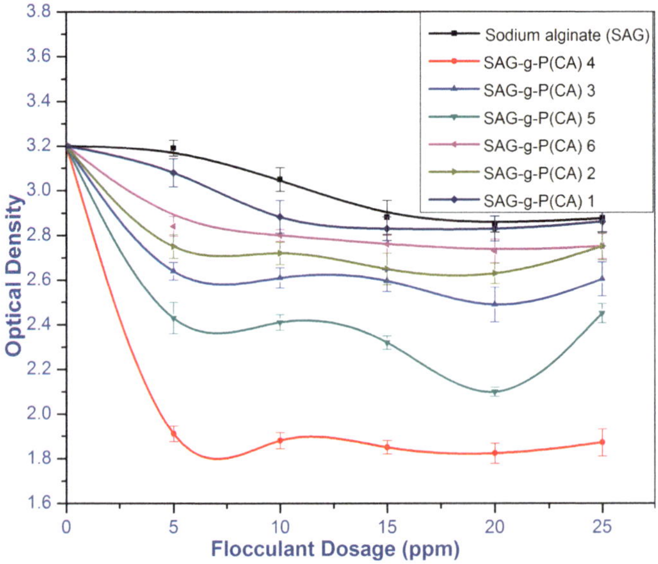

FIGURE 13.6b Flocculation characteristics of SAG and synthesized grades of SAG-g-P (CA) by determination of optical density.

13.3.1 SETTLING TEST OF SUGARCANE JUICE

The test involves a 100 mL stoppered measuring cylinder having the cane suspension.[24] Concentrated solution of flocculant (SAG-g-P (CA) or SAG) was added to affect the optimized dosage (as evaluated by the jar test procedure above). The contents were mixed properly by turning the cylinder upside down for at least 10 times. After that, the cylinder was set in upright position and the height of the clear sugarcane interface was traced with respect to time. The settling rate was computed in each case. The result has been graphically sketched as interfacial height versus settling time and presented in Figure 13.6c.

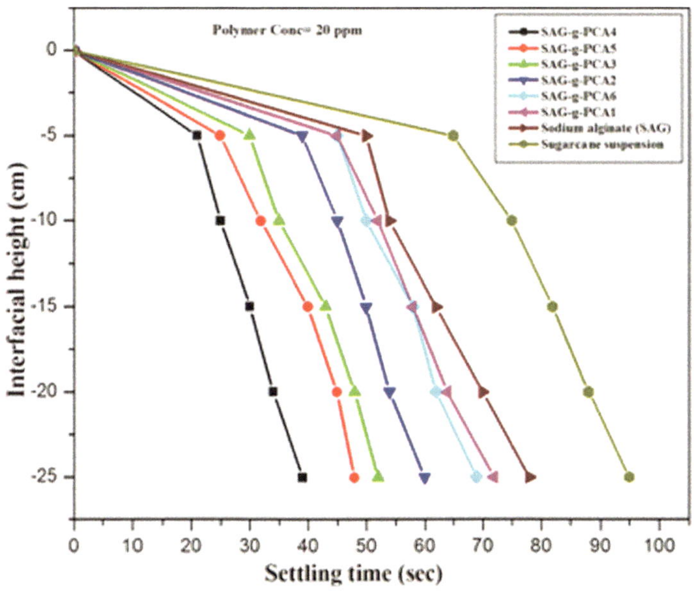

FIGURE 13.6c Correlation between interfacial height and time.

13.3.2 CHARACTERIZATION OF CANE JUICE SUSPENSION

13.3.2.1 DETERMINATION OF ZETA POTENTIAL AND FLOC SIZE

The zeta potential and average floc size of cane juice suspension were measured before and after flocculation (Fig.13.7a–c), by DLS analysis (Model: Zeta Sizer Nano ZS, Malvern, UK). Zeta potential is a measure of degree of electrostatic attraction or repulsion and constitutional parameters established to cause flocculation. The data have been given in Table 13.3 and the correlation between percentage grafting, zeta potential and floc size has been graphically presented in Figure 13.7d.

TABLE 13.3 Relation Between Settling Velocity, Average Floc Size, Zeta Potential, and Degree Brix of the Various Grades of (SAG-g-P (CA)). The Results Are Mean (N = 3) Including Standard Deviation.

Polymer grades	Settling velocity (cm/s)	Average floc size (nm)	Zeta potential value (mV)	Degree Brix value
Control	0.263	396	−12.9	1.3645 ± 0.00025
SAG	0.320	431	−12.0	1.3617 ± 0.00030
SAG-g-P (CA) 1	0.347	549	−11.8	1.3641 ± 0.00035
SAG-g-P (CA) 2	0.416	690	−9.04	1.3631 ± 0.00020
SAG-g-P (CA) 3	0.480	720	−8.88	1.3624 ± 0.00041
SAG-g-P (CA) 4	**0.641**	**1189**	**−7.08**	1.3606 ± 0.00052
SAG-g-P (CA) 5	0.520	820	−8.23	1.3596 ± 0.00035
SAG-g-P (CA) 6	0.362	566	−10.9	1.3634 ± 0.00025

Note: The bold values represent the characteristics of best grade of graft copolymer.

FIGURE 13.7a Sugarcane juice before flocculation.

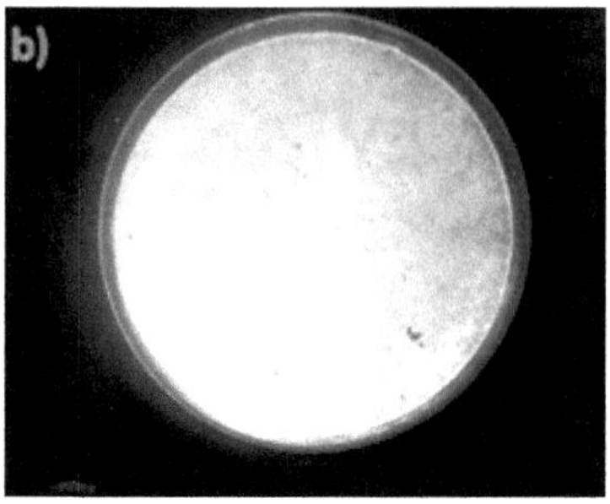

FIGURE 13.7b Sugarcane juice after flocculation with sodium alginate.

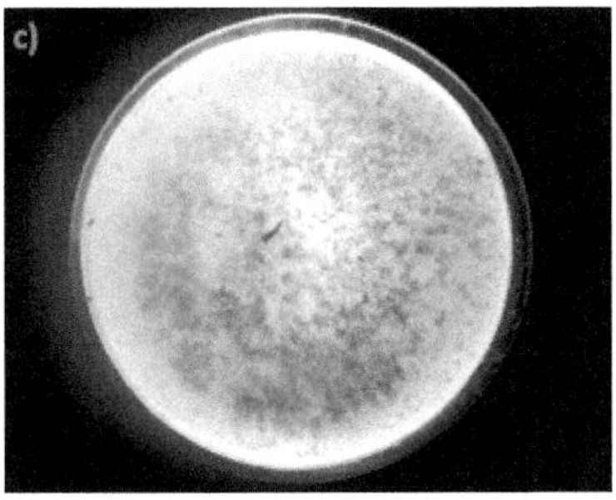

FIGURE 13.7c Sugarcane juice after flocculation with SAG-g-P (CA).

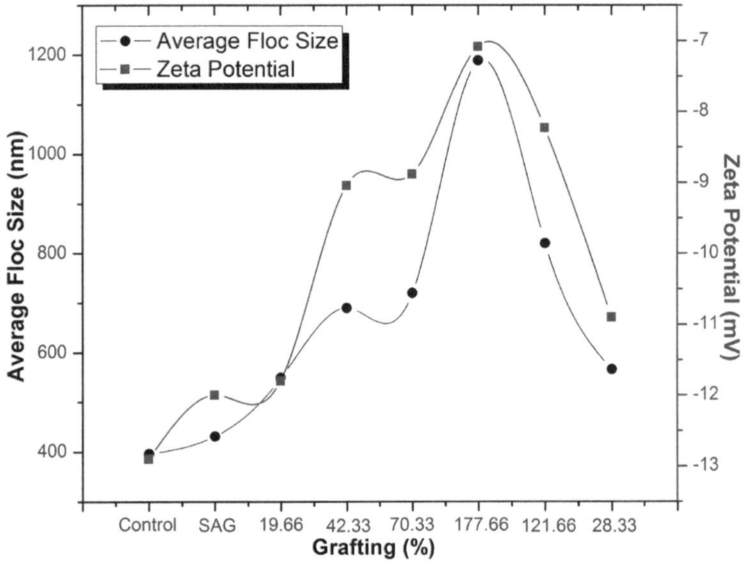

FIGURE 13.7d Correlation between percentage grafting, zeta potential, and floc size.

13.3.2.2 DETERMINATION OF DEGREE BRIX OF SUGARCANE JUICE SUSPENSION

Degree brix (°bx) of cane juice depends upon the amount of dissolved solids and can, therefore, serve as a measure of sugar content. The °bx of juice samples was measured before and after treatment with various grades of flocculant using a digital Abbe Refractometer (Model: LMAR-1317, Chennai, India) at ambient temperature and reported as a mean of three readings in Table 13.3.

13.4 RESULT AND DISCUSSION

13.4.1 SYNTHESIS OF SAG-G-P(CA) BY MICROWAVE-ASSISTED PROCESS

Various grades of the graft copolymer were prepared by altering the concentration of CA and CAN in presence of microwave radiation. The optimized grade has been determined by its highest percentage grafting and consequently through highest intrinsic viscosity which is directly proportional to molecular weight as shown in Table 13.1.

Being electron deficient molecule CAN abstract electrons from alcoholic oxygen of SAG and built a contemporary new bond with Ce. This new bond breaks quickly under the effect of microwave irradiation due to being more polar than original O—H bond and develops free radical sites onto the backbone of parent polymer.[25] The grafted chains of monomer get attached at these sites through initiation, propagation, and termination, leading to the successful synthesis of modified graft copolymer (SAG-g-P (CA)).

13.4.1.1 EFFECT OF INITIATOR CONCENTRATION

It has been observed that %G as well as intrinsic viscosity increases with increasing CAN concentration from 0.2 to 0.4 g, above this quantity, it decreases. This may be explained by the fact that at low concentration of catalyst/initiator, fewer grafting sites develops, which result in longer monomer chains. On the other hand, at high concentration of CAN, larger number of grafting sites develops which makes the typical grafted chains shorter. Hence, by grafting CA chains onto SAG backbone, two probabilities are there

1. Incorporation of small number of lengthy monomer chains.
2. A large number of short CA chains.

In the first case, due to the existence of longer grafted chains, the compressed structure of the graft copolymer would be modified. This results in larger hydrodynamic volume, and hence increased intrinsic viscosity. On the other hand, a large number of short CA chains (second case) do not convert the original structure and hence reduces the hydrodynamic volume (i.e., intrinsic viscosity)[26] shown in Table 13.1 and Figure 13.8a.

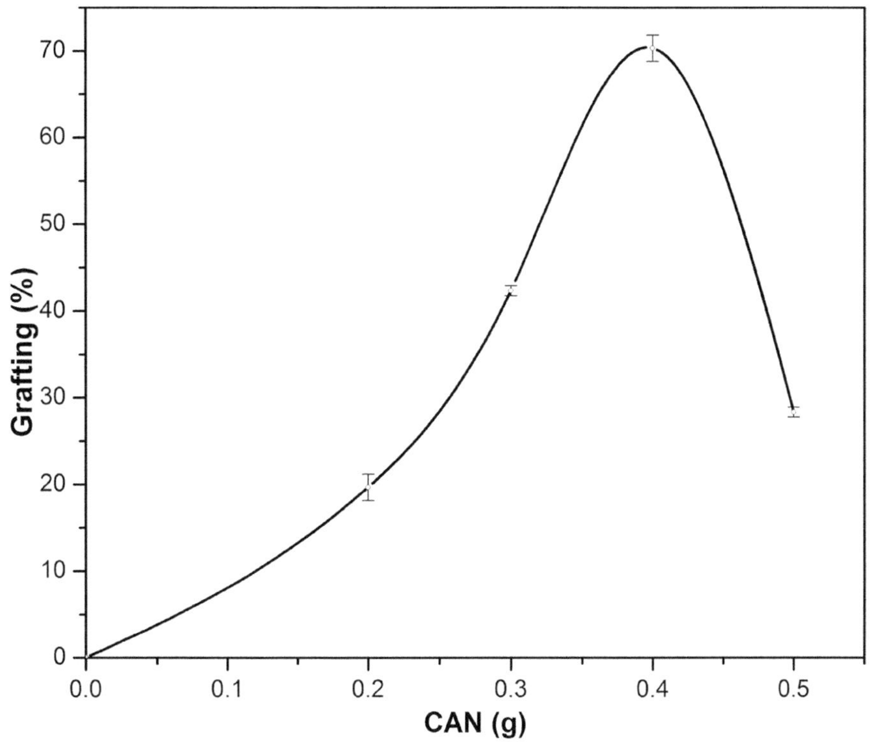

FIGURE 13.8a Interaction between CAN concentration and percentage grafting.

13.4.1.2 EFFECT OF MONOMER CONCENTRATION

Percentage grafting increases with increasing the concentration of monomer (from 2.5 to 7.5 g) and attains maximal at concentration of 7.5 g in the reaction mixture. Subsequently, the percentage grafting reduces. This may be illustrated by the reason that an increase in monomer concentration results in the addition of monomer molecules adjacent to the SAG backbone. The reduction in percentage grafting after optimization may be attributed to the unavailability in the active sites on the polymer backbone as graft copolymerization progresses. Apart from this, the homopolymer formation predominates in the presence of increased monomer concentration, leading to reduction in percentage grafting, and intrinsic viscosity as presented in Table 13.1 and Figure 13.8b.

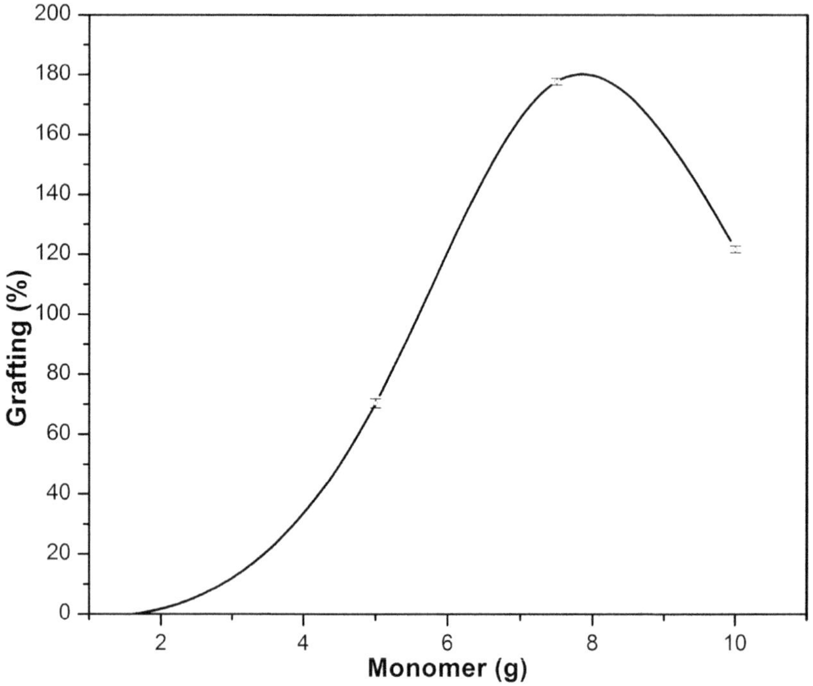

FIGURE 13.8b Interaction between monomer concentration and percentage grafting.

13.4.2 CHARACTERIZATION

13.4.2.1 EVALUATION INTRINSIC VISCOSITY

The intrinsic viscosity of SAG and distinct grades of SAG-g-P (CA) has been given in Table 13.1. From results, it is evident that the intrinsic viscosity of all grades of graft copolymer is higher than that of SAG. This may be described by the increased hydrodynamic volume of SAG-g-P (CA) as a result of grafting of CA chains which increase intrinsic viscosity by two manners:

1. By straightening of the polysaccharide chain.
2. By providing their individual hydrodynamic volume.

Further, it is in excellent compliance with Mark–Houwink–Sakurada relationship (intrinsic viscosity $\eta = KM^{\alpha}$ where K and α are constants, both related to stiffness of the polymer chains),[27] which demonstrates the increment in intrinsic viscosity as a result of increase in molecular weight (*M*) due to the attachment of monomer chains. The interaction between intrinsic viscosity, molecular weight, and percentage grafting has been illustrated in Figure 13.1.

13.4.2.2 SOLUBILITY OF GRAFT COPOLYMER IN POLAR AND NONPOLAR SOLVENT

The solubility profile of SAG and distinct grades of SAG-g-P (CA) in polar as well as in nonpolar solvent have been shown in Figure 13.2. Though SAG is soluble in water, its solubility increases after grafting with cinnamic acid, that is, all grades of graft copolymer are more soluble in water than parent polymer due to the incorporation of polar group on to the main chain of parent polymer. Consequently, the best grade of SAG-g-P (CA) showed highest solubility. This may be described by the fact that higher the percentage grafting higher is the number of polar monomer chains attached which in turn increases the solubility in water.

But in case of nonpolar solvent (i.e., n-hexane) the trend reversed due to the hydrophilic nature of graft copolymer in (SAG-g-P (CA)) as shown in Figure 13.2.

13.4.2.3 INSTRUMENTAL ANALYSIS

13.4.2.3.1 Number Average Molecular Weight of Graft Copolymers

As apparent from Table 13.1 and Figure 13.1, number average molecular weight of distinct grades of SAG-g-P (CA) is greater than that of parent polymer, that is, SAG. As explained by Mark–Houwink–Sakurada relationship (intrinsic viscosity $\eta = KM^{\alpha}$ where K and α are constants), higher the percentage grafting, higher is the intrinsic viscosity and thus higher the number average molecular weight by virtue of grafting of CA chains.[27]

13.4.2.3.2 Elemental Analysis

The results of elemental analysis for SAG, cinnamic acid, and that of the best grade of SAG-g-P (CA) 4 are given in Table 13.2. As expected, the elemental composition of the grafted polymers is intermediate of its constituents (SAG and CA). This is an indication of grafting of cinnamic acid onto the backbone of parent biopolymer.

13.4.2.3.3 FTIR Spectroscopy

As apparent from Figure 13.3a, SAG has a peak at 3342 cm^{-1} which belongs to O—H stretching. The peaks at 2887 and 1600 cm^{-1} are associated with C—H stretching and C—O—C stretching, respectively. Two COO^- symmetric stretching peaks are apparent at 1421 and 1325 cm^{-1}.

The FTIR spectrum (Fig. 13.3b) of SAG-g-P (CA) showed in addition to the above peak, an additional peak at 1681 cm^{-1}, due to the stretching vibration of C=O bond of the grafted CA chains. Peak at 3062 cm^{-1} is associated with C—H stretching vibrations of the benzenoid rings in CA. The existence of these peaks finally approves the grafting of CA chains onto the backbone of SAG.

13.4.2.3.4 Scanning Electron Microscopy Analysis

A closure inspection of SEM micrographs of SAG Figure 13.4a and that of SAG-g-P (CA) 4 Figure 13.4b, reveals that after grafting the granular structure of inherent SAG has been deformed and transformed to fibrillar structure, due to grafting of CA monomers. This observation confirms the grafting and indicates that grafting of CA influences the morphological characteristics of SAG.

13.4.2.3.5 Thermal Gravimetric Analysis

The thermogram of SAG Figure 13.5a acquires three specific sector of weight loss. The initial weight loss was at (25–110°C) owing to the loss of moisture present. The second sector (225–380°C) appeared as a result of decomposition of COO^- groups and resultant removal of carbon dioxide (decarboxylation). The third sector of weight loss (380–800°C) was due to complete deterioration of the polysaccharide backbone.

SAG-g-P (CA) 4 (Fig. 13.5b) in addition to the above sector of weight loss, had a fourth sector of weight loss (350–620°C), due to the degeneration of CA chains grafted on SAG moiety.

13.4.3 CLARIFICATION OF CANE JUICE BY FLOCCULATION USING GRAFT COPOLYMER

The dangling, lengthy chains of monomer connected on the rigid backbone of SAG form branched comb like structure. This customized architecture exaggerates hydrodynamic volume and consequently, the radius of gyration of graft copolymer. These grafted chains forms bridges[24] between suspended and colloidal impurities due to increase in their approachability[28,29] and every particle may hook up two or more such type of links. The impurities get entrapped at the surface of this revised structure, that is, migrated from liquid phase to solid phase, leaving the cane juice more clean and clear. In other words, polymeric chains get adsorbed onto the surface of one or more particles. In this way, the particles forming impurities get adjacent to each other and concentrated in the mode of floc which undergoes gravity sedimentation.[30]

For all grades of graft copolymers and parent polymer, there is an optimal dosage which shows the highest flocculation efficacy (i.e., the optical density and turbidity of the supernatant collected is minimum) (Fig. 13.6a,b). Above this dosage, the flocculation efficacy reduces as a result of destabilization of flocs developed by excess of flocculant. This characteristic of the flocculation curve precisely supports the bridging mechanism.[24] The optimal dosage of SAG-g-P (CA) 4 as flocculant in sugarcane juice was at 20 ppm.

13.4.3.1 SETTLING TEST OF SUGARCANE JUICE SUSPENSION

For this test the dosage of flocculant was maintained at 20 ppm (i.e., the optimized dosage as determined by jar test described above). Figure 13.6c exhibits the settling attributes in cane juice treated with SAG and various grades of SAG-g-P (CA). The settling rate of the various grades follows the same order as in jar test results (Fig. 13.6a,b). As expected the best grade of synthesized graft copolymer (SAG-g-P (CA) 4) has the highest settling rate. Thus, it is another positive sign of flocculation competency, which is directly related to the radius of gyration. The greater the settling rate of the floc, the greater will be its flocculation performance. As indicated from Table 13.3, with increase in percentage grafting there is an increase in the settling rate of flocs and SAG-g-P (CA) 4 has the highest settling velocity. The settling time was plotted against the interfacial height (Fig. 13.6c).

13.4.3.2 DETERMINATION OF ZETA POTENTIAL AND FLOC SIZE

Zeta potential and Floc size of cane juice suspension have been determined before and after the flocculation. Table 13.3 indicates that the average floc size increases with increasing percentage grafting. This indicates the larger accumulation of impurities from cane juice as expected, due to the larger structure of graft copolymer to combine impurities. As indicated from Figure 13.7a–c, the floc is very prominent in case of flocculation by SAG-g-P (CA) 4 than by SAG and in cane juice (control).

The attraction between colloidal particles exceeds repulsion and the dispersion of particles breaks and flocculate. This significantly reduces the magnitude of zeta potential of the cane juice from −12.9 to −7.08 mV. The details are given in Table 13.3 and Figure 13.7d. The virtue of the graft copolymer reduces the magnitude of zeta potential accredits its higher potential.

13.4.3.3 DEGREE BRIX DETERMINATION OF SUGARCANE JUICE SUSPENSION

The most important precaution to be observed in clarification of cane juice is minimum or no loss of sucrose content. From Table 13.3, it is evident that there is no significant change in °bx, that is, percentage of sugar content in juice suspension before and after flocculation. This may be explained by the fact that flocculation is a surface phenomenon and act only upon suspended colloidal particles and does not interfere with dissolved solids. Therefore, the composition of dissolved solids (concentration of sucrose content) remains unaltered.

13.5 CONCLUSIONS

A novel method for clarification of sugarcane juice by SAG-g-P (CA) assisted flocculation was developed. This technique is capable of producing the clarified, sparkling sugar cane juice without loss of sugar content which is a desirable quality from consumer end and thus increases the market value of the final product. It is worthwhile to note that this technology may be further useful for clarification of other fruit and vegetable juices, alcoholic, and nonalcoholic beverages, etc.

ACKNOWLEDGMENTS

Author "Pinki Pal" gratefully acknowledge the research grant from Department of Science and Technology (DST), New Delhi, India (sanction order no. SR/WOS- A/ET- 13/2014).

KEYWORDS

- **sugar cane**
- **coagulation**
- **cinnamic acid**
- **viscometer**
- **biopolymer**
- **microwave-assisted method**
- **graft copolymer**
- **flocculation**

REFERENCES

1. Honig, P. *Principles of Sugar Technology;* Elsevier Publishing Company: New York, NY, 1953.
2. Mohammed, H.; Solomon, W. K.; Bultosa, G. Optimization of Phosphate and Anionic Polyacrylamide Flocculant (APF) Level for Sugar Cane Juice Clarification Using Central Composite Design. *J. Food Process. Preserv.* **2016,** *40*, 67–75.
3. Schoonees-Muir, B. M.; Gwegwe, B. M. M. The Use of Polyaluminium Coagulants for the Removal of Color During Clarification. *Int. Sugar J.* **2009,** *111* (1324), 246–249.
4. Zhang, Z.; Mu, G.; Peng, J.; Zhao, M.; Wen, H.; Zhang, X. Study on Decolourisation of Sugar Cane Juice Using Activated Carbon. *Adv. Mat. Res.* **2012,** *529*, 376–379.
5. Lutin, F.; Guerif, G.; Herz, G.; Tani, Y. In *Ion Exchange Membranes: An Industrial Reality*, Proceedings of Euromembrane, Bath, UK, 1995; 2, p 75.
6. Thampy, S. K.; Narayanan, P. K.; Trevedi, G. S.; Gohil, D. K.; Indusekhar, V. R. In *Demineralization of Sugar Cane Juice by Electrodialysis*, Abstract, Proceeding of Twelfth Conference, IMS, Bhavnagar, India, 1994.
7. Himachi, M.; Gupta, B. B.; Ben Aim, R. Ultrafiltration: A Mean for Decolorization of Cane Sugar Solution. *Sep. Purif. Technol.* **2003,** *30*, 229–239.
8. Patricia, P.; Roberto-Herminia, M. Study of Clarification Process of Sugar Cane Juice for Consumption. *Cienc. Technol. Aliment. Champinas.* **2010,** *30* (3), 776–783.
9. Gonzalez-Munoz, M. J.; Pena, A.; Meseguer, I. Role of Beer as a Possible Protective Factor in Preventing Alzheimer's Disease. *Food Chem. Toxicol.* **2008,** *46*, 49–56.

10. Bourzutscky, H. C. C. Color Formation and Removal-options for Sugar and Sugar Refining Industries: A Review. *Zuckerindustrie.* **2005,** *130*, 545–553.
11. Gyura, J.; Seres, Z.; Vatai, G.; Molnar, E. B. Separation of Non-sucrose Compound from the Syrup of Sugar-beet Processing by Ultra and Nanofiltration Using Polymer Membrane. *Desalination* **2002,** *148*, 49–56.
12. Jacob, S.; Jaffrin, M. Y. Purification of Brown Cane Sugar Solutions by Ultrafiltration with Ceramic Membrane: Investigation of Membrane Fouling. *Sep. Sci. Technol.* **2000,** *35*, 989–1010.
13. Jegatheesan, V.; Phong, D. D.; Shu, L.; Ben Aim, R. Performance of Ceramic Micro and Ultrafiltration Membrane Treating Limed and Partially Clarified Sugar Cane Juice. *J. Membr. Sci.* **2009,** *327*, 69–77.
14. Ghosh, A. M.; Balakrishnan, M. Pilot Demonstration of Sugar Cane Juice Ultrafiltration in an Indian Sugar Factory. *J. Food Eng.* **2003,** *58*, 143–150.
15. Pal, P.; Pandey, J. P.; Rahul, R.; Sen, G. A Novel Biodegradable Cinnamic Acid Grafted Carboxymethyl Cellulose Based Flocculant for Water Treatment. In *Advanced Functional Materials: Properties and Applications;* Inamuddin, Al-Ahmad, A., Eds.; Trans Tech Publications: Switzerland, 2016; pp 156–166.
16. Mishra, S.; Mukul, A.; Sen, G.; Jha, U. Microwave Assisted Synthesis of Polyacrylamide Grafted Starch (St-g-PAM) and Its Applicability as Flocculant for Water Treatment. *Int. J. Biol. Macromol.* **2011,** *48*, 106–111.
17. Pal, S.; Sen, G.; Ghosh, S.; Singh, R. P. High Performance Polymeric Flocculants Based on Modified Polysaccharides—Microwave Assisted Synthesis. *Carbohydr. Polym.* **2012,** *87*, 336–242.
18. Zhang, X.; Meng, H.; Di, Y. Synthesis and Characterization of Cinnamic Acid-grafted Poly (Vinylidene Fluoride) Microporous Membranes. *Energy Procedia.* **2012,** *17*, 1850–1857.
19. Dua, H.; Zhang, J. The Synthesis of Poly (Vinyl Cinnamates) with Light-induced Shape Fixity Properties. *Sens. Actuators A.* **2012,** *179*, 114–120.
20. Fanta, G. F. Synthesis of Graft and Block Copolymers of Starch. In *Block and Graft Copolymerization;* Ceresa, R. J., Ed.; Wiley Interscience: New York, NY, 1973; p 1.
21. Fanta, G. F. Properties and Applications of Graft and Block Copolymers of Starch. In *Block and Graft Copolymerization;* Ceresa, R. J., Ed.; Wiley Interscience: New York, NY, 1973; p 29.
22. Collins, E. A.; Bares, J.; Billmeyer, F. W. *Experiments in Polymer Science;* John Wiley & Sons: New York, NY, 1973.
23. Rong, Y.; Sillick, M.; Gregson, C. M. Determination of Dextrose Equivalent Value and Number Average Molecular Weight of Maltodextrin by Osmometry. *J. Food Sci.* **2009,** *74*, C33–C40.
24. Ruehrwein, R. A.; Ward, D. W. Mechanism of Clay Aggregation by Polyelectrolytes. *Soil Sci.* **1952,** *73*, 485–492.
25. Galema, S. A. Microwave Chemistry. *Chem. Soc. Rev.* **1997,** *26*, 233–238.
26. Nayak, B. R.; Singh, R. P. Synthesis and Characterization of Grafted Hydroxypropyl Guar Gum by Ceric Ion Induced Initiation. *Eur. Polym. J.* **2001,** *37*, 1655–1666.
27. Rani, P.; Sen, G.; Mishra, S.; Jha, U. Microwave Assisted Synthesis of Polyacrylamide Grafted Gum Ghatti and Its Application as Flocculant. *Carbohydr. Polym.* **2012,** *89*, 275–281.
28. Singh, R. P. Advanced Turbulent Drag Reducing and Flocculating Materials Based on Polysaccharides. In *Polymers and Other Advanced Materials, Emerging Technologies*

and Business Opportunities; Prasad, P. N., Mark, J. E., Fai, T. J., Eds.; Plenum Press: New York, NY, 1995; p 227–249.
29. Singh, R. P.; Nayak, B. R.; Biswal, D. R.; Tripathy, T.; Banik, K. Biobased Polymeric Flocculant for Industrial Effluent Treatment. *Mat. Res. Innovation* **2003,** *7*, 331–340.
30. Brostow, W.; Pal, S.; Singh, R. P. A Model of Flocculation. *Mat. Lett.* **2007,** *61*, 4381–4384.

CHAPTER 14

FISH MINCE AND SURIMI PROCESSING: NEW TRENDS AND DEVELOPMENT

L. N. MURTHY[1*], G. P. GIRIJA[1], C. O. MOHAN[2], and C. N. RAVISHANKAR[2]

[1]*Mumbai Research Centre of Central Institute of Fisheries Technology, CIDCO Admin Building, Sector 1, Vashi, Navi Mumbai 400703, Maharashtra, India*

[2]*ICAR–Central Institute of Fisheries Technology, Wellington Island, Matsyapuri P.O., Cochin 682029, India*

**Corresponding author. E-mail: murthycift@gmail.com*

CONTENTS

ABSTRACT

Fish mince is comminuted meat separated from bone, whereas surimi is stabilized myofibrillar protein obtained from mechanically deboned fish flesh that is washed with water and blended with cryoprotectant. A variety of products can be custom-made using the surimi as base material to meet the customers' demands. Japanese are the pioneers in surimi technology which later spread to world over making it one of the highly appreciated technology. Although Alaska Pollock was most sought after fish for surimi, many other fish varieties are also being used. A brief historical perspective, production process for mince and surimi, and equipments needed are discussed in this chapter.

14.1 INTRODUCTION

Fish and seafood constitute an important food component for a large section of world population which is widely consumed as a source of proteins.[206] They come after meat and poultry as staple animal protein foods where fish forms a cheap source of protein.[206] Fish consumption is increasing day by day and it is not always possible to consume fish in the fresh form. Today, even more people are turning to fish as a healthy alternative to real meat.[1] Fish-processing industries are important in order to increase the consumption of fish and for increasing the contribution to the earnings through foreign exchange. The fish processing operations contribute to develop commercial value-added fish products from fish and help in reducing the postharvest losses. Indian planners ensuring food security for the increasing population has accorded the utilization of aquatic resources for human consumption.

14.2 IMPORTANCE OF FISH IN A BALANCED DIET

Fish and seafood play a pivotal role from nutritional point of view. As a rich source of nutrients, fish provide a good balance of protein, vitamins, and minerals and a relatively low caloric content. In addition, fish is one of the excellent sources of omega-3 polyunsaturated fatty acids which appear to have beneficial effects in reducing the risk of cardiovascular diseases and are linked with positive benefits in many other pathological conditions particularly, certain types of cancer and arthritis. Coronary heart disease, hypertension, cancer, obesity, iron deficiency, protein deficiency, osteoporosis, and arthritis are contemporary health problems for which fish provide a number of nutritional advantages and some therapeutic benefits. Nutritional factors of importance are calories, proteins, lipids, cholesterol, minerals, and vitamins. Proteins form one of the most important constituents of fish muscle. Fish protein is an excellent source of lysine and sulfur-containing amino acids which are lacking in plant proteins.[1]

14.2.1 PROTEINS FROM FISH MEAT

The knowledge on the proteins of fish meat is mainly derived from proteins of other muscle systems. Muscle proteins are composed of both sarcoplasmic as well as myofibrillar protein (MFP) fractions.[167] The sarcoplasmic fractions are mainly composed of enzymes, membranous tubular structures, and some lipoprotein constituents. The MFP fractions are composed of myosin, actin, actomyosin, tropomyosin, troponin, and actinin.[192] The fish muscle consists of bundles of fibers referred as myofibrils.[64,186] The intra-muscular connective tissues are mainly composed of "collagen" which constitutes about 3–5% of total proteins in fish meat.[103]

14.2.1.1 SARCOPLASMIC PROTEINS

The soluble proteins in sarcoplasm are referred as sarcoplasmic proteins. The major constituents of sarcoplasmic protein are enzymes of the glycolytic pathway and are water-soluble proteins.[208] Sarcoplasmic proteins constitute about 25–30% of total muscle protein.[192] The sarcoplasmic proteins from fish are more or less like those from land animals, that is, they include myoglobin, a number of enzymes, and other albumins.[192] The content of sarcoplasmic proteins is generally higher in pelagic fishes such as sardine and mackerel and lower in demersal fishes like plaice and snapper.

The sarcoplasmic proteins are separated into four different components based on their sedimentation velocity in differential centrifugation.[6] They include nuclear, mitochondrial, microsomal, and cytoplasmic fractions.[14] They can be obtained as pellets after centrifugation at 1000 × *g* (nuclear), 10,000 × *g* (mitochondrial), 100,000 × *g* (microsomal) and from the supernatant (cytoplasmic). These protein fractions make up more than 100 different sarcoplasmic proteins including most of the enzymes involved in energy metabolism.[215] Despite their diversity, sarcoplasmic proteins share many common physicochemical properties. For instance, most are of relatively low molecular weight components and have high globular or rod-shaped structures. These structural characteristics may be partially responsible for the high solubility of these proteins in water or dilute salt solutions.[215]

Nutritionally, sarcoplasmic proteins are equivalent to the MFPs of muscle. However, they are less functional than the MFP and therefore, have not been highly valued as a separate food material although they have been considered as a potential source of enzymes for food and industrial processing.[50] More recent studies have shown that the sarcoplasmic protein fractions might actually supplement the gelling of the MFPs.[86,113–115,122] The addition of sarcoplasmic protein improves thermal gelation, has a promotive effect on suwari (gel setting) and a restrictive effect on modori (gel-breaking phenomenon).[86]

14.2.1.2 MYOFIBRILLAR PROTEINS

Based on physiological functions in muscle, MFPs are further classified into two subgroups namely, contractile proteins such as myosin and actin and regulatory proteins such as tropomyosin, troponin, α-actinin, and c-protein.[7] The protein myosin and actin, which are involved in the contraction and relaxation of living muscle, are termed contractile proteins. Myosin, actin, and their conjugated form actomyosin impart various functionalities to the fish meat system.[105]

14.2.1.2.1 Myosin

Myosin contributes 55% of muscle proteins and forms the thick filament in the myofibril.[117] It is the largest MFP molecule with a molecular weight of 520 kDa.[164] It possesses three important functions, namely, ATPase activity, actin-binding ability, and filament forming ability.[65] The role of myosin in contributing to different functional properties is well established.[215]

Fish myosin, similar to myosin of other vertebrates, is a hexameric protein consisting of two heavy chains (MW–210 kDa each) and four light chains (MW–15.25 kDa each) with a total molecular weight of 520 kDa.[178] The N-terminal portion of the heavy chain forms a globular head, which contains the actin, nucleotide, and light chain binding sites. The c-terminal portion of the two heavy chains associates to form a coiled-coil rod, which is involved in the filament formation under physiological ionic condition.[141] To each globular head are attached two light chains by non-covalent interaction which can be easily dissociated from the head by denaturing agents like urea, alkali or guanidine hydrochloride.[192] One of the light chains is called regulatory light chain (RLC) and the other is essential light chain (ELC).[141]

When trypsin or chymotrypsin digests myosin, it is cleaved into two components, a rapid sediment component called heavy meromyosin (HMM) and a slow sediment component called light meromyosin (LMM). When HMM is treated with trypsin, it is cleaved into a head portion called myosin subfragment 1 (S_1) or "myosin head" and a neck portion called myosin subfragment 2 (S_2) or "myosin neck."[192] The LMM and the S_2 constitute the tail or rod of the myosin molecule and the S_1 constitutes the myosin head.[215] S_1 is responsible for most of the activities of myosin molecule. It bears the nucleotide (ATPase), actin, and light chain binding sites. The S_1 has a molecular size of 95 kDa consisting of a 25 kDa N-terminal domain, a 50 kDa central domain, and a 20 kDa c-terminal domain. The 25 kDa domains form the ATPase site where ATP binds for hydrolysis to ADP and P_1. The actin-binding site, where actin filaments bind to form the actomyosin cross bridge is formed by the 50 and 20 kDa domains. The two light chain-binding sites are located on the 25 kDa domain.[66,141] The tail or rod of myosin molecule, which consists of the LMM and the S_2 of HMM of the two heavy chains, has a length of 1700 A° and a diameter of 20 A°.[190]

14.2.1.2.2 Actin

Actin is the major constituent of thin filaments of myofibril. It makes up 15–20% of muscle protein by weight.[117] In solutions of low ionic strength, it remains as a monomer called G-actin because of its globular shape.[190] It has a molecular weight of 42 kDa. As ionic strength increased to physiological level, G-actin polymerizes non-covalently into an insoluble double helical fibrous form, F-actin, which closely resembles the thin filaments of intact muscle. Actin, like myosin, has an ATPase enzyme activity. However, ATP hydrolysis by actin does not power the contraction of living muscle; rather, the ATP-ADP cycle of actin participates in filament assembly and disassembly.[190] G-actin has three binding sites: (1) a nucleotide (ATP or ADP) binding site, (2) a divalent cation (calcium) binding site, and (3) a myosin-binding site.[215] F-actin is formed by the polymerization of G-actin monomers when salt concentration is raised. Each monomer binds one molecule of ATP, which is hydrolyzed to ADP on polymerization.[154] F-actin complexes with myosin S_1 in absence of ATP resulting in arrowhead-shaped filaments.[112]

14.2.1.2.3 Actomyosin

In actomyosin, myosin and actin are associated via non-covalent bonds, which can be easily dissociated by ATP or at high ionic strength. Electrostatic interactions, through phosphorus groups play a major role in stabilizing the actomyosin complex.[215] The actomyosin complex is biochemically and physico-chemically similar to myosin in many aspects. For instance, actomyosin retains most of the myosin ATPase activity and thermally induced gelation properties typically observed on myosin. However, structural changes in the myosin heavy chain (MHC) brought about by cross-linking with actin do alter some physical and functional characteristics of myosin. Some functional properties of myosin such as emulsifying capacity are lost when actomyosin is formed.[43] In general, actomyosin does not exhibit any physico-chemical or functional attributes of F-actin.[215]

14.3 MEANING OF MINCEMEAT AND SURIMI

Fish mince is nothing but mechanically deboned fish meat. Due to the mincing operations, fish mince becomes more susceptible to quality deterioration when compared with whole fish which in turn lead to fat oxidation and autolytic changes at faster rate. Mince quality depends on various factors such as season of catching, species, handling and processing methods, etc.[8] Fish mince can be stored up to 6 months under frozen storage conditions.[26] Minced meat is generally stored at −18°C in the form of frozen blocks of 1–2 kg. In order to prevent lipid oxidation and protein denaturation during frozen storage of mince, spices, cryoprotectants, and hydrocolloids may be incorporated into the mince.[74,76] An important use of fish mince is in the preparation of surimi, an intermediate product which, because of its characteristic ability to form gels, can be used to develop a variety of products conforming to consumer fancies.

Surimi is mechanically deboned and water-washed-minced meat from fish added with cryoprotectants and frozen[89,142] which is originated in Japan. Although primarily used to prepare *kamaboko* in Japan, it finds application in the manufacture of various mince-based products such as fish paste products, sausages. Minced meat technology has received considerable attention in the recent years in order to develop imitated analog products such as crab sticks and shrimp analogues.[45] Since the beginning of the 1980s much interest in surimi has been generated throughout the seafood and food industry. The popularity of the surimi-based products especially crab analog have paved way for rapid development of surimi industry. Surimi is generally frozen stored in the form of blocks typically frozen in plate freezers.

14.4 HISTORICAL PREVIEW

Japanese started production of surimi hundreds of years ago as a means of fish meat preservation.[145] Surimi was originally developed as a raw material for *kamaboko* that is produced by taking advantage of gel-forming ability of MFPs. The industrialized surimi-making process was refined in 1969 by Nishitani Yosuke of Japan's Hokkaido Fisheries Experiment Institute.[216] The *kamaboko* type gels made from Walleye Pollock surimi by treating with 2–3% sodium chloride and by a two-step heating procedure including setting prior to final cooking.[178]

Before 1960s, manual operations for fish mince production was practiced on small scale which was shifted to mechanized mode on larger scale for fish mince production till the year 2000. Afterwards, there is a shift toward the maximum utilization and efficiency in operations till 2015 and now everyone is thinking of sustainable and responsible utilization of resources. Since last 900 years, Japan is practicing the minced meat technology with technological development. Initial family-based production of mince-based products turned into commercial scale from the year 1603 with progressive changes till the date such as use of sugar as cryoprotectants and frozen storage of fish mince and mince-based products. Nishiya Kyosuke and Takeda Fumio discovered frozen surimi.[144]

Surimi is a popular food not only in Japan but also in other countries due to the unique textural attributes and nutritional value.[145] The popularity of surimi-based products among consumers favors the development of surimi manufacturing plants in many countries.[146]

14.5 MINCED MEAT AND SURIMI PRODUCTION PROCESS

14.5.1 RAW MATERIAL SUITABLE FOR MINCED MEAT AND SURIMI PRODUCTION

For production of minced meat and surimi, any fish species can be used as a raw material and has fairly long shelf life at subzero temperature.[192] Fish species having low contents of fat, white color meat and good gel-forming ability on heating can be used as a raw material for surimi production. The species most commonly used in the manufacture of surimi are Alaska pollock or walleye pollock (*Theragara chalcogramma*).[49,60]

There are several types of fish that are commonly used as surimi raw material which are different for temperate and tropical region. Alaska pollock, Pacific whiting, arrow tooth flounder (ATF), blue whiting, mackerel, menhaden, bigeye snapper, threadfin bream, lizardfish, croaker, and tilapia are commonly utilized for surimi production.[11,21,49,151,163] In tropical countries, fishes such as croaker, threadfin bream, soles, haddock, and other pelagic fishes are also considered suitable for surimi production. Fish species used for surimi production in tropical countries are listed in Table 14.1. Fatty fish such as sardines, herrings, and mackerels are difficult to process into surimi due to higher content of dark muscle, high lipid content, and poor gel-forming ability.[183] Factors that affect the composition of the fish such as seasonal variation, feeding, pH of habitat, water, adaptation temperature, sex, and spawning have profound influence on the quality of surimi produced from each species. Fish caught during active feeding season yield surimi of the highest quality. A drop in moisture follows increase in lipid content of meat. Fish of low moisture content yield surimi of high gelling ability.[45] Small pelagic fish having high water-soluble proteins and dark muscle are now used for surimi production after suitable modification in the process line.[156] Surimi is graded into based on the characteristics of meat such as gelation properties of proteins, whiteness of meat, etc. Gopakumar[45] graded tropical varieties of fish into various grades as very good, good, and poor wherein; croakers, Alaska pollock and threadfin bream are few species categorized as very good. Freshwater fishes like carps and tilapia, marine species like ribbonfish, sciaenids, and yellowfin tuna, and elasmobranchs like sharks and rays are categorized as good for surimi production. Sardine, mackerel, anchovies, and shrimps are categorized as poor grade species for surimi production.

The points which need to be considered regarding raw material include freshness of fish as a principal factor as it affects gel-forming ability.[39] Surimi can be prepared from pre-rigor or from post-rigor fish. Physical and chemical properties of fish muscle undergo major postmortem changes that significantly alter functional properties. Surimi attains maximum gel strength when fish is processed immediately after death, while "in" rigor fish is difficult to handle and surimi made tends to leave fishy odor. It is better to prepare surimi just after rigor mortis is resolved.[39]

TABLE 14.1 Tropical Fish Species Suitable for Surimi Production and Trade Names.

Common name	Scientific name	Trade name of surimi product
Threadfin breams (Five lined threadfin bream, Rosy threadfin bream, Redspine threadfin bream, Japanese threadfin bream, Double whip threadfin bream, Slender threadfin bream, Rosy threadfin bream, and Red spine threadfin bream)	*Nemipterus* spp. (*Nemipterus tambuloides, Nemipterus peronii, Nemipterus nemurus, Nemipterus japonicus, Nemipterus metatophorus,* and *Nemipterus zysron*)	*Itoyori*
Lizard fish (Shortfin lizardfish, Brush toothed lizardfish, and Wanieso lizardfish)	*Saurida* spp. (*Saurida micropectoralis, Saurida undosquamis,* and *Saurida wanies*)	*Eso*
Big eye (Purple-spotted big eye, and Red big eye)	*Priacanthus* spp. (*Priacanthus tayenus,* and *Priacanthus macracanthus*)	*Kintoki-dai*
Croaker (Big eye croaker, White croaker, Big head pennah croaker, and Sharp nose hammer croaker)	*Pennahia spp.* and *Johnius* spp. (*Pennahia macrocephalus* and *Johnius borneensis*)	*Guchi*
Goatfish (Gold band goatfish)	*Upeneus* spp. (*Upeneus moluccensis*)	*Himeji*
Pike conger	*Muraenesox spp.*	*Hamo*
Pony fish (Slip mouth Orange fin ponyfish, and Striped ponyfish)	*Leiognathus* spp. (*Leiognathus bindus, Leiognathus fasciatus*)	*Shirosaki*
Sardine	*Sardinella* spp.	*Mamakari*
Snapper (Bigeye snapper)	*Lutjanus* spp. (*Lutjanus lineolatus*)	*Fuedai*

14.5.2 PREPARATION OF MINCED MEAT FROM WHOLE FISH

Finely ground paste of fish meat is referred as fish mince. For the production of fish mice, fish is forced against a screened/perforated surface. The flesh passing through the screened surface is collected.

14.5.2.1 PREPROCESSING OF FISH BEFORE MINCING

Gutting and beheading are carried out before mincing. The portion of backbone just above the belly cavity is removed in order to avoid discoloration caused by blood along the backbone and spoilage.

14.5.2.2 MEAT PICKING/BONE SEPARATION

Meat picking machine/meat bone separators are commonly used for meat mincing operation. Different types of meat and bone separators are available. Simple kind of bone separators is used for fish mince production. Meat separator or meat picker is fed with fish from the hopper in random manner. Performance improvement is possible by placing fish with skin side against the belt and cut surface against the drum in order to remove the skin cleanly and to avoid clogging of separator.

One common type of which is belt and drum type wherein, fish is allowed to compress on the perforated surface which allows separation of meat without bones. The fishes are generally beheaded, gutted, and passed between counter-rotating belt and a perforated drum. The fish pieces are supplied into the feed hopper from where a conveyor belt of strong highly elastic material transfers the material to a perforated drum. The meat is pressed through the perforations into the interior of the drum while bones and other solid parts are retained on the outer drum shell from where it is removed by a scraper. The bone-free meat is delivered to one side by means of a screw conveyor in the perforated drum. The perforated drum rotates slightly faster than the conveyor. The perforations in the drum are usually 3–7 mm in dia. Pressure applied by the conveyor to the drum can be regulated depending on the type and size of raw material and on the diameter of perforations.[79] A picture of a meat bone separator is given in Figure 14.1.

Minced meat of fish is generally packed added with cryoprotectants and antioxidants, frozen in plate freezers and stored at −18°C in cold storage. Mincing of fish helps to improve yields of flesh when compared with filleting alone. It is widely acceptable due to the absence of bones in flesh as like in fillets. It has better control on the processing parameters too and it is possible to produce molded forms also using fish mince.

14.5.3 SURIMI PROCESSING

Surimi processing operation involves minced fish meat which is water washed and added with cryoprotectants. It involves various steps[146] which include:

1. Beheading and gutting
2. Mincing/deboning

3. Water washing and dewatering
4. Refining
5. Dehydration using screw press
6. Addition of cryoprotectants
7. Packing, freezing, and storage.

FIGURE 14.1 Mechanical meat bone separator.

Surimi processing line and few types of equipment used in surimi processing are represented in Figures 14.2 and 14.3, respectively.

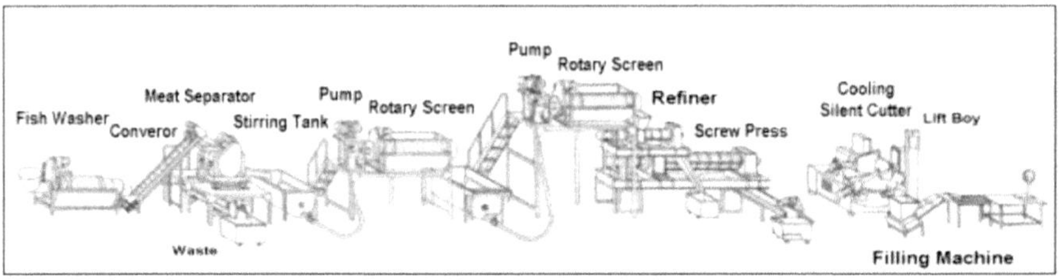

FIGURE 14.2 Schematic presentation of surimi processing line.

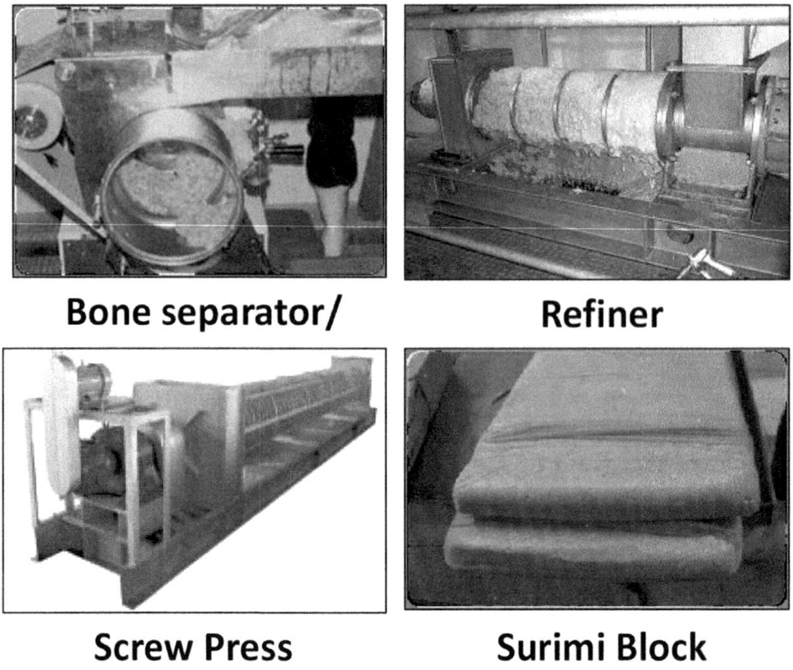

FIGURE 14.3 Meat bone separator, refiner, and screw press used in surimi processing.

14.5.3.1 BEHEADING AND GUTTING

Generally, two methods are used for gutting before deboning.[153] In one method, fish is beheaded, gutted, and belly walls are thoroughly washed before deboning. In the other method, initially fish are cut into fillets prior to deboning. The former method can better retain the meat. The filleting prior to deboning reduces the yield but produces enhanced quality of minced meat.

14.5.3.2 DEBONING AND MINCING

Fish is generally cut along the backbone and evisceration is carried out in order to maintain the quality of mince. Fish with white meat and freshness are generally selected as it decides the quality of mince and mincing yield. Various types of meat bone separators such as Baader type are commonly used with varying capacities with 50–60% minced meat yield. Deboned flesh of fish that is fish mince is subjected to further processing. For the preparation of surimi, the drum of the deboner should have a medium orifice of size 3–4mm for better quality and yield.[93,195]

14.5.3.3 WATER WASHING

Water washing of separated fish flesh in surimi production is very critical. The objectives of the water washing the fish flesh are to reduce lipid content, undesirable color, and soluble protein fraction from the muscle.[199] The repeated washing of mince removes water-soluble nitrogenous matter and flavor compounds. Washing also enhances the gel-forming capacity of the structural proteins. The washing technique used is an important key in determining the quality of the surimi. Minced fish flesh is rapidly washed with chilled water (5–10°C), as low-temperature water helps preserve the freshness of the raw material. This process removes undesirable matter such as blood, pigments, and other impurities, leaving the myofibril protein. The maximum amount of myofibril protein extracted is desirable because it influences the gel-forming ability of surimi. MFPs which are salt soluble in nature plays an important role in gel formation,[62,146] and the gels contained concentrated MFP due to the three cycle washing process. The breaking force (gel strength) of sardine and mackerel surimi-like material produced using surimi-washing process (using water or NaCl washing) was greater than that of materials produced using the alkaline solubilization process with or without prewashing as reported by Chaijan et al.[22]

The water-soluble fraction includes sarcoplasmic proteins, enzymes (primarily proteases), inorganic salts, and low-molecular organic substances such as trimethylamine oxide. The removal of sarcoplasmic proteins facilitates an increase in the concentration of MFPs, which are primarily responsible for gel formation.[199] The non-proteinaceous substances are known to accelerate the denaturation of muscle protein during frozen storage and are removed during water washing.[132] The rate at which these undesirable soluble substances are removed from the minced meat is a function of several factors, including the water temperature, the degree of agitation, and the contact time between water and meat particles.[48] The number of washing cycles required depends upon the type, composition, and freshness of the fish to be processed. The majority of the soluble components are freely and rapidly removed in the first washing cycle primarily by dilution of the free solubles. However, a definite residence time is required for the elution of water-soluble components from the minced meat.[199] The extraction of water-soluble components from the mince at a given washing cycle appears to be a function of agitation time which independent of the meat to water ratio. The amount of protein leached out increases markedly with washing time extended up to 9–12 min, and thereafter, it levels off at all ratios of meat to water studied.[95] This indicates that a 9–12 min. agitation is adequate for washing. If the residence time is unduly prolonged, the fish meat will absorb an excess amount of water and subsequent dewatering becomes difficult. Although the optimal time varies with the freshness of the raw material, water temperature, and size of the meat particles, 15–20 min for the entire washing process is usually regarded as appropriate for a commercial operation.[95] Lower gelling ability of washed tilapia meat as a result of extensive dehydration of MFPs due to higher washing cycles was reported by Murthy et al.[118]

The temperature of water used for washing separated fish flesh will be in the range of 5–10°C.[93] The temperatures need not to be kept unnecessarily low; the type of fish and fish species, specifically the thermostability of actomyosin ATPase enzyme should determine it. It has been found that warm water fish can tolerate a higher water temperature than cold-water fish without reduction in protein functionality.[4]

In manual batch process, the volume of water for each washing will be 5–10 times that of fish and at least three washing cycles are required.[93] The modern surimi plant adapts continuous washing process. The washing tanks will have paddles for agitation and further leads to rotary screen for dewatering processing.[199]

Generally, with continued washing, the gel strength increases up to two washing cycles and thereafter levels off according to a study conducted at a 3:1 (water-to-meat) ratio.[94] Such increased gel strength reflects increased MFP content and decreased sarcoplasmic protein content. However, the leveling off of gel strength after two washings suggests that it no longer behaves as a function of MFP concentration once a certain level is reached. Additional washing beyond this point is unnecessary from the standpoint of enhancing gel strength.[95]

The quality of water used for washing is very important. Principal factors determining the effectiveness of washing are ionic strength or concentration of various inorganic salts (hardness), metal ions, pH, and temperature. Since fish meat tends to swell more readily in the latter part of the repeated washing process, it is good practice to add a small amount of salt in the final washing cycle in order to increase ionic strength. The minimum hydrophilic condition occurs at ionic strength between 0.005 and 0.1.

Soft water having very low levels of Ca^{++}, Mg^{++}, Fe^{++}, Mn^{++}, and Zn^{++} ions is recommended for washing the fish mince. Hard water produces textural deterioration and color change during processing and subsequent frozen storage. Demineralized water, using ion exchange resins or reverse osmosis, is suitable for use in surimi plants.[45]

A single washing at 3:1 (water:flesh) ratio and washing under acidic conditions at pH 5.0–5.3 were found to be most efficient and effective in removing lipid and trimethylamine oxide.[143] By this method water requirements was reduced by 80% and yield was improved by up to 34% over conventional processing. However, the texture of gels was poorer when prepared from surimi produced under acidic conditions.

14.5.3.4 INTERMEDIATE DEWATERING

The intermediate dewatering is normally carried out to improve the efficiency of final dewatering process. The diameter of the holes in screens is about 0.5 mm and maximum dewatering occurs within the first sections of the rotary screens.[199] Fine particles of the meat lost through screens may account for about 8% of the starting mincemeat weight.[95] A significant portion of these particles are normally recovered and recycled.

14.5.3.5 REFINING

Partially dewatered mince is refined to remove any connective tissue, skin scales, or other undesirable inclusions. The most common type of refiner consists of cylindrical screen (Perforation size is typically 1–3 mm) and a screw-shaped rotor. The washed mince is fed into machine and the soft mass is selectively forced through the perforation under the compressive force generated by the rotor.[199]

The quality consideration involved in the refining processes include the problem of removal of the impurities effectively while obtaining a high yield and to prevent temperature raise in the mincemeat resulting from the high shear condition.[95] This condition calls for appropriate water content in the washed meat (about 90%) and maintenance of temperature of the meat below 10°C.

14.5.3.6 FINAL DEWATERING

Final dewatering is carried out using a screw press. The moisture content of the washed meat is reduced to about 80–84% by screw press.[199] The screw press consists of a rotating screw and a cylindrical screen. The compression ratio (difference between the capacities of inlet and outlet) of the screw permits the application of pressure and thus removal of water from the meat; the higher the compression ratio, the greater is the dewatering power. At high pressures, mince is forced out of the screen along with water and the yield will decrease. Higher pressure also leads to protein denaturation due to temperature rise and higher shear condition.[199] Ordinarily a compression ratio 2.5–3.5 is used. The dewatering power of screw press also depends upon size of holes in the screen. Generally, the hole's diameter ranges from 0.5 to 1.0 mm at the inlet and from 1.0 to 2.0 mm at the outlet. The screw speed is determined by the degree of dewatering intended for the finished mince. A slow speed is employed to achieve maximum dewatering.[199]

Surimi processing involves various steps. In brief, raw material is washed using 5–10°C water, gutted, beheaded, and washed again with 5–10°C water. Raw material is subjected to meat separation to get minced fish meat. Minced meat is then subjected to 2–3 cycles of water washing with intermediate dewatering and finally, water-washed meat is subjected to straining and dehydration followed by blending with additives such as polyphosphate, sugar, salt, etc., and stuffed. Water-washed meat added with cryoprotectants is finally frozen at −40°C, packed in master cartons and stored in the cold storage at −30°C.[45] Though marine fish is more often used for surimi preparation, considerable attempts have been done to utilize freshwater fishes also. Many authors have eloborated equipments needed for processing freshwater fish surimi.[20]

14.6 FUNCTIONAL PROPERTIES OF MINCED MEAT AND SURIMI

Since surimi is a wet concentrate of MFP, all the functional properties exhibited by surimi has been attributed to MFP. Among various functional protein properties, gelation, surface-active property, hydration, and viscosity are important in terms of commercial significance.

14.6.1 *HYDRATION CAPACITY*

Apparently, proteins are surrounded by a "loose" hydration shell composed of several layers of water, an "innermost" layer consisting of water (10–20 molecules of H_2O per molecule of protein) tightly bound to specific sites on or in protein molecules (i.e., absorbed water) and in another layer of water, about 102–105 molecules are more loosely bound, covering the immediate surface of protein molecules (i.e., adsorbed, non-freezable water). The second, third and additional layers of water have the properties similar to bulk water, that is, essentially non-structured water surrounding the adsorbed layer.[88,221]

Gerald et al.[44] have suggested that all changes in water holding capacity (WHC) are due to changes in the volume of myofibrils. Water binding capacity varies with protein source, composition, and presence of carbohydrates (i.e., hydrophilic polysaccharides), lipids, pH, and salts.[59]

According to Briskey[17] and Hermansson[57] the degree of protein hydration and viscosity of particular food systems are interrelated and these are directly influenced by pH, ionic strength, temperature, and protein concentration. Studies of water absorption over a range of conditions and concentrations are useful in assessing potential application of new proteins. About 4–5% of muscle water is directly bound to the protein molecules as hydration water, which has a highly ordered structure and different from those of free water. The water of hydration remains tightly bound even during the application of severe mechanical or other physical force. It is not greatly influenced by changes in structure or charge of meat proteins.[54] The term WHC collectively refers to the ability of meat to retain water. The common changes in WHC are related to the water other than hydration water.[55]

Hamm[53] demonstrated the importance of the spatial arrangement of myofilament. He suggested that at a pH away from isoelectric point of the proteins, the electrostatic repulsion between the protein molecules is high with the result network of the myofilament is enlarged and more water is immobilized between them. Conversely, at a pH near isoelectric point of proteins, the tightening of myofilament is facilitated, which results in decrease of WHC.

Heat causes a decrease in myofibrillar WHC and a large part of muscle water becomes freely movable and is released from tissue. Shamasundar and Prakash[180] reported the WHC of freeze-dried and powdered prawn meat as 4.96 ± 0.3 g of water/g of dried material. This value of WHC is relatively higher compared to many other protein sources of both vegetable and animal origin.

14.6.2 GELATION

Gelation is a process wherein native structure of protein molecules are denatured and undergo ordered aggregation resulting in three-dimensional network with entrapment of water.[152,215] Gel is an endpoint wherein textural properties are exhibited. Commercially typical protein gel products are *kamaboko* (fish meat), tofu (soybean curd), and fu (wheat gluten).[110]

Structurally, a gel is a form of matter intermediate between a solid and a liquid, consisting of strands or chains cross-linked to create a continuous network immersed in a liquid medium.[197] Gel may be defined by their ability to immobilize a liquid, their macromolecular structure, or their textural or rheological properties.[82] Rheologically gel is a substantially diluted system, which exhibits no steady-state flow.[38] Thus, collectively, a gel is a continuous network of macroscopic dimensions immersed in a liquid medium and exhibiting no steady-state flow.[220] Gels can be classified into four types, on the basis of structural criteria.[41]

1. Well-ordered lamellar structures including gel mesophases.
2. Covalent polymeric networks; completely disordered.
3. Polymer network formed through physical aggregation; predominantly disordered, but with local regions of order.
4. Particular disordered structures.

Most proteins form thermally irreversible gels.[220] A thermally irreversible gel is a visco-elastic solid formed during heating, which does not reverse to a viscous liquid on reheating.

A theoretical framework for protein gelation suggested by Ferry[37] and later modified by Schmidt[176] as given below:

native protein → denatured protein (a)

denatured protein → soluble aggregate (b)

soluble aggregate → insoluble aggregate (c)

insoluble aggregate→ gel (d)

The process of protein gelation was considered in detail by Ferry[37] He proposed that two steps are involved in the heat-induced gelation process. In the first step, proteins were thermally denatured with concomitant conformational changes. In the second step, the denatured proteins get aggregated. The two steps are:

$$ {}_{X}P_{N} \xrightarrow{\text{Ist step}} XP_{D} \xrightarrow{\text{IInd step}} (P_{D})_{X} $$

where $_X$ is the number of protein molecules; P_N is the native protein and P_D is the denatured protein.

Many mathematical models describing gelation from a mechanical or rheological perspective have been reviewed,[28,92] including Flory-Stockmayer theory,[40,188] mean field theory,[28] and percolation theory.[187]

The Flory-Stockmayer theory describes gelation as a sudden event that occurs when the degree of cross-linking between polymers reaches a critical value called the "gel point" at which viscosity diverges to infinity. This aggregation process is ultimately responsible for the formations of three-dimensional structure requiring the participating molecule to interact at specific point. The percolation theory assumes that monomers form small aggregates and at a critical threshold of bonding, a gel is reached after which the aggregates cross link throughout the percolation lattice.

Gelation is a controlled process in which the final network displayed some degree of order as opposed to coagulation in which structure is totally random.[58,198] The degree of order in gelation is controlled by number of factors and one such factor is thermal treatment. A high heating rate allows insufficient time for molecules to suitably orient themselves resulting in discontinuous matrix.[116] The other factors contributing to the nature of the gel include protein species, concentration, pH, and ionic environment.[84]

The model proposed by Foegeding and Hamann,[42] the two-step process of Ferry[37] to more completely include aspects of some theoretical models by dividing aggregation into a series of steps leading to the final gel structures as given below:

	Heat		Time		Time		Time		Time	
Native protein	→	Protein unfolding	→	Protein-protein interaction	→	GEL POINT	→	Primary matrix	→	Equi-librium matrix

With respect to fish muscle proteins, gelation has an important role in the production of surimi-based products. Fish proteins have good gelling ability than other muscle proteins.[89] To produce a gel product with the required elasticity and firmness, the two basic requirements must be met.[89,93,192]

1. MFPs must initially be solubilized in a salt solution.
2. On heating to form a gel, the proteins must be denatured in such a way that they form a regular network structure capable of immobilizing the water present.

When surimi is ground with sodium chloride, it dissolves the MFPs. Simultaneously, myosin combines with actin to yield macromolecular actomyosin. Both myosin and actomyosin have dominant roles in surimi gelation and show species specificities with regard to gelation properties.[182] Actin binding has been shown to modify the gelation characteristics of myosin of other species,[217] but tropomyosin seems to have no effect on gelation.[172] The gelation properties of actomyosin are mainly derived from the myosin.[126]

Among MFPs, myosin is responsible for the thermal gelation of fish meat. The thermal aggregation of myosin molecules is a crucial process for developing the elastic gel.

When fish meat is ground with sodium chloride, the MFPs get dissolved to produce a viscous "Sol." The sol can be converted to gel in two ways, either by heating or by application of high pressure. Heat-induced gelation is the conventional method of preparing fish gel products like *kamaboko*, fish sausage and seafood analogs. Pressure-induced gelation is of recent origin and is still under experimentation. In the heat-induced gelation, protein–protein interaction such as disulfide bonding, hydrophobic interaction, and hydrogen bonding are most common.[125] The kinetics of the sol–gel transition of surimi during heating has been studied by Iso et al.[67] The change in the viscoelastic property (P) of meat products during cooking follows a quadratic or exponential model:

$$Pr = C_1 T + C_2 T \tag{14.1}$$

or

$$Pr = \exp\,(C_1 T + C_2 T) \tag{14.2}$$

where C_1 and C_2 are constants and T is the absolute temperature. Based on reaction kinetics and Eyring's absolute reaction rate theory,[36] a relationship between Pr and thermodynamic parameters has been established by Correia and Mittal[31] as follows:

$$\frac{dPr}{dt} = \frac{KT(Pr_m - Pr_i)}{hq} \frac{exp\Delta s^{\neq}}{R} \frac{\Delta H^{\neq}(1-X)^n}{RT} \quad (14.3)$$

where subscripts i and m denote initial and minimum or maximum value of Pr during cooking, q is the temperature gradient (Kmin^{-1}), R is the gas constant, h is plank's constant, K is Boltzmann's constant, $\Delta S^{\neq}$ is the entropy change of activation, $\Delta H^{\neq}$ is the free energy change of activation, n is the order of the reaction, and X is the degree of cooking.

To elucidate the heat-induced gelation mechanism, actomyosin, myosin, and myosin subfragments have been successfully used to study protein–protein interactions.[10,135,136,174,212] Myosin is the most important component for adequate gel formation in fish gel products. Gels prepared from myosin alone have higher gel strength and elasticity than those prepared from natural actomyosin.[174] Observation of the dynamic viscoelastic behavior and differential scanning calorimetry (DSC) analysis revealed that gelation of carp actomyosin occurred at two stages: at a temperature range of 30–41°C and 51–80°C.[174] Differential shear modulus studies also revealed two transitional temperature ranges for hake natural actomyosin, 36–38°C and 48°C.[10] Sano et al.[173] proposed that the first stage of elasticity development was due to interaction among the tail portion of myosin molecules. The second stage was attributed to hydrophobic interaction among the head portions of myosin molecules because the protein conformation changes during heating such that the hydrophobic amino acid residues, which are found mainly in the head portion, become exposed on the surface. Ziegler and Foegeding[220] summarized that myosin from mammals generally undergoes gelation by losing its non-covalently stabilized α-helical structure as a result of heating, followed by increased turbidity caused by intermolecular association. Myosin then forms a rigid structure that is stabilized by covalent disulfide bonds and non-covalent interactions. The important chemical forces involved in the protein gelation phenomena are hydrophobic interaction, hydrogen bond, electrostatic interaction, and disulfide bonds.

The thermal gelation of fish muscle is influenced by two unique phenomena not occurring in the gelation of mammalian muscle. These are setting (suwari) and gel degradation (Himodori or Modori). Ribbonfish had moderate gel-forming ability, which could be enhanced by the addition of either 9% corn or tapioca starch.[33]

Fish flesh mince when ground with salt, a sol is formed. Commercially this sol is incubated at low temperature or at 40°C for a period of time resulting in a soft, cohesive, translucent gel called as "suwari" in Japanese and the phenomenon is known as "setting."[89] Without salt, such a paste is not obtained even if fish mince is kept for long time.[124] This phenomenon of setting has been effectively applied to the manufacture of gel (*kamaboko*) type of products.[75,80,189] Setting influences molding process and fiber development process in surimi products; hence control of setting is very important for quality control of gel products.[100,150]

Setting phenomenon is temperature dependent and has been categorized into low-temperature setting (4–10°C) and high-temperature setting (40°C).[212] Both setting methods have been applied in the commercial manufacture of fish gel products and each appears to produce unique textural characteristics in the final gel products.

The principal protein constituent in setting phenomenon is myosin molecule. It is MHC, which takes part in setting process resulting in cohesive gel.[78] The mode and rate of crosslinking reaction of MHC of the salt ground meat depends notably on the pH, a larger decrease in the amount of MHC was accompanied by the cross-linked MHC and unidentified components. In contrast, the decrease in MHC with concomitant formation of cross-linked MHC was supported under alkaline pH.

Setting is a gel-forming phenomenon, which occurs when salted meat paste is incubated below 40°C. It can be achieved within a short period (2–4 h) near 40°C (high-temperature setting) or following an extended period (12–24 h) at lower temperatures of 0—40°C (low-temperature setting).[212] Setting enhances surimi gelling properties to a great extent.[78,178]

Upon heating at higher temperatures above 70°C, a stronger, more elastic gel with higher WHC results as compared to a gel cooked without setting.[178]

Effect of setting time (0, 6, and 24 h), heating temperature (70°C, 80°C, and 90°C), heating time (10–60 min), storage time (0 or 36 days at −20°C), and addition of polysaccharides (1% W/W) on textural characteristics of emulsified sausage made from washed or unwashed tilapia mince was studied and it was found that all textural characteristics of the sausages decreased after 36 days of frozen storage.[25]

Setting and thermal treatment effects on texture and color of tropical tilapia surimi gels were compared to Alaska pollock and Pacific whiting gels. Pollock gels were generally the strongest and whitest, whereas tilapia gel quality was generally second to pollock gels.[85,137] investigated the mechanism of conformational changes of fish actomyosin from tilapia, lemon sole, lingcod, and rockfish during setting by using Laser Raman Spectroscopy.

They observed that, a slow unfolding of α-helix and exposure of hydrophobic amino acid residue occurring in long-time incubation at 40°C, thereby forming hydrophobic interactions among AM molecules. On the other hand, tilapia AM did not form a gel with heating at 40°C; its α-helical structures in the myosin being far more stable than that of the other species. The heat resistance of the coiled α-helix, may prevent the gelation of tilapia AM. It is therefore, likely that the unfolding of the α-helix of myosin is a prerequisite for the gelation of AM during setting.

The modori or himodori is a gel degradation phenomenon. It occurs when salted fish mince is heated at 50–70°C. It results in a considerable degradation of the gel to nonelastic state. The mechanism of modori is still poorly understood. However, three hypotheses have been described.[106] The first is based on "thermal coagulation" of MFPs during heating near 60°C. At this temperature (near 60°C) hydrophobic interaction of proteins are thought to be maximum, resulting in the liberation of water and gel weakening. The second mechanism is based on the participating of nonenzymatic modori inducing proteins. Two types of modori-inducing proteins with different molecular weights (44 and 50 kDa), which have no proteolytic activity, are known to reduce gel strength of surimi gels.[69] The third mechanism involves the degradation of formed gels by a heat-activated proteolytic enzyme, which degrades MHC.[3,69,218] This group of proteinases which degrade MHC with a maximum activity at 55°C resulting in modori is called Modori-Inducing Proteinases (MIPs).[200] MIPs are a group of serine proteinase.

MIPs are grouped into two classes based on their extractability from myofibrils. The easily extractable ones are called sarcoplasmic MIP (Sp-MIP) and the hardly extractable ones as myofibrillar MIP (Mf-MIP). Further, each class is classified into 50°C–MIP and 60°C–MIP according to their optimum temperature of 50°C and 60°C, respectively. Thus, the four types of MIPs are sp-50°C–MIP, Sp-60°C–MIP, Mf-50°C–MIP, and Mf-60°C–MIP. To date only Sp-60°C–MIP has been purified from threadfin bream.[81] It is a glycoprotein having monomer structure and its molecular weight is 77 kDa.

The important technique used in the study of protein gelation is electron microscopy and rheology. As gelation is influenced by the partially denatured state of protein molecules, protein denaturation during gelation is studied by DSC, circular dichroism, fluorescence, and UV absorption differential spectra.[51]

Rheology is the study of flow and deformation of water.[9] The rheological study of surimi-based products is categorized into two groups: (1) the small deformation test and (2) the large deformation tests. The small deformation test is very useful for monitoring sol–gel transition and for characterizing the visco-elastic behavior of gels in the "linear-region" in which the amplitude of the stress and strain are adjusted sufficiently to low values where stress is proportional to strain.[170] The stress–strain relationship as a function of different processing condition during gelation phenomenon can be conveniently measured for better understanding of gel structure and process.

Small deformation tests probe viscoelastic parameters, whereas large deformation tests normally probe elastic parameter. Two independent parameters are obtained from small deformation tests: the storage modulus (G') describes the amount of energy that is stored elastically in the structure which measures the elastic component and loss modulus (G") is a measure of energy loss which indicates viscous response[170] The ratio of G"/G' is the phase angle delta, is a measure of how much the stress and strain are "out" of phase with each other end.

$$\tan \delta \frac{G''}{G'} \tag{14.4}$$

The tan δ values at any given temperature can be measured continuously and the sol–gel transition temperature can be obtained.

The large deformation test is used for obtaining the rheological parameters corresponding to the food processing conditions and sensory properties.[27]

Changes in rheological properties of ATF myosin during thermal scanning were studied by Visessanguan et al.[205] Of the rheological parameters assessed, storage modulus (G') and phase angle were used to evaluate gel formation. An increase in storage modulus (G') is a measure of energy recorded per cycle of sinusoidal shear deformation indicated the increase in rigidity of the sample associated with formation of elastic gel structure.[35] Changes in dynamic viscoelastic properties indicated that ATF myosin formed a gel in three different stages as shown by the first increase in G' at 25°C, followed by a decrease at 35°C and second increase at 42°C.[205]

The use of phase angle to evaluate network characteristics has the advantage of incorporating the contributions of both G' and G" into a single parameter to evaluate the final network.[35] A change in phase angle reflects the transition of the viscous myosin sol to the elastic myosin gel, which correlates to the changes in G' during heating.

Noguchi[128] studied the viscoelastic changes of surimi during thermal gelation by using a low-frequency sinusoidal wave. The changes in storage modulus values were observed at, at least eight characteristic regions; four structure forming regions and four critical or unstable regions appeared alternatively on increasing the temperature from 5°C to 80°C.

An elastic modulus of the resulting gel increased with an increase in the binding amount of the arylating reagent but this increase was suppressed by the addition of sucrose.[127]

Rathnakumar and Shamasundar[162] investigated the viscoelastic properties of pink perch (*Nemipterus japonicus*) meat during setting and gelation using Controlled Stress Rheometer under oscillation mode. The increase in elastic component was maximum at 50°C in set meat and at 70°C in fresh meat. The transition from sol–gel state as indicated by tan δ value was at 50°C for set meat and 36.7°C and 56.8°C for fresh meat. Murthy et al[120]reported moderate gel-forming ability of tilapia meat.

The number of transitions and transition temperature of G' are influenced by pH, protein solubility, and ionic conditions and vary among proteins.[211,213] Linear or gradient heating (thermal scanning) usually produces a single G' or shear modulus peak for myosin suspended in low ionic strength ($\mu < 0.03$) solutions.[34] However, as ionic strength is increased to 0.5 or above, the peak becomes less pronounced. The magnitudes of G' transitions as well as the shape of G' curve for actomyosin were closely related to the myosin to actin ratio.[210] Storage modulus (G') is a good index for the gel-forming ability of food proteins. The higher the G' valve, the greater is the gel-forming ability of protein system. Assuming that the protein gel is made up of a highly viscoelastic network.[38] Gelation of food proteins is a function of the nature of the proteins and processing conditions such as pH, ionic strength, binding agent, and heating regimes.[209] There was evidence that under certain specific conditions, oxidation could promote functional performance of muscle proteins such as in surimi from fish[223]. Whether or not oxidation can lead to enhanced gelation, emulsification, and other physicochemical properties of muscle proteins seems to be determined by the extent of oxidative modification as well as type of change in proteins.[209] It has been hypothesized that limited oxidative modification in proteins would facilitate balanced protein and protein–solvent interaction conducive to gelation. Excessive oxidation would suppress protein functionality due to the formation insoluble protein aggregates and destruction of some functional side chains.[214] According to Lee and Lanier,[96] gelation of surimi protein involves both covalent and non-covalent bonds. The major covalent bonds are disulfide and glutamyl-lysine covalent linkage and the major noncovalent bonds are hydrophobic interactions, hydrogen bonds, and ionic bonds. Formation of such intermolecular bonds could theoretically be driven by oxidative stress and hence, affect the gelation process of surimi.

14.7 PROPERTIES OF PROTEINS DURING FROZEN STORAGE

Though, freezing and frozen storage have been recognized as one of the best methods of long-term preservation of fish and fishery products, the process itself has some deteriorative effect on the final eating quality. The change in the properties of proteins from fish during frozen storage has been a subject of investigation for the last six decades. It has been documented well that eating quality of fish meat is mainly attributed to major protein components, myosin, actin, and actomyosin. As a consequence of freezing and frozen storage these proteins will undergo series of alteration leading to changes in physico-chemical and functional properties which will have bearing on final eating quality.[184]

The rate of loss in eating quality is very much dependent on species, method of freezing, time, and temperature of frozen storage. The physiological differences between fish and mammalian or avian counterparts make fish muscle less stable at low temperatures. The changes in the protein properties during frozen storage is also affected by lipid content of fish as lipids are highly susceptible for oxidation because of high content of unsaturated fatty acids.[104] The various adverse effects during frozen storage on the proteins from fish will lead to alteration of the secondary and tertiary structure of proteins due to cleavage of the bonds that contribute to stability of native molecule.

A denaturing factor brings about a partial or total deconformation of native molecule depending upon the extent of free energy caused by its action. Denaturation, which involves the rupture of confirmation of the native proteins, brings about an exposure of hydrophobic groups to the bulk water. This results in alterations in the nearby water structures and increases pK value of carboxyl and amino groups of polypeptide chain. A denatured protein in the form random coil with all its functional groups exposed is highly reactive and can be involved in the formation of various new intra and intermolecular covalent crosslinks.[103,184]

Physical and chemical characteristics such as various hydrodynamic, optical, and enzymatic properties can follow the denaturation of proteins.

Minced fish meat is prone to undesirable biochemical reaction during frozen storage. The alteration in the physicochemical properties of MFP will lead to altered functional properties like decrease in hydration ability; gel-forming ability and other surface-active properties.[89] The frozen storage stability of mince is influenced by type of species, preprocessing condition, and storage of fishes.[56] Among MFP, it is the myosin molecule, which is most susceptible for biochemical reaction leading to decreased functionality.[32]

Frozen storage results in aggregation of carp myosin and actomyosin and the number of cross bridges between protein molecules increased during storage. The cross bridges formed during the storage attributed to formation of ionic (electrostatic) bonds, hydrogen bonds, nonpolar (hydrophobic) bonds, and disulfide (S—S) bonds.[201]

Upon freezing and frozen storage of carp actomyosin in solution or in suspension, it was found that myosin was dissociated from actin filaments and formed amorphous masses and further actin filament become entangled which contains masses of aggregated myosin.[138–140,201]

The denaturation of MFPs during frozen storage which leads to loss of solubility occurs due to proteins aggregation resulting from formation of disulfide bonds,[160] hydrogen bond and hydrophobic interactions.[19]

The formation of insoluble, high molecular weight protein aggregates in myosin solutions increased as the temperature decreased to below the freezing point and reached a maximum near the eutectic point (−11°C) of the myosin KCl water solution.[18] Solubility is important functional characteristics related to protein hydration and is a prerequisite for many other functional attributes of muscle proteins in processed meats.[52] Hence, factors such as structure of proteins, the pH and ionic strength of the medium, and the various other intrinsic and extrinsic parameters could influence protein solubility. MFPs are susceptible to freezing treatment and their solubility usually decreases during frozen storage.[215]

The immediate consequence of protein denaturation is formation of soluble aggregates; in a second step the aggregates become insoluble and the protein is not extractable.[207] The actomyosin from frozen stored fillets of post-spawned hake remained soluble (85%) up to 240 days. However, its hydrodynamic property indicated that it was highly affected by frozen storage.[111]

Another way to follow protein aggregation is to monitor changes in reactive —SH groups.[91]

A split plot design was applied to study the effect of frozen storage on the quality attributes of surimi.[61] Gel strength of surimi products was shown to be significantly affected by storage and its three-way or lower order interactions with washing, grinding, setting, and heating process.

Surimi gel cohesiveness, an index of gel deformability, has been referred as the most sensitive parameter to describe the quality or functionality of surimi proteins.[89] Decrease in gel deformability during frozen storage of surimi signified reduced gel-forming ability of proteins and have been associated with freeze-induced denaturation of MFPs.[148,193]

Surimi gel firmness as a function of frozen storage increased up to 4 weeks and after 4 weeks gel firmness decreased. Sych et al.[94] associated initial increase in gel firmness with decrease in protein WHC. Gel expressible moisture, an index of protein WHC, correlated well with firmness of cod surimi gel.[194]

The gel-forming ability of surimi made from fresh (1–2 days old) fish in good condition does not change significantly up to one year when held at constant temperature below −20°C.[68,70] However, when the surimi is stored at −10°C, the gel-forming ability gradually decreases, and the surimi becomes useless after 3 months, this is attributed to a decrease in extractable actomyosin.[70]

Loss of gel-forming ability in frozen surimi is attributed to denaturation and aggregation of MFPs.[192] Frozen storage reduced the gel-forming ability of MFPs from red hake, amber fish, mackerel, and hoki.[72,102] During storage of Alaska Pollock surimi at −20°C, MFPs were gradually denatured, resulting in lowering of extractability and gel-forming ability and changes in ultra-centrifugal pattern.[179] At −30°C denaturation of MFPs preceded much more slowly than at −20°C. In contrast, no or little denaturation of MFPs occurred below −40°C the most reasonable temperature for a short form storage is −30°C, whereas −40°C is preferable for a long-term storage.[179]

Gel-forming ability of pink perch meat as a function of frozen storage at −20°C was evaluated by large strain and small strain test.[161] A progressive reduction in gel strength with storage demonstrates the inability of MFPs to orient itself for proper network formation. The loss in gel strength is mainly due to the loss of solubility due to the formation of insoluble aggregates. This was demonstrated in protein from herring surimi.[23]

MFPs have different functional properties depending upon intrinsic, environmental, and processing factors. Among these properties, viscosity is important because of its involvement in many foods manufacturing process. Viscosity further provides information on physico-chemical interaction among protein molecules.[82,169] Viscosity has been used to determine the degree of protein denaturation and aggregation, during frozen storage.[30] It is considered a more reliable index of fish protein quality than protein solubility or emulsifying capacity.[30] Changes in apparent viscosity (η app) of muscle homogenates are hypothesized to be related to change in actomyosin.[15]

The viscosity of proteins decreases markedly during frozen storage of surimi and changes in viscosity can be taken as a more reliable index of protein quality.[30] A high correlation among different functional properties like viscosity and solubility was obtained frozen storage of fish.[29] The measurement of viscosity can be made use of to determine any change in shape or state of protein molecule.[155,181] The reduced viscosity of 1% protein solution of prawn increased with increase in frozen storage period indicating the process of aggregation.[181]

An association between loss of solubility and the decrease in —SH reactive groups during frozen storage has been reported for myosin[24] and actomyosin.[73] Total —SH groups decreased continuously during 15 day frozen storage of fish myosin to reach 33% of the initial valve, confirming the importance of disulfide bonds in frozen-induced aggregation of fish myosin in solution.[157] A decrease in total —SH groups during frozen storage of myosin solution in 0.6 M KCl has been reported.[24] This was accompanied a major decrease of ATPase enzyme activity.

Lin and Park[99] studied the relationship between solubility and conformational changes of myosin from salmon at high ionic strength and pH during frozen storage. The loss of solubility of proteins during frozen storage is mainly related to their denaturation resulting from the changes in the properties of myosin, actin, and actomyosin systems.[99] Preprocessing conditions,[165] freezing, and storage temperature[73] have been known to influence the solubility of MFPs.

During frozen storage (−20°C) of pre-rigor and post-rigor actomyosin from rohu (*Labeo rohita*) it was observed that there was greater decrease in solubility in the case of post-rigor actomyosin than pre-rigor actomyosin.[158,159] Shenoy and James (1972)[222] reported that extractability of proteins from tilapia decreased during frozen storage at −18°C. The salt soluble proteins decreased at a faster rate during the first week of frozen storage and thereafter the change took place at a slower rate.

Loss of ATPase activity is not necessarily synonymous with aggregations because it is possible to have no aggregation but lose 100% of the activity by denaturation of active site.[157]However, relations between frozen induced aggregation of myosin (determined as loss of solubility) and loss of ATPase activity have been reported.[24] ATPase activity decreased by 89% of its initial activity during 15th day frozen storage of fish myosin[157] suggesting that myosin heads could be involved in the fish myosin aggregation.

Proteolytic degradation of tropical tilapia surimi was biochemically and rheologically characterized to identify a group of proteinases responsible for its textural degradation. Proteolysis of tilapia surimi occurred as the temperature increased and attained the highest activity at 65°C. MHC completely disappeared when incubated at 65°C for 4 h. Soybean trypsin inhibitor (SB) and leupeptin (LE) significantly inhibited proteolysis. Storage modulus (G^1) of surimi gels mixed with either SB or LE was higher, indicating that serine type proteinase(s) were involved in proteolysis of tropical tilapia surimi.[219]

Dynamic viscoelastic behavior of actomyosin from green mussel (*Perna viridis*) revealed the ability to form network during heating and strength of gel was found to be weak as indicated by frequency sweep.[13]

In comminuted meat products[209] when the fat particles are coated with surface-active salt-soluble proteins, their stability is expected to increase sharply due to a reduction in interfacial tension.[46] The ability to stabilize fat particles in the form of hydrodynamic emulsions is an important property of muscle proteins. Emulsifying ability of muscle proteins depends on the balance of solubility and hydrophobicity.[97] Protein with high solubility and hydrophobicity had excellent emulsifying ability[209] and emulsifying capacity decreased significantly with frozen storage temperature and time.

The emulsion capacity of total proteins from pink perch meat registered a decrease of 53% from its initial value at the end of 300 days of frozen storage.[161] The reduction is attributed to the formation of protein aggregates during frozen storage.

14.7.1 CRYOSTABILIZATION OF PROTEINS IN SURIMI

The dewatered fish flesh, which is nothing but a concentrate of MFP, will be mixed with certain additives, referred as cryoprotectants, prior to freezing and frozen storage. The added cryoprotectants will minimize the rate of denaturation of proteins during frozen storage and thus maintaining the functionality of the mince. The effectiveness of cryoprotectants depends on fish species and quality prior to surimi preparation and method of processing into surimi.[192] In the absence of cryoprotectants, the washed mince will lose its important functional properties like gelling ability, hydration, and binding ability.[108]

The use of cryoprotectants to prevent denaturation of muscle protein of Alaska pollock during frozen storage was initially found by Nishiya's group.[123,196] The cryostabilization technique used in surimi manufacture effectively minimizes freeze denaturation of proteins.

In early surimi manufacturing process as proposed by Nishiya's group 10% sucrose and 0.3% polyphosphates were used as cryoprotectants.[123] Several pioneering studies were done in selecting the most suitable compounds and determining their optimum mixture ratios.[123,130]

The cryoprotective effect of sugars, namely glucose, galactose, fructose, lactose, and sorbitol was studied by Nishiya's group.[196] Hexoses (glucose and fructose and disaccharides (sucrose and lactose) were found to be most effective cryoprotectants.[133] Sodium glutamate was the first amino acid studied for its cryoprotective effect on frozen surimi.[129]

Based on several studies, a list of chemical attributes that seem to be characteristic of cryoprotective substances was established[107,131] and they are;

1. The molecule has to possess one essential group, either —COOH or —OH, and more than one of the following supplementary groups: —COOH, —OH, —SH, $-NH_2$, $-SO_3H$, and $-OPO_3H_2$.
2. The functional groups (both essential and supplementary ones) must be suitably spaced and oriented relative to each other.
3. The molecule must be comparatively small.

The latter requirement however, seems to be violated by the positive cryoprotective effect of polydextrose (R).[148] Oligosaccharides have been widely used as a food ingredient due to their favorable properties such as high WHC, low-sweetness, low viscosity, and low calories.[148] Cryoprotective effects of lactitol dihydrate at 2%, 4%, 6%, and 8% levels in cod surimi were investigated and compared to industrial control, 1:1 mixture of sucrose: sorbitol, 8% (W/W).[193] Data revealed the excellent cryoprotective properties of lactitol in maintaining various functional properties.

Surface tension and water binding capacity of these compounds are responsible for preventing the protein denaturation.[18,109,171,191] Addition of sucrose or sorbitol increases the stability of MFPs of frozen surimi prepared from pacific pomfret muscle.[134] The most common surimi formula is 92% washed mincemeat, 4% sugar and sorbitol and this product can be stored up to a year without loss of gelling properties.[168] A mixture of sucrose (4%), sorbitol (4%), and polyphosphates (0.3%) was fairly effective in surimi.[93] Park et al.[149] reported that a mixture of 5–6% sucrose and sorbitol (1:1) has an effective cryoprotective property. Extended frozen storage of surimi is made possible by the addition of 4% sucrose, 4% sorbitol, and 0.2% polyphosphate which inhibit the freeze denaturation of MFPs.[2,131]

The most commonly used cryoprotectants in the surimi industry have been low molecular weight sugar and polyols such as sucrose and/or sorbitol. They are added to 8% (W/W) alone or in mixture of 1:1 to washed fish muscle.[101] These carbohydrates were chosen because of their relative low cost, availability, and less tendency to cause Maillard browning in white gel products. Sucrose imparts a considerable sweet taste to surimi, which consumers sometimes found objectionable for certain applications. Non-sweet additives with high molecular weight such as polydextrose (R) and maltodextrin were found to have cryoprotective effect.[147,148,193]

The protection of MFPs, especially myosin, from denaturation during frozen storage was related to type of sugar and concentration.[87] Prevention of protein denaturation by sugars can be explained by their ability to increase the surface tension of water,[5] as well as the amount of bound water, which prevents withdrawal of water molecule from the protein, thus stabilizing the protein.[18,191]

The osmolytic cryoprotectants such as sugars and polyols function by stabilizing lattice structure of water, thus increasing surface tension of water and stabilize hydration shells and protect against aggregation by increasing the molecular density of the solution without changing dielectric constant.[175] The high molecular weight polymers such as dextran functions by increasing the molecular density and solvent viscosity and thus lowering protein aggregation. The amino acids function by weak electrostatic interaction with protein and increase surface tension of water. The ionic compounds such as citrates and phosphates function by shielding the charges at low concentration and stabilize; at higher concentration causes aggregation by competing for water.[175] The forces responsible for interaction of polymers (maltodextrin, carrageenan, xanthan, and carboxymethylcellulose) with proteins and their stabilizing roles are electrostatic in nature. Other interactions being hydrogen binding, hydrophobic or covalent bonds may also contribute to their stabilizing role.[12]

Alternatively, Lim[98] has proposed that diffusion controlled process in frozen system like surimi can be influenced by the presence of glassy states. Significant retardation of protein aggregation was observed below the glass transition temperature (Tg') of polymers like carboxymethyl cellulose (−10.5°C), maltodextrin (−5.5°C), and gums (−5.7°C), which is higher than sucrose (−33.2°C). Their protective effects on muscle protein might be explained by the presence of glassy states. Sucrose is an excellent cryoprotectant, preventing protein aggregation, but did not delay other chemical reaction at temperature below its (Tg'). Sucrose functions to protect protein by a different mechanism, that is, by solute exclusion.

The salts and amino acids present a potential application due to their charged state, as electrostatic interactions are possible with charged sites on proteins.

14.7.2 STABILIZING WITH CRYOPROTECTANTS

In surimi production, the addition of cryoprotectant plays a significant role in retaining functional properties during freezing and frozen storage.[101] The addition of cryoprotectants is important to ensure maximum functionality of frozen surimi since freezing induces protein denaturation and aggregation. The cryoprotectants also help to retain WHC and prevent drip loss.

Generally, a composition of 4% sugar, 4% sorbitol, and 0.3% polyphosphates is used as cryoprotectants. Recently enzyme inhibitors, such as beef plasma protein, egg white, or potato extracts have been used in conjunction with cryoprotectants, gel enhancers, and color enhancers. Enzyme inhibitors are commonly formulated with sucrose, sorbitol, sodium tripolyphosphate, tetrasodium pyrophosphate, calcium carriers, sodium bicarbonate, and partially hydrogenated canola oil. Trehalose, a disaccharide with 45% sweetness of sucrose is a newly introduced cryoprotectant which effectively replaced sucrose and/or sorbitol.[63] Cryoprotectants were originally incorporated into the dewatered mince using a kneader. At present, silent cutters are used which uniformly distribute the cryoprotectants faster and the temperature increase is minimum during chopping. The temperature during mixing should not exceed 10°C since higher temperatures may adversely affect protein functionality. Addition of egg white, soy protein isolates, and potato starch enhanced functional properties of common carp (*Cyprinus carpio*) surimi gel.[71] Cryoprotectants were found to be less effective in minimizing deteriorative changes in tilapia meat during frozen storage.[119]

14.7.2.1 FREEZING, PACKING, AND STORAGE

In commercial applications, surimi is formed in a standard 10 kg block in a plastic bag of 3–7 mm thickness which is placed on a stainless steel tray. The trays are placed in a contact plate freezer and held approximately for 2.5 h or until the core temperature reaches −25°C. After inspecting the frozen surimi blocks with a metal detector, two 10 kg blocks are packed in cardboard cartons and frozen stored at −20°C.

14.8 SURIMI TYPES AND PROCESSING YIELD

The processing operation generates 18–36% surimi with generation of 50–70% solid wastes. Generally, two types of surimi are produced one is salted in which 2.5% salt is used and it is used for preparation of *kamaboko* and other one is non-salted surimi.

14.9 QUALITY OF SURIMI

Gel strength, water content, and color are important parameters when quality of surimi is considered.[90]

14.9.1 GEL STRENGTH

Gel strength can be measured either by folding test or using compression response technique. Surimi gel is prepared by addition of 2–3% salt to thawed surimi and the resultant gel is filled in casings which are sealed at both ends, heated at 90°C for half an hour, immediately cooled and tested.

In folding test, 3 mm thick slices of cooked surimi are cut off and folded into quadrants. Cracks developed on folding are observed in order to grade surimi samples. The grading is:

AA—No crack after folding twice
A—No crack after first folding but crack occurs on second folding
B—Crack occurs gradually on first folding
C—Crack occurs immediately on first folding
D—Breaks by finger press even before folding.

Texture profile analyser is generally used to measure gel strength of surimi using compression technique.

14.9.2 WATER CONTENT

Water content is measured with an infrared moisture meter.

14.9.3 COLOR EVALUATION

Surimi color is generally determined using Hunters Scale Colorimeter wherein "L" indicates whiteness or darkness; "a" describes red or green whereas "b" indicates yellow or blue.

Surimi samples are also evaluated for presence of impurities by spreading.

14.10 MINCE AND SURIMI-BASED FISH PRODUCTS

Surimi is a primary material used for gelling foods such as *kamaboko* and fish balls. Minced meat can be used for the production of battered and breaded products such as fish cutlet, burgers, etc. The textural properties of the products prepared from surimi are mainly contributed by MFPs.[45] Among MFPs, it is myosin and actin which contribute to a great extent to many of the eating qualities like texture, flavor, and mouthfeel.[83]

Surimi is popular due to excellent gelling and elastic properties. It is possible to prepare imitated products for lobster tails, shrimps, and crab legs which are expensive wherein, appearance, flavor, and texture can be imitated. Elastic nature of surimi can be modified by use of polyphosphates, starch, egg white, etc. Natural seafood extracts can be used for flavor and different molds can be used for attaining the appearance and desired shape. Emulsified products such as sausages can be made using fat and few other ingredients with surimi.

Traditional products from Japan include *Chikuwa* (tube-shaped fish paste), *kamaboko* (boiled fish paste), *Satsuma age* (fried fish paste product), and *Hampen* (floating type boiled fish paste). Artificial/imitated surimi products include *kanikama* (artificial crab legs), hampen (in cheese sandwiched form), *kamaboko* (in easy to eat form), and *satsumage* with hampen paste which are marketed in Japan.

14.10.1 KAMABOKO

It is oldest paste product from Japan. Surimi is the raw material used for *kamaboko* preparation. Surimi is put into silent cutter with salt, seasonings, starch, etc., and fish paste is prepared. It is then transformed into desired shape, steam cooked, and packed.[45]

14.10.2 FISH SAUSAGE

Surimi is mixed with salt, spices, seasoning, starch, and fat and ground well to make paste in silent cutter. It is then stuffed in casings, sealed, steam cooked, and stored at cold temperatures.

14.10.3 FISH HAM

It is fish sausage added with few pieces of fish and lard.[45] It is stuffed in casings and cooked in hot water.

High nutritional value fish crackers were developed from mince of different low-value species containing low fat and were expanded by microwave cooking.[121] The mince from different species could be combined to prepare composite fillets.[202] Variety of value-added products can be prepared from surimi which include fish patties, fish fingers, fish balls, fish wafers, fish fingers, loaves, dehydrated mince, and cutlet can be prepared from surimi and minced meat,[47,77,166,203,204] fish salads and fish pakodas,[177] fish noodles,[16] fish sausages,[185] etc., are further developments in the range of products. Surimi-based analog products are given in Figure 14.4.

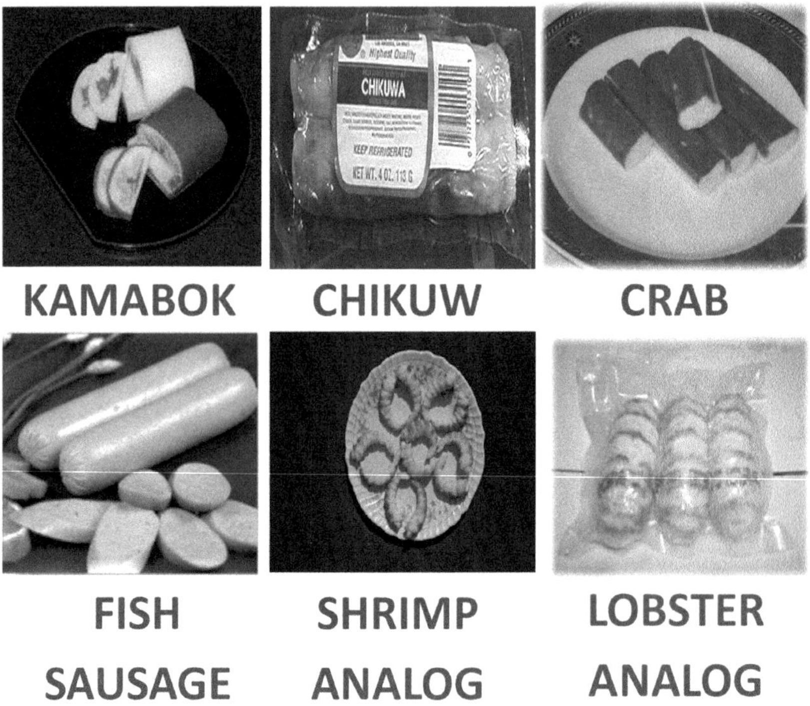

FIGURE 14.4 Surimi-based products.

Surimi has potential for utilization in pasta as a source of protein supplement. It can be used in pet foods for preparing various extruded formed products. In sausages and restructured products also, surimi can be utilized. It can be used as protein supplement in health drinks too.

14.11 CONCLUSION

Surimi which is originated from Japan is popular throughout the world due to its gelation behavior. Surimi is water washed minced meat added with cryoprotectants for longer storage life under frozen conditions. Surimi could be able to add a new aspect in seafood production. Water washing and gelation characteristics of MFPs play an important role in surimi production. Surimi processing allows fish proteins to retain for a longer time. It also enhances functionalities of fish meat in order to prepare gel-based surimi products. The popular among them are *kamaboko* and sausage. It is possible to achieve consumption throughout the year by surimi processing. Preparation of surimi is influenced by various factors such as raw material used for surimi production, gel-forming ability of the species selected, maintenance of process parameters such as temperature, etc. Surimi seafood is popular ingredient in seafood preparations. An innovative approach is necessary for improvement in surimi seafood production.

KEYWORDS

- **surimi**
- **actomyosin**
- **sarcoplasmic proteins**
- **myofibrillar proteins**
- ***kamaboko***

REFERENCES

1. Adebayo-Tayo, A. C.; Odu, N. N.; Michael, M. U.; Okonko, I. O. Multi-Drug Resistant (MDR) Organisms Isolated from Sea-foods in Uyo, South-Southern Nigeria. *Nat. Sci.* **2012,** *10* (3), 61–70.
2. Akahane, T.; Chihara, S.; Yashida, Y.; Tsuchiya, T.; Noguchi, S.; Ookami, H.; Matsumoto, J. J. Application of Differential Scanning Colorimetry to Food Technological Study of Fish Meat Gels. *Nippon Suisan Gakk.* **1981,** *47*, 105–111.
3. An, H.; Peters, M. Y.; Seymour, T. A. Roles of Endogenous Enzymes in Surimi Gelation. *Trends Food Sci. Technol.* **1994,** *7*, 321–327.
4. Arai, K.; Kawamura, K.; Hayashi, C. The Relative Thermal Stabilities of Actomyosin ATPase from a Dorsal Muscles of Various Fish Species. *Bull. Jap. Soc. Sci. Fish.* **1973,** *30*, 1077–1085.
5. Arakawa, T.; Timasheff, S. N. Stabilization of Protein Structure by Sugars *Biochemistry* **1982,** *21*, 6536–6544.
6. Asghar, A.; Pearson, A. M. Influence of Ante-and Post-mortem Treatments upon Muscle Composition and Meat Quality. *Adv. Food Res.* **1980,** *26*, 53–213.

7. Asghar, A.; Samejima, K.; Yasui, T. Functionality of Muscle Protein Gelation Mechanisms of Structured Meat Products. *CRC Crit. Rev. Food Sci. Nutr.* **1985,** *22* (1), 27–106.
8. Babbitt, J. K. Suitability of Seafood Species as Raw Materials. *Food Technol.* **1986,** *40*, 97–100.
9. Barnes, H. A.; Hutton, J. F.; Walters, K. *An Introduction to Rheology*; Elsevier Science Publishers: Amsterdam, the Netherlands, 1989; pp 199.
10. Beas, V. E.; Crupkin, M.; Trucco, R. E. Gelling Properties of Actomyosin from Pre- and Post-spawning Hake (*Merluccius hubbsi*). *J. Food Sci.* **1988,** *53*, 1322–1326.
11. Benjakul, S.; Visessanguan, W.; Thongkaew, C.; Tanaka, M. Effect of Frozen Storage on Chemical and Gel-forming Properties of Fish Commonly Used for Surimi Production in Thailand. *Food Hydrocoll.* **2004,** *19*, 197–207.
12. Bernal, V. M.; Smajda, C. H.; Smith, J. L.; Stanley, D. W. Interaction in Protein/Polysaccharide/Calcium Gels. *J. Food Sci.* **1987,** *52*, 1121–1125.
13. Binsi, P. K.; Shamasundar, B. A.; Dileep, A. O. Some Physico-chemical, Functional and Rheological Properties of Actomyosin from Green Mussel (*Perna viridis*). *Food Res. Int.* **2006,** *39*, 992–1001.
14. Bodwell, C. F.; McClain, P. E. Chemistry of Animal Tissue. In *The Science of Meat and Meat Products;* Price, E. F., Scheweighert, B. S., Eds.; Freeman: San Francisco, CA, 1971; p 78.
15. Borderias, A. J.; Colemenero, J. F.; Tajeda. M. Viscosity and Emulsifying Ability of Fish and Chicken Muscle Protein. *J. Food Technol.* **1985,** *20*, 31–42.
16. Brillantes, S. Fish Noodles Using Indian Carp. *Asian Food.* **2000,** *J7*, 137.
17. Briskey, E. J. Functional Evaluation of Protein in Food Systems. In *Evaluation of Novel Proteins Products;* Bender., Ed.; Pergamon Press: London, 1970; p 303.
18. Buttkus, H. Accelerated Denaturation of Myosin in Frozen Solution. *J. Food Sci.* **1970,** *35*, 558–562.
19. Buttkus, S. H. On the Nature of Chemical and Physical Bonds Which Contribute to Some Natural Properties of Protein Foods: A Hypothesis. *J. Food Sci.* **1974,** *39*, 484–489.
20. Bykowski, P.; Dutkiewicz, D. Freshwater Fish Processing and Equipment in Small Plants. Food and Agriculture Organization of the United Nations: Rome, 1996; No. 905, p 59.
21. Campo-Deano, L.; Tovar, C. The Effect of Egg Albumen on the Viscoelasticity of Crab Sticks Made from Alaska Pollock and Pacific Whiting Surimi. *J. Food Hydrocoll.* **2009,** *23*, 1641–1646.
22. Chaijan, M.; Benjakul, S.; Visessanguan, W.; Faustman, C. Physicochemical Properties, Gel Forming Ability and Myoglobin Content of Sardine (*Sardinella gibbosa*) and Mackerel (*Rastrelliger kanagurta*) Surimi Produced by Conventional Method and Alkaline Solubilisation Process. *Eur. Food Res. Technol.* **2006,** *222*, 58–63.
23. Chan, J. K.; Gill, T. A.; Thompson, J. W.; Singerd, S. Herring Surimi During Low Temperature Setting, Physico-chemical and Textural Properties. *J. Food Sci.* **1995,** *60*, 248–253.
24. Chen, C. S.; Hwang, D. C.; Jiang, S. T. Effect of Storage Temperature on the Formation of Disulfides and Determination of Milk Fish Myosin (*Chanos chanos*). *J. Agri. Food Chem.* **1989,** *37*, 1228–1231.
25. Cheng, C. C.; Tsai, J. S.; Chang, C. Textural Characteristics of Tilapia Emulsified Sausage. *Chemistry* **1996,** *265*, 8347–8350.
26. Ciarlo, A. S.; Boeri, R. L.; Giannini, D. H. Storage Life of Frozen Blocks of Patagonian Hake *(M. hubbsi)* Filleted and Minced. *J. Food Sci.* **1985,** *50*, 723–738.

27. Clark, A. H.; Lee-Tuffnell, C. D. Gelation of Globular Proteins. In *Functional Properties of Food Macromolecules;* Mitchell, J. R., Ledward, D. A., Eds.; Elsevier: Amsterdam, the Netherlands, 1986; p 203.
28. Clark, A. H. Gels and Gelling. In *Physical Chemistry of Foods;* Schwartzkerg, H. G., Hartel, R. W., Eds.; Marcel Dekker: New York, NY, 1992; pp 263–305.
29. Colmenero, J. F.; Borderias, A. J. A Study of the Effects of Frozen Storage on Certain Functional Properties of Meat and Fish Protein. *J. Food Technol.* **1983,** *18*, 731–737.
30. Colmenero, J. F.; Tejada, M.; Borderias, A. J. Effect of Seasonal Variations on Protein Functional Properties of Fish During Frozen Storage. *J. Food Biochem.* **1988,** *12*, 159–170.
31. Correia, R.; Mittal, G. S. Kinetics of Hydration Properties of Meat Emulsions Containing Various Fillets During Smoke House Cooking. *Meat Sci.* **1991,** *29*, 335–351.
32. Davis, J. R.; Ledwar, D. A.; Bardsley, R. G.; Poulter, R. G. Species Dependence of Fish Myosin Stability to Heat and Frozen Storage. *Int. J. Food Sci. Tech.* **1994,** *29*, 287–230.
33. Dileep, A. O.; Shamasundar, B. A.; Binsi, P. K.; Badii, F.; Howell, N. K. Effect of Ice Storage on the Physicochemical and Dynamic Viscoelastic Properties of Ribbonfish (*Trichiurus* spp) Meat. *J. Food Sci.* **2005,** *70* (9), E537–E545.
34. Egelandsdai, B.; Frethem, K.; Samejima, K. Dynamic Rheological Measurement of Heat Induced Myosin Gels: Effect of Ionic Strength, Protein Concentration and Addition of Adenosine Tri Phosphate or Pyrophosphate. *J. Sci. Food Agri.* **1986,** *37*, 915–926.
35. Egelandsdai, B.; Martisen, B.; Actiu, K. Rheological Parameters as Predictor of Protein Functionality. A Model Study Using Myofibrils of Different Fibre-type Composition. *Meat Sci.* **1995,** *39*, 97–111.
36. Eyring, H. The Activated Complex in Chemical Reaction. *J. Chem. Phys.* **1935,** *3*, 107–115.
37. Ferry, J. D. Protein Gels. *Adv. Prot. Chem.* **1948,** *4*, 1–78.
38. Ferry, J. D. *Viscoelastic Properties of Polymers;* Wiley: New York, NY, 1980.
39. Flick, J.; Barua, M. A.; Enriquoz, L. G. Processing Finfish. In *The Seafood Industry;* Martin, R. E., Flick, J. F., Eds.; Osprey Book: Oxford, London, 1990.
40. Flory, P. J. Molecular Size Distribution in Three Dimensional Polymers. I. Gelation. *J. Am. Chem. Soc.* **1941,** *63*, 3083–3091.
41. Flory, P. J. Introductory Lecture. *Faraday Discuss. Chem. Soc.* **1974,** *57*, 7–18.
42. Foegeding, A. E.; Hamann, D. D. Physicochemical Aspects of Muscle Tissue Behavior. In *Physical Chemistry of Foods;* Schwartsberg, H. G., Hartel, R. W., Eds.; Marcel Dekker; New York, NY, 1992; pp 423–441.
43. Galluzzo, S. G.; Regenstein, J. M. The Role of Chicken Breast Muscle Protein in Meat Emulsion Formation: Myosin, Actin and Synthetic Actomyosin. *J. Food Sci.* **1978,** *43*, 1761–1765.
44. Gerald, O.; David, R.; John, T. *Recent Advances in the Chemistry of Meat;* ARC Meat Research Institute Longford; Allen., Ed.; The Royal Society of Chemistry Burlington House: Mayfair, London, 1984.
45. Gopakumar, K. *Tropical Fishery Products;* Oxford & IBH Publishing Co. Pvt. Ltd.: New Delhi, India, 1997.
46. Gordon, A.; Barbut. S. Mechanism of Meat Better Stabilization: A Review. *CRC Crit. Rev. Food Sci. Nutri.* **1992,** *32*, 299–332.
47. Grantham, G. J. Fisheries Technical Paper 216, *Minced Fish Technology—A Review;* FAO: Rome, Italy, 1981.

48. Green, D. P. Leaching Soluble Nitrogenous Components from Atlantic Menhaden Muscle in Surimi Manufacturing. Ph.D. Thesis, North Carolina State University, Raleign, NC, 1989.
49. Guenneugues, P.; Morrissey, M. T. Surimi Resources, In *Surimi and Surimi Seafood*, 2nd ed.; Park, J. W., Ed.; CRC Press, Taylor & Francis Group: Boca Raton, FL, 2005; p 923.
50. Haard, N. F. Biochemical Reactions in Fish Muscle During Frozen Storage. In *Sea Food Science Technology;* Bligh, E. G. Ed.; M. A. Blackwell Scientific Publishers: Cambridge, London, 1992; pp 176–209.
51. Hall, G. M., Ed.; Basic Concepts. In *Methods of Testing Protein Functionality*; Blackie Academic and Professional: New York, NY, 1996; pp 1–10.
52. Halling, P. J. Protein-stabilized Foams and Emulsions. *CRC Crit. Rev. Food Sci. Nutri.* **1981,** *15*, 155–203.
53. Hamm, R. Biochemistry of Meat Hydration. In *Advances in Food Research;* Chichester, C. O., Ed.; Academic Press: Cambridge, MA, 1960; Vol. 10, pp 356–463.
54. Hamm, R. The Importance of Water Binding Capacity of Meat in Frankfurter-type Sausage. *Der Flies Wirtschaft.* **1973,** *53*, 73.
55. Hamm, R. On the Rheology of Minced Meat. *J. Texture Stud.* **1975,** *6*, 281–296.
56. Hasting, R. J.; Rodger, G. W.; Park, R.; Matthews, A. D.; Anderson, E. M. Differential Scanning Calorimetry of Fish Muscle: The Effect of Processing and Species Variation. *J. Food Sci.* **1985,** *50*, 503–506.
57. Hermansson, A. M. Functional Properties of Proteins for Foods. *Labensm. Wiss. Technol.* **1972,** *5*, 24–29.
58. Hermansson, A. M. Physico Chemical Aspects of Soil Protein Structure Formation. *J. Texture Studies*. **1978,** *9*, 33–58.
59. Hermansson, A. M.; Servik, B.; Skjoldebran, M. D. C. Functional Properties of Protein: Swelling of Fish Protein Concentrates. *Chem. Alas*. **1972,** *76*, 981–994.
60. Holmes, K. L.; Noguchi, S. F.; Macdonald, G. A. The Alaska Pollock Resource and Other Species Used for Surimi. In *Surimi Technology;* Lanier, T., Lee, C. M., Eds.; Marcel Dekker: New York, NY, 1992; pp 41–78.
61. Hsu, S. Y. Effect of Frozen Storage and Other Processing Factors on the Quality of Surimi. *J. Food. Sci.* **1990,** *55*, 661–664.
62. Hultin, H. O.; Kristinsson, H. G.; Lanier, T. C.; Park, J. W. Process for Recovery of Functional Protein by pH Shifts. In *Surimi and Surimi Seafood*; Park, J. W., Ed.; CRC Press: Boca Raton, FL, 2005; pp 107–139.
63. Hunt, A.; Park, J. W.; Zoreb, H. Trehalose as Functional Cryprotectant for Fish Proteins, Abstract # 99–6, IFT Annual Meeting, Anaheim, CA, June 16–19, 2002.
64. Huxley, H. E.; Hanson, J. Changes in the Cross Striation of Muscle During Contraction and Stretch and Their Structural Interpretation. *Nature* **1954,** *173*, 973–976.
65. Huxley, H. E. Sliding Filaments and Molecular Motile Systems. *J. Biol. Chem.* **1990,** *265*, 8347–8350.
66. Imai, J.; Harayama, Y.; Kikuchi, K.; Kakinuma, M.; Watabe, S. DNA Cloning of Myosin Heavy Chain Iso Forms from Carp Fast Skeletal Muscle and Their Gene Expression Associated with Temperature Acclimation. *J. Exp. Biol.* **1997,** *200*, 27–34.
67. Iso, N.; Mizuno, H.; Ozawa, H.; Mochizuki, Y. Kinetic Analysis of the Sol–gel Transition of Fish Paste (Surimi). *Fisheries Sci.* **1994,** *60* (2), 163–164.
68. Iwata, K.; Okada, M.; Fujii, Y.; Miyamoto, K. Influence of Storage Temperature on Quality of Alaska Pollock Surimi. *Refrigeration* **1968,** *43*, 1145–1148.

69. Iwata, K.; Kann, K.; Okada, M. Kamaboko Formation in Mackerel and Red Sea Bream Myosin. *Bull. Jap. Soc. Sci. Fish.* **1977,** *43*, 237.
70. Iwata, K.; Kobashi, K.; Hase, J. Studies on Muscle Alkaline Protease. II. Some Enzymatic Properties of Carp Muscle Alkaline Protease. *Bull. Jap. Sci. Fish.* **1971,** *40*, 189.
71. Jafarpour, A.; Hajiduon, H. A.; Rezaie, M. A Comparative Study on Effect of Egg White, Soy Protein Isolate and Potato Starch on Functional Properties of Common Carp (*Cyprinus carpio*) Surimi Gel. *J. Food Process. Technol.* **2012,** *3*, 190. doi:10.4172/2157-7110.1000190
72. Jiang, S. T.; Ho, M. L.; Lee, T. C. Optimization of the Freezing Conditions on the Mackerel and Amber Fish for Manufacturing Minced Fish. *J. Food Sci.* **1985,** *50*, 727–732.
73. Jiang, S. T.; Hwang, D. C.; Chen, C. S. Denaturation and Change in S H Group of Actomysin from Milkfish (*Chanos chanos*) During Storage at –20°C. *J. Agric. Food. Chem.* **1988,** *36*, 433–437.
74. Jiang, S. T.; Lan, C. C.; Tsau, C. Y. New Approach to Improve the Quality of Minced Fish Products from Freeze–Thawed Cod and Mackerel. *J. Food Sci.* **1986,** *51* (2), 310–312.
75. Joseph, D.; Lanier, T. C.; Hamann, D. D. Temperature and pH Effects of Transglutaminase Catalyzed "setting" of Crude Fish Actomyosin. *J. Food Sci.* **1994,** *59*, 1018–1023.
76. Joseph, J.; George, C.; Perigreen, P. A. Effect of Spices on Improving the Stability of Frozen Stored Mince. *Fish Technol.* **1992,** *29* (1), 30–34.
77. Joseph, J.; Perigreen, P. A.; Thampuran, N. Preparation & Storage of Cutlet from Low Priced Fish. *Fish Technol.* **1984,** *21*, 70–74.
78. Kamath, G. G.; Lanier, T. C.; Foegeding, A. E.; Hamann, D. D. Non-disulphide Covalent Cross Linking of Myosin Heavy Chain in Setting of Alaska Pollock and Atlantic Croaker Surimi. *J. Food Biochem.* **1992,** *16*, 151–172.
79. Keay, J. N. *Minced Fish. Ministry of Agriculture, Fisheries and Food, Torry Research Station, Torry Advisory Note. 79, Food and Agriculture Organization* ; FAO Document Repository: Rome, 2001. http://www.fao.org/wairdocs/tan/x5950e/x5950e00.htm
80. Kimura, I.; Sugimoto, M.; Toya, D. A.; Seki, N.; Arai, K.; Fujita, T. A Study of the Cross-linking Reaction of Myosin in Kamaboko, "Suwari" Gels. *Nippon Suisan Gakk.* **1991,** *57*, 1389–1396.
81. Kinoshita, M.; Toyohara, H.; Shimizu, Y. Proteolytic Degradation of Fish Gel (Modori-phenomenon) During Heating Process. In *Chilling and Freezing of New Fish Products;* Aberdeen, Scotland, 1990.
82. Kinsella, J. E. Functional Properties of Proteins in Foods: A Survey. *CRC Crit. Rev. Food Sci. Nutri.* **1976,** *7*, 219–280.
83. Kinsella, J. E. Relationship between Structure and Functional Properties of Food Protein. In *Food Proteins;* Fox, P. F., Condon, J. J., Eds.; Applied Science Publishers: New York, NY, pp 51–103.
84. Kinsella, J. E. Functional Properties in Food Proteins. Thermal Modification Involving Denaturation and Gelation. In *Research in Food and Nutrition;* Mclongllin, J., Ed.; Book Press: Dublin, Ireland, 1984; p 226.
85. Klesk, K.; Yongsawatdigul, J.; Park, J. W.; Viratchakul, S.; Virhulakul, P. Gel Forming Ability of Tropical Tilapia as Compared with Alaska Pollock and Pacific Whiting Surimi. *J. Aqua. Food Product Technol.* **2000,** *9* (3), 91–104.
86. Ko, W. C.; Hwang, M. S. Contribution of Milk Fish Sarcoplasmic Protein to the Thermal Gelation of Myofibrillar Protein. *Fish. Sci.* **1995,** *61*, 75–78.

87. Kumazawa, Y.; Oozaki, Y.; Iwani, S.; Matsumoto, T.; Arai, K. Combined Preventive Effect of Inorganic Phosphate and Sugar on Freeze Denaturation of Carp Myofibrillar Protein. *Bull. Jap. Soc. Sci. Fish.* **1990,** *56*, 105–113.
88. Kuntz, I. D.; Kauzmann, W. Hydration of Protein and Polypeptide. *Adv. Protein Chem.* **1974,** *29*, 239–245.
89. Lanier, T. C. Functional Properties of Surimi. *Food Technol.* **1986,** *97*, 107–124.
90. Lanier, T. C. Measurement of Surimi Composition and Functional Properties, In *Surimi Technology;* Lanier, T. C., Lee, C. M., Eds.; Marcel Dekker Inc.: New York, NY, 1992; pp 138–144.
91. LeBlanc, E. L.; LeBlanc, R. J. Determination of Hydrophobicity and Reactive Groups in Protein of Cod (*Gadus morhua*) During Frozen Storage. *Food Chem.* **1992,** *43*, 3–4.
92. Ledward, D. A. Gelation of Gelation. In *Functional Properties of Food Macromolecules;* Mitchell, J. R., Ledward, D. A., Eds.; Elsevier Applied Science Publisher: New York, NY, 1986; p 171.
93. Lee, C. M. Surimi Process Technology. *Food Technol.* **1984,** *38* (11), 69–80.
94. Lee, C. M. A Pilot Plant Study of Surimi Making Properties of Red Ham (Urophycis chuss). In *Symp. On Engineered Sea Foods Including Surimi;* Martin, R., Collette, R., Eds.; Nakonal Fisheries Institute: Washington, DC, 1986; pp 225–243.
95. Lee, C. M. Surimi Manufacturing and Fabrication of Surimi-based Products. *Food Technol.* **1986,** *40* (3), 115–124.
96. Lee, H. G.; Lanier, T. C. The Role of Covalent Cross Linking in the Texturising of Muscle Protein Sols. *J. Muscle Foods.* **1995,** *6*, 125–138.
97. Li-Chan, E.; Kawai, L.; Nakai, S. Physico-chemical and Functional Properties of Salt-extractable Proteins from Chicken Breast Muscle Deboned after Different Post-mortem Handling Times. *Can. Inst. Food Sci. Technol. J.* **1986,** *19*, 241–248.
98. Lim, M. H. Studies of Reaction Kinetics in Relation to Thermal Behavior of Solutes in Frozen Systems. Ph.D. Thesis, University of California, Davis, 1989.
99. Lin, T. M.; Park, J. W. Solubility of Salmon Myosin as Affected by Conformational Changes at Various Ionic Strengths and pH. *J. Food Sci.* **1998,** *63*, 215–218.
100. Ma, L.; Grove, A.; Barbosa, Canovas, G. V. Viscoelastic Characterization of Surimi Gel: Effect of Setting and Starch. *J. Food Sci.* **1996,** *61*, 881–884.
101. Macdonald, G. A.; Lanier, T. C. Carbohydrates as Cryoprotectants for Meats and Surimi. *Food Technol.* **1991,** *45*, 150–159.
102. Macdonald, G. A.; Lelievre. J.; Wilson, D. C. N. Effect of Frozen Storage on the Gel-Forming Properties of Hoki (*Macruronus novaelzelancliae). J. Food Sci.* **1992,** *57*, 69–71.
103. Mackie, I. M. Advances in Analytical Methods for Evaluating the Effects of Freezing on the Properties and Characteristics of Fish Proteins. *Proc. M. O. C. A.* **1984,** *II*, 160–169.
104. Mackie, I. M. *The Effect of Freezing on Flesh Proteins*; Torry Document No., 1992, 2534, 1–59.
105. Mackie, I. M. Methods of Identifying Species of Raw and Processed Fish. In *Fish Processing Technology;* Hall, G. M., Ed.; Springer: New York, NY, 1997; pp 160–163.
106. Makinodan, Y.; Toyohara, H.; Niwa, E. Implication of Muscle Alkaline Proteinase in the Textural Degradation of Fish Meat Gel. *J. Food. Sci.* **1985,** *50*, 1351–1355.
107. Matsumoto, J. J.; Noguchi, S. F. *Control of Freeze Denaturation of Fish Muscle Proteins by Chemical Substance*, Proceedings XIII International Congress of Refrigeration. Washington, DC, 1971, pp 237–241.

108. Matsumoto, J. J.; Noguchi, S. F. Cryostabilisation of Protein in Surimi. In *Surimi Technology;* Lanier, T. C., Lee, C. M., Eds.; Marcel Decker Inc.: New York, NY, 1992; pp 357.
109. Matsumoto, J. J. Denaturation of Muscle Proteins During Frozen Storage. In *Proteins at Low Temperature;* Fennema, O., Ed.; ACS Symposium Series 180; American Chemical Society: Washington, DC, 1979; pp 205–224.
110. Matsumura, Y.; Mori, T. Gelation. In *Methods of Testing Protein Functionality;* Hall, G. M., Ed.; Blackie Academic and Professional: New York, NY, 1996; pp 76–109.
111. Montecchia, C. L.; Roura, S. I.; Rolandan, H.; Perej-Borla, O.; Crupkin, M. Biochemical Properties of Actomyosin from Frozen Pre-and Post Spawned Hake. *J. Food Sci.* **1997,** *62*, 491–495.
112. Moore, P. B.; Huxley, H. E.; DeRosier, D. J. Three-dimensional Reconstruction of F-actin, Thin Filaments and Decorated Thin Filaments. *J. Mol. Biol.* **1970,** *50* (2), 279–295.
113. Morioka, K.; Kurashima, K.; Shimizu, Y. Heat Gelling Properties of Fish Sarcoplasmic Protein. *Nippon Suisan Gakk.* **1992,** *58*, 767–772.
114. Morioka, K.; Nishimura, T.; Obatake, A. Changes in the Strengths of Heat Induced Gels from Myofibrils in Combination with Sarcoplasmic Proteins from Lizard Fish and Pacific Mackerel. *Fish. Sci.* **1998,** *64* (3), 503–504.
115. Morioka, K.; Shimizu, Y. Contribution of Sarcoplasmic Proteins to Gel Formation of Fish Meat. *Nippon Suisan Gakk.* **1990,** *56*, 929–933.
116. Mulvihill, D. M.; Kinsella, J. E. Gelation Characteristics of Whey Proteins and Beta Lactoglobulin. *Food Tech.* **1987,** *41*, 102–111.
117. Murray, R. K.; Granner, D. K.; Mayes, P. A.; Rodwell, V. W. *Harpers Biochemistry*; Prantice-Hall International Inc.: Englewood Cliffs, NJ, 1993.
118. Murthy, L. N.; Rajanna, K. B. Effect of Washing on Composition and Properties of Proteins from Tilapia (*Oreochromis mossambicus*) Meat. *Fish. Tech.* **2011,** *48* (2), 125–132.
119. Murthy, L. N.; Panda, S. K.; Rajanna, K. B. Effect of Cryoprotectants on the Functional Properties of Proteins from Tilapia (*Oreochromis mossambicus*) During Frozen Storage. *Fish. Tech.* **2012,** *49*, 155–160.
120. Murthy, L. N.; Panda, S. K.; Shamasundar, B. A. Physicochemical and Functional Properties of Proteins of Tilapia (*Orechromis mossambicus*). *J. Food Proc. Eng.* **2011,** *34*, 83–107.
121. Neiva, C. R. P.; Machado, T. M.; Tomita, R. Y.; Furlan, E. F.; Lemos-Neto, M. J.; Bastos, D. H. M. Fish Crackers Development from Minced Fish and Starch: An Innovative Approach to a Traditional Product. *Ciênc. Tecnol. Aliment. Campinas.* **2011,** *31* (4), 973–979.
122. Ninomiya, K.; Ookawa, T.; Tsuchiya, T.; Matsumoto, J. J. Concentration of Fish Water Soluble Protein and Its Gelation Properties. *Nippon Suisan Gakk.* **1990,** *56*, 1641–1645.
123. Nishiya, K.; Takeda, F.; Tamoto, K.; Tanaka, O.; Fukumi, T.; Kifabayashi, T.; Aizawa, S. Studies on Freezing of Surimi (Fish Paste) and Its Applications (IV): On Freezing Surimi of Atka Mackerel Meat, Mon. Rep. Hokkaido Municipal. *Fish. Exp. Stn.* **1961,** *18*, 391–397.
124. Niwa, E. Chemistry of Surimi Gelation. In *Surimi Technology;* Lanier, T. C., Lee, C. M., Eds.; Marcel Dekker Inc.: New York, NY, 1992; p 389.
125. Niwa, E.; Mastubara, Y.; Hamada, I. Hydrogen and Other Polar Binding in Fish Flesh Gel and Setting Gel. *Bull. Jap. Sci. Fish.* **1982,** *48*, 667–670.
126. Niwa, E.; Nakayama, T.; Hamada, I. State of Water in Fish Flesh Determined by IR Technique. *Nippon Suisan Gakk.* **1980,** *46*, 1147–1150.

127. Niwa, E.; Nakayama, T.; Hamada, I. Effects of Arylation for Setting of Muscle Proteins. *Agric. Biol. Chem.* **1981,** *45*, 341.
128. Noguchi, F. F. Dynamic Viscoelastic Changes of Surimi (Minced Fish Meat) During Thermal Gelation. *Bull. Jap. Soc. Sci. Fish.* **1986,** *52*, 1661–1270.
129. Noguchi, S.; Matsumoto, J. J. Studies on the Control of the Denaturation of Fish Muscle Properties During Frozen Storage-I. Preventive Effect of Na-glutamate. *Bull. Jap. Soc. Sci. Fish.* **1970,** *36*, 1078–1087.
130. Noguchi, S.; Matsumoto, J. J. Studies on the Control of Denaturation of Fish Muscle Proteins During Frozen Storage-IV. Preventive Effect of Carboxylic Acids. *Bull. Jap. Soc. Sci. Fish.* **1975,** *41* (3), 329–335.
131. Noguchi, S. The Control of Denaturation of Fish Muscle Protein During Frozen Storage. Doctoral Thesis, Sophia University, Japan, 1974.
132. Noguchi, S. Science of Frozen Surimi I, II and III. In *Practical Technical Handbook for Kneaded Seafoods;* Nippon Shokuhinkeizaisha: Tokyo, 1982; pp 40–60.
133. Noguchi, S.; Osawa, K.; Matsumoto, J. J. Studies on the Control of Denaturation of Fish Muscle Proteins During Frozen Storage-VI. Preventive Effect of Carbohydrates. *Bull. Jap. Soc. Sci. Fish.* **1976,** *42*, 77–82.
134. Numakura, T.; Nakamura, T.; Takama, K. F.; Fujie, T.; Arai, K. Preparation of Frozen Surimi from Pacific Pomfret Muscle. *Bull. Jap. Soc. Sci. Fish.* **1983,** *49* (12), 1863–1870.
135. Ogawa, M.; Ehara, T.; Tamiya, T.; Tsuchiya, T. Thermal Stability of Fish Myosin Comp. *Biochem. Physiol.* **1993,** *106B*, 517–521.
136. Ogawa, M.; Kanamaru, J.; Miyashija, H.; Tamiya, T.; Tsuchiya, T. Alpha-helical Structure of Fish Actomyosin: Change During Setting. *J. Food Sci.* **1995,** *60*, 297–299.
137. Ogawa, M.; Nakamura, S.; Horimoto, Y.; An, H.; Tsuchiya, T.; Nakai, S. Raman Spectroscopic Study of Changes in Fish Actomyosin During Setting. *J. Agric. Food Chem.* **1999,** *47*, 3309–3318.
138. Oguni, M.; Invove, N.; Ohi, K.; Shinano, H. Denaturation of Carp Myosin B During Frozen and Super Cooled Storage at −8 °C. *Bull. Jap. Soc. Sci. Fish.* **1987,** *53*, 789– 794.
139. Ohnishi, M.; Tsuchiya, T.; Matsumoto, J. J. Electron Microscope Study of the Cryoprotective Effect on Frozen Denaturation of Carp Actomyosin. *Bull. Jap. Soc. Sci. Fish.* **1978,** *44*, 755–762.
140. Ohnishi, M.; Tsuchiya, T.; Matsumoto, J. J. Kinetic Study on the Denaturation Mechanism of Carp Actomyosin During Frozen Storage. *Bull. Jap. Soc. Sci. Fish.* **1978,** *44*, 27–37.
141. Ojima, T.; Kawashima, N.; Inoue, A.; Amauchi, A.; Togashi, M.; Watabe, S.; Nishita, K. Determination of Primary Structure of Heavy Meromyosin Region of Walleye Pollock Myosin Heavy Chain by cDNA Cloning. *Fisheries Sci.* **1998,** *64*, 812–819.
142. Okada, M.; History of Surimi Technology of Japan. In *Surimi Technology;* Lanier, T. C., Lee, C. M., Eds.; CRC Press: Boca Raton, FL, 1992.
143. Pacheo-Aguilar, R.; Crawford, D. L.; Lampila, L. E. Procedures for the Efficient Washing of Minced Whitting (*Merluccius Products*) Flesh for Surimi Production. *J. Food Sci.* **1989,** *54*, 248–252.
144. Park, J. W. Surimi Technology–Past, Present and Future. Keynote Lecture at 8th World Fisheries Congress, 23–27th May, 2016, Busan, Korea. http://surimischool.org/wfc2016 (accessed June 18, 2016).
145. Park, J. W.; Morrissey, M. T. Manufacturing of Surimi from Light Muscle Fish. In *Surimi and Surimi Seafood*; Park, J. W., Ed.; Marcel Dekker Inc.: New York, NY, 2000; pp 23–58.

146. Park, J. W.; Lin, T. M. J. Surimi: Manufacturing and Evaluation. In *Surimi and Surimi Seafood*, 2nd ed.; Park, J. W., Ed.; CRC Press, Taylor & Francis Group: Boca Raton, FL, 2005; p 923.
147. Park, J. W.; Lanier, T. C. Combined Effects of Phosphates and Sugar or Polyols on Protein Stabilization of Fish Myofibrils. *J. Food Sci.* **1987,** *52*, 1509–1513.
148. Park, J. W.; Lanier, T. C.; Green, D. P. Cryoprotective Effects of Sugars, Polyols, and/or Phosphates on Alaska Pollock Surimi. *J. Food Sci.* **1988,** *53*, 1–3.
149. Park, J. W.; Lanier, T. C.; Keeton, J. T.; Hamann, D. D. Use of Cryoprotectants to Stabilize Functional Properties of Pre Rigor Salted Beef During Frozen Storage. *J. Food Sci.* **1987,** *52*, 537–542.
150. Park, T.; Yongasawatdigul, J.; Lin, T. M. Rheological Behavior and Potential Cross-Linking of Pacific Waiting (*Merluccius*) Surimi Gel. *J. Food Sci.* 1994, *59*, 773–776.
151. Perez-Mateos, M.; Lanier, T. C. Comparison of Atlantic Menhaden Gels from Surimi Processed by Acid or Alkaline Solubilization. *J. Food Chem.* 2006, *101*, 1223–1229.
152. Philips, L. G.; Whitehead, D. M.; Kinsella, T. *Structure-function Properties of Food Proteins.* Academic Press Inc.: New York, NY, 1994; p 271.
153. Pigott, G. M. Surimi: The "High Tech" Raw Material from Minced Fish Flesh. *Food Rev. Int.* **1986,** *2*, 213–246.
154. Pollard, T. D. Actin. *Curr. Opinion Cell Biol.* **1990,** *2*, 33–40.
155. Prakash, V. Partial Specific Volume and Interaction with Solvent Component of α-globulin from Seseamum L. in Urea and Guanidine Hydrochloride. *J. Biosci.* **1982,** *5*, 347–359.
156. Putro, S. Surimi-prospects in Developing Countries. *Infofish Intl.* **1989,** *5*, 29–32.
157. Ramirez, J. A.; Martin, P. L. M.; Bandman, E. Fish Myosin Aggregate as Affected by Freezing and Initial Physical State. *J. Food Sci.* **2000,** *65*, 556–560.
158. Rao, S. B. Denaturation of *Labeo rohita* (Rohu) Actomyosin on Frozen Storage- preservative Effect of Carbohydrates. *Fish Technol.* **1983,** *20*, 29–33.
159. Rao, S. B. Denaturation of *Labeo rohita* (Rohu) Actomyosin on Frozen Storage- preservative Effect of Carboxylic Acids. *Fish. Technol.* **1984,** *21*, 29–33.
160. Rao, S. B. Changes in *Labeo rohita (*Rohu) Actomyosin During Frozen Storage: Protein Denaturation or Coagulation. *Fish. Technol.* **1998,** *35*, 80–83.
161. Ratnakumar, K. Changes in the Properties of Major Protein Fraction from Pink Perch *(Nemipterus japonicus*) Meat During Processing. Ph. D. Thesis, University of Agricultural Sciences, Bangalore, 1999; pp 170.
162. Ratnaumar, K.; Shamasundar, B. A. Visco-elastic Properties of Pink Perch (*Nemipterus japonicus*) Meat During Setting and Gelation. In *Advances and Priorities in Fisheries Technology;* Balachandran, K. K., Iyer, T. S. G., Madhavan, P., Joseph, J., Perigreen, P. A., Raghunath, M. R., Varghese, M. D., Eds.; Society of Fisheries Technologists: Cochin, India, 1998; pp 251–256.
163. Rawdkuen, S.; Sai-Ut, S.; Khamsorn, S.; Chaijan, M.; Benjakul, S. Biochemical and Gelling Properties of Tilapia Surimi and Protein Recovered Using an Acid Alkaline Process. *J. Food Chem.* **2009,** *112*, 112–119.
164. Rayment, I.; Rypniewski, W. R.; Schmidt-Base, K.; Smith, Tomchick, D. R.; Benning, M. M.; Winkelmann, D. A.; Wesenberg, G.; Holden, H. M. Three-dimensional Structure of Actin-myosin Complex and Its Implications for Muscle Contraction. *Science* **1993,** *261*, 58–65.
165. Reddy, G. V. S.; Srikar, L. N. Pre-processing Ice Storage Effects on Functional Properties of Fish Mince Protein. *J. Food Sci.* **1991,** *56*, 965–968.

166. Regenstein, J. M. Total Utilization of Fish. *Food Technol.* **2004,** *58* (3), 28–30.
167. Regenstein, J. M.; Regenstein, C. E. *Food Protein Chemistry: An Introduction for Food Scientists.* Academic Press: New York, NY, 1984.
168. Regenstein, J. M.; Regenstein, C. E. Special Processing Procedure, Surimi. In *Introduction to Fish Technology;* Van Nostrand Rainhold: New York, NY, 1991; pp 139–147.
169. Rha, C.; Pradipsena, P. Viscosity of Proteins. In *Functional Properties of Food Macromolecules;* Mitecell, J. R., Ledward, D. A., Eds.; Elsevier Applied Science Publishers: London, 1986; pp 79–120.
170. Rose-Murphy, S. B. Rheological Methods. *Crit. Rep. Appl. Chem.* **1984,** *5*, 103–137.
171. Salde, L.; Levina, H. Beyond Water Activity: Recent Advances on an Alternative Approach to the Assessment of Food Quality and Safety. *Cri. Rev. Food Sci. Nutr.* **1991,** *30*, 115–360.
172. Samejima, K.; Ishioroshi, M.; Yasui, T. Heat Induced Gelling Properties of Actomysin. Effect of Tropomyosin and Troponin. *Agric. Biol. Chem.* **1982,** *46*, 535–540.
173. Sano, T.; Noguchi, S. F.; Matsumoto, J. J.; Tsuchiya, T. Thermal Gelation Characteristics of Myosin Sub Fragments. *J. Food Sci.* **1990,** *55*, 55–58.
174. Sano, T.; Noguchi, S. F.; Tsuchiya, T.; Matsumoto, J. J. Dynamic Viscoelastic Behavior of Natural Actomyosin and Myosin During Thermal Gelation. *J. Food Sci.* **1988,** *53*, 924–928.
175. Schein, C. H. Solubility as a Function of Protein Structure and Solvent Components. *Biotechnology.* **1990,** *16*, 295–301.
176. Schmidt, R. H. Gelation and Coagulation. *ACS Symp. Ser.* **1981,** *147*, 131–148.
177. Sehgal, H. S.; Sehgal, G. K. Aquacultural and Socio-economic Aspects of Processing Carps into Some Value-added Products. *Biores. Technol.* **2002,** *82*, 291–293.
178. Seki, N.; Nakahaa, C.; Jakeda, H.; Mayana, N.; Nozawa, H. Dimeization Site of Carp Myosin by Endogenous Transglutaminase. *Fish. Sci.* **1998,** *64* (2), 314–319.
179. Shaban, O.; Ochiai, Y.; Watabe, S.; Hishimoto, K. Quality Changes in Alaska Pollock Meat Paste (Surimi) During Frozen Storage. *Bull. Jap Soc. Sci. Fish.* **1985,** *51*, 1853–1858.
180. Shamasundar, B. A.; Prakash, V. Physicochemical and Functional Properties of Proteins from Prawn (*Metapenaeus dobsoni*). *J. Agric. Food Chem.* **1994,** *42*, 169–174.
181. Shamasundar, B. A.; Prakash, V. Effect of Sodium Acetate on Physico-Chemical Properties of Proteins from Frozen Prawn (*Metapenaeus dobsoni*). *J. Agric. Food Chem.* **1994,** *42*, 175–180.
182. Shimizu, Y.; Nishioka, F.; Machida, R.; Shien, C. M. Gelation Characteristics of Salt Added Myosin Sol. *Bull. Jap. Soc. Sci. Fish.* **1983,** *49*, 1239–1240.
183. Shimizu, Y.; Toyohara, H.; Lanier, T. C. Surimi Production from Fatty and Dark Fleshed Fish Species. In *Surimi Technology;* Lanier, T. C., Lee, C. M., Eds.; Marcell Dekker Inc.: New York, NY, 1992; pp 181–208.
184. Sikorski, Z.; Olley, J.; Kostuch, S. Protein Changes in Frozen Fish. *CRC Crit. Rev. Food Sci. Nutr.* **1976,** *8*, 97–129.
185. Sini, T. K.; Santhosh, S.; Joseph, A. C.; Ravishankar, C. N. Changes in the Characteristics of Rohu Fish (*L. rohita*) Sausage During Storage at Different Temperatures. *J. Food Process. Pres.* **2008,** *32* (3), 429–442.
186. Squire, J. M. *The Structural Basis of Molecular Contraction;* Plenum Press: London, 1981.
187. Stauffer, D.; Coniglio, A.; Adam, M. Gelation and Critical Phenomena. *Adv. Polm. Sci.* **1982,** *44*, 103–158.

188. Stockmayer, W. H. Theory of Molecular Size Distribution and Gel Formation in Branched-chain Polymers. *J. Chem. Phys.* **1943,** *11*, 45–55.
189. Stone, A. P.; Stanley, D. W. Mechanism of Fish Muscle Gelation-review Paper. *Food Res. Intr.* **1992,** *25*, 381–388.
190. Stryer, L. *Biochemistry*; W. H. Freeman and Company: New York, NY, 1995; p 1064.
191. Sun, C. T.; Wang, H. H. Cryoprotective Mechanism of Polyols and Monosodium Glutamate in Frozen Fish Mince. *3rd Intl. Cong. Eng. Foods Dublin*, 1984.
192. Suzuki, T. *Fish and Krill Protein: Processing Technology.* Applied Science Publishers: Barking, Essex, 1981.
193. Sych, J.; Lacroix, C.; Carrier, M. Determination of Optimal Level of Lactitol for Surimi. *J. Food Sci.* **1991,** *56*, 285—290.
194. Sych, J.; Lacroix, C.; Adambounu, L. T.; Castaigne, F. Cryoprotective Effects of Some Materials on Cod-surimi Proteins During Frozen Storage. *J. Food Sci.* **1990,** *55*, 1222–1227.
195. Takeda, F. Technological History of Frozen Surimi Industry. *New Food Ind.* **1971,** *13*, 27–31.
196. Tamoto, T.; Tanaka, O.; Takeda, F.; Fukumi, T.; Nishiya, K. Studies on Freezing of Surimi (Fish Paste) and Its Application (IV). The Effect of Sugar upon the Keeping Quality of Frozen Alaska Pollack Meat. *Bull. Hokkaido Reg. Fish. Res. Lab. Fish. Agency* **1961,** *23*, 50–60.
197. Tanaka, T. Gels. *Sci. Am.* **1981,** *244*, 124–138.
198. Tombs, M. P. Gelation of Globular Proteins. *Faraday Discuss Chem. Soc.* **1974,** *57*, 158–164.
199. Toyoda, K.; Kimura, I.; Fujita, T.; Naguchi, F. S.; Lee, C. M. The Surimi Manufacturing Process. In *Surimi Technology;* Lanier T. C., Lee, C. M., Eds.; Marcel Dekker Inc.: New York, NY, 1992; pp 99–112.
200. Toyohara, H.; Kinoshita, M.; Shimizu, Y. Proteolytic Degradation of Thread Fin Bream Meat Gel. *J. Food Sci.* **1990,** *55*, 259–260.
201. Tsuchiya, Y. The Nature of Cross Bridge Constituting Aggregates of Frozen Stored Carp Myosin. In *Advances in Fish Science and Technology;* Connell, J. J., Ed.; Fishing News Books Ltd.: UK, 1980; pp 434–438.
202. Venugopal, V. Mince and Mince-based Products, In *Seafood Processing: Adding Value Through Quick Freezing, Retortable Packaging and Quick Chilling;* CRC Press, Taylor & Francis Group: Boca Raton, FL, 2006; p 485.
203. Venugopal, V.; Ghadi, S. V.; Nair, P. M. Value Added Products from Fish Mince. *Asian Food J.* **1992,** *7*, 3–12.
204. Venugopal, V.; Shahidi, F. Value Added Products from Under-utilized Fish Species. *Crit. Rev. Food Sci. Nutr.* **1995,** *35* (5), 431–453.
205. Visessanguan, W.; Ogawa, M.; Nakai, S.; An, H. Physiochemical Changes and Mechanisms of Heat-induced Gelation of Arrow Tooth Flounder Myosin. *J. Agric. Food Chem.* **2000,** *48*, 1016–1023.
206. Wafaa, M. K.; Bakr, Walaa, A.; Hazzah, A.; Abaza, F. Detection of Salmonella and Vibrio Species in Some Seafood in Alexandria. *J. Am. Sci.* **2011,** *7* (9), 663–668.
207. Wagner, J. R. Denaturalization de Proteinas de Musculo Bovino Durante La Congelacion y el Almacenamiento Congelat - D. Thesis, Unversided Nacional de La Plata, Agrentina, 1986.

208. Wahyuni, M.; Ishizaki, S.; Tanaka, M. Improvement of Thermal Stability of Fish Water Soluble Proteins with Glucose-6-phosphate Through the Mail Lord Reaction. *Fish. Sci.* **1998,** *64*, 973–978.
209. Wang, B.; Xiong, Y. L. Functional Stability of Antioxidant Washed, Cryoprotectants Treated Beef Heart Surimi During Frozen Storage. *J. Food Sci.* **1998,** *63*, 293–298.
210. Wang, S. F.; Smith, D. M. *Heat Induced Gelation of Chicken Breast Actomyosin as Influence by Weight Ratio of Actin to Myosin, Paper No. 846*, Presented at the 53rd Annual Meeting of Inst, Food Technologist, Chicago, Il., July 10–14, 1993.
211. Wang, S. F.; Smith, D. M.; Stello, J. F. Effect of pH on the Dynamic Rheological Properties of Chicken Breast Salt Solution Proteins During Heat-induced Gelation. *Poultry Sci.* **1990,** *69*, 2220–2227.
212. Wu, M. C.; Akahane, M. T.; Lanier, T. C.; Hamann, D. D. Thermal Transition of Actomyosin and Surimi Prepared from Atlantic Croaker as Studied by Differential Scanning Calorimetry. *J. Food Sci.* **1985,** *50*, 10–13.
213. Xiong, Y. L. A Comparison of Rheological Characteristics of Different Fractions of Chicken Myofibrillar Proteins. *J. Food Biochem.* **1993,** *60*, 217–227.
214. Xiong, Y. L. Impacts of Oxidation on Muscle Protein Functionality. *Recip. Meat Conf. Proceed.* **1996,** *49*, 79–86.
215. Xiong, Y. L. Structure Function Relationships of Muscle Proteins. In *Food Proteins and Their Applications;* Damodaran, S., Paraf, A., Eds.; Marcel Dekker Inc.: New York, NY, 1997; pp 341–386.
216. Yasothai, R.; Giriprasad, R. Surimi Washing Process and Salting in and Salting out Method of Protein Extraction. *Int. J. Sci. Environ. Technol.* **2015,** *4* (1), 161–163.
217. Yasui, T.; Ishioroshi, M.; Samejima, K. Heat Induced Gelation of Myosin in the Presence of F-actin. *J. Food Biochem.* **1980,** *4*, 61–78.
218. Yongsawatdigul, J.; Park, J. W.; Dagga, Y. A.; Kolbe, E. Ohmic Heating Maximizes Gel Functionality of Pacific Whiting Surimi. *J. Food Sci.* **1995,** *60*, 10–14.
219. Yongsawatdigul, J.; Park, J. W.; Virulhakul, P.; Viratchakul, S. Proteolytic Degradation of Tropical Tilapia Surimi. *J. Food Sci.* **2000,** *65* (1), 129–133.
220. Ziegler, G. R.; Foegeding, E. A. The Gelation of Proteins. In *Advances in Food and Nutrition Research;* Kinsella J. E., Ed.; Academic Press Inc.: San Diego, CA, 1990; Vol. 34, p 204.
221. Zimm, B. H.; Lundberg, J. L. Sorption of Vapors by High Polymers. *J. Phys. Chem.* **1956,** *60*, 425–428.
222. Shenoy, A. V.; James, M.A. Freezing Characteristics of Tropical Fishes II. Tilapia (*Tilapia mossambica*). *Fish Technol.* **1972,** *9*(1), 37–41.
223. Srinivasan, S.; Hultin, H. O. Chemical, Physical, and Functional Properties of Cod Proteins Modified by a Nonenzymic Free-radical Generating System. *J. Agric. Food Chem*, **1997,** *45*, 310–320.

CHAPTER 15

HIGH-PRESSURE APPLICATIONS FOR PRESERVATION OF FISH AND FISHERY PRODUCTS

BINDU J.* and SANJOY DAS

ICAR–Central Institute of Fisheries Technology, Wellington Island, Matsyapuri P.O., Cochin 682029, India

*Corresponding author. E-mail: bindujaganath@gmail.com

CONTENTS

ABSTRACT

The desire for alternative technologies or minimal processing and preservation technologies that are environment friendly, low in cost, and able to preserve fresh quality attributes of the food has led to the introduction of many novel nonthermal technologies. Thermal pasteurization and thermal sterilization have resulted in making safe product with extended shelf life than its raw counterparts. But, the thermal process often resulted in significant changes in its sensorial attributes and its nutritional quality and ends with overprocessed food. Of the different nonthermal technologies in the vogue, only high-pressure (HPP) processing has commercial application in seafood. HPP had a significant effect on vegetative bacteria, virus, and bacterial endospores. Its application is varied depending on the end product and end users and in spite of its initial high-capital investment, the benefits of high pressure are recognized. So, HPP products find a niche market segment where unique high ended products are placed.

15.1 INTRODUCTION

The application of very high pressures (up to 87,000 psi, 6000 bar, or 600 MPa) for preservation of food substances in combination with or without heat is known as high-pressure processing (HPP). It is also known as high hydrostatic pressure processing or ultra high-pressure processing. HPP can be undertaken at ambient or refrigerated temperature. The technology is highly beneficial for heat sensitive products. Among the different nonthermal techniques in vogue, HPP is gaining in popularity because of its food preservation capability and products having fresher taste, better appearance, texture, and nutritional value when compared to thermal processing. The first HPP line was introduced in Japan for jam manufacture in 1990s and has since been upgraded to several food products. A number of HPP products have been commercialized in Japan, Spain, North America, Europe, and China. Machines are now available with operating pressures in the range 400–900 MPa and capacities ranging up to 900 kg per batch.

The consumers demand safe food but at the same time, they also prefer minimally processed foods in which the nutrients are not destroyed during food processing. HPP is considered as one of the very effective tool of destroying both pathogenic and spoilage microorganisms in food. Due to non-involvement of heat-treatment, HPP is very often referred as cold pasteurization of food. HPP treatment for preservation of food was initiated long ago by Hite,[23] who showed that pressure treatment of 600 MPa for 1 h at room temperature increases the shelf-life of raw milk by 4 days. But it is only in mid-1980s, when research was initiated for finding potential of application of HPP as an alternative method of food preservation. First high pressure processed food was marketed in Japan by Meidi-ya, who marketed jams, jelly, and sausages.[60] Due to its very low impact on covalent bonds, food does not under many chemical changes during HPP and thus it maintains the texture and sensory characters of food.

The two main principles of direct relevance to the use of high pressures in foods are the Le Chatelier's Principle and the Isostatic Principle. Le Chatelier's applies to all physical processes and states that when a system at equilibrium is disturbed the system responds in a way that tends to minimize the disturbance.[48] This means that HP stimulates reactions that result in a decrease in volume but opposes reactions that involve in an increase in volume. Second, the Isostatic means that pressure will be transmitted instantaneously and uniformly throughout the sample which may or may not be in direct contact with the pressure-transmitting medium. The time necessary for pressure processing is limited and independent of the sample size and quantity. As such the total process time is minimum.

The food product to be treated is generally packed in a flexible or semi-flexible container or pouch and placed in a pressure vessel capable of sustaining the required pressure. The product is submerged in pressure-transmitting medium, which is a liquid. Water is commonly used as the pressure-transmitting medium. Other liquids include castor oil, silicone oil, sodium benzoate, ethanol, glycol, etc., in various combinations with water or separately. The pressure-transmitting fluid should be able to protect the inner vessel from corrosion and liquids base on the manufacturer's specification is usually used. The process temperature range and the viscosity of the fluid under pressure are some of the factors involved in selecting the medium. Each processing cycle consists of an initial pressurization period wherein the pressure is build up and processing is undertaken with or without external application of heat to the product. The hydraulic fluid is pressurized with a pump, and this pressure is transmitted uniformly throughout the packaged food. HPP is independent of size and geometry of the food and acts instantaneously, thereby reducing the total processing time. The process is most suitable for liquid foods and solids which contain a certain amount of moisture. Since the pressure is transmitted uniformly and simultaneously in all directions, food retains its shape even at extreme pressures. Once the pressure is build up to the desired level the product is held at this pressure for a few minutes and then decompression or pressure release takes place. Once there is a fall in pressure the product temperature falls below that of the initial product temperature. During the pressurization process, adiabatic heating takes place and there is an increase in the temperature of the food product which is again dependent on the pressure transmitting fluid, product, pressurization rate, temperature, and pressure. In the case of water, the increase in temperature due to adiabatic heating is 3°C for every 100 MPa increase in pressure.

15.2 MAJOR ADVANTAGES OF THE TECHNOLOGY

1. It does not break covalent bonds; therefore, the development of flavors unacceptable to the product quality is prevented and the natural qualities of products are maintained.
2. It can be applied at room temperature, thus reducing the amount of thermal energy required unlike conventional processing.
3. Since HPP is uniform throughout the food, the preservation treatment is evenly distributed throughout the medium.
4. High pressure acts instantaneously and is not dependent on size or shape and therefore there is reduction in processing time.
5. The process is environment friendly since it requires only electric energy and there are no waste products.[60]

15.2.1 EFFECT ON VEGETATIVE FORM OF BACTERIA

Bacteria being the prokaryote are more pressure resistant than eukaryotes.[45] High pressure treatment causes many changes in bacterial cells, including alteration of cell wall and cell membrane, inhibition of key enzymes and inhibition of protein synthesis. The cell membrane is considered as primary targets of HP treatment and research in this regard has showed the relationship between pressure resistance and membrane fluidity.[58] The temperature of the medium or food matrices is also very much important in the microbial destruction point of view. The microorganisms generally show less pressure resistance at their optimal growth temperature than higher or lower temperature. This is because of the fact that the membrane fluidity is disturbed while placed in higher or lower temperature than the optimum temperature. The HP treatment also disrupts the ribosome of bacterial cell thus leading to inhibition of protein synthesis.[42] But the sensitivity of different microbes to HP treatment varies to a great extent among different bacterial species and strains and it also depends upon the different food matrices. For example, the microorganisms are more resistant to high-pressure treatment in milk as compared to meat and seafood. The presence of more carbohydrates, proteins, and lipid in the food render the microorganisms more resistant to pressure treatment.[56] HP treatment also causes condensation of nucleic acid within bacterial cell. The high hydrostatic pressure also causes alteration of morphology of bacterial cell including cell lengthening, pore formation, and also separation of cell wall from cell membrane.[6] Calcium, magnesium, and sucrose also act as protective agent against pressure destruction of micro-organisms.[8] The high pressure sensitivity of microorganisms also depends upon the water activity (a_w) and generally microorganisms are more pressure sensitive when water activity is more.[57] The presence of cholesterol makes the organisms more pressure sensitive. As per the pH is concerned, the microorganisms are more pressure sensitive, when the pH of the food is low.[1] Among bacterial species, Gram-positive bacteria generally show more resistance to HP treatment than Gram-negative bacteria. This is due to presence of thick cell wall in Gram-positive bacteria and more complex nature of Gram-negative bacterial cell wall.[52] In different study, it was also observed that the cocci groups of bacteria are more resistant to pressure treatment than rod-shaped bacteria.[57] The high pressure sensitivity of the bacteria also depends upon the growth phase during pressure treatment. Generally, bacteria in exponential phase of growth are more pressure sensitive compared to stationary phase.[20]

High-pressure destruction curves generally show a nonlinear curve with a tailing. Among nonspore-forming bacteria, generally heat-resistant organisms are more pressure resistance, but there are exceptions too. *Salmonella senftenberg* is the most heat resistant *Salmonella*, but this particular strain is more pressure sensitive than *Salmonella typhimurium.*[37] Among seafood-borne bacterial pathogens, *Listeria monocytogenes*, *Staphylococcus aureus*, and *Escherichia coli* O157:H7 are considered as more high pressure resistant than the other seafood borne pathogens. Pressure treatment of 600 MPa for 3 min can reduce the population of *L. monocytogenes* only by 3.85–4.35 $\log_{10}$ cfu/g in ready-to-eat (RTE) sliced ham and turkey meat[70] and it also found that to achieve 5 $\log_{10}$ reduction of this organisms in frankfurters cheese, 700 MPa pressure treatment for 9 min is required.[69] The pressure destruction kinetics of *L. monocytogenes* in Indian white prawn was studied and the pressure D value of this organism at 250, 300, 350, and 400 MPa pressure level was found 34.521, 11.806, 5.92, and 5.099 min, respectively.[10] It has been observed that 250 MPa pressure treatments for 20 min on cold smoked salmon could not significantly reduce the count of *L. monocytogenes*, but this pressure treatment was sufficient to increase the length of Lag phase of this organism.[32] *S. aureus* is also a barotolerant organism, which can withstand a very high hydrostatic pressure. Gervilla et al.[15] found that 500 MPa pressure treatment at 50°C for 15 min could reduce the population of *S. aureus* in milk by around 7.3 $\log_{10}$. The pressure D value of this organism at 450 MPa pressure level was found 20 and 16.7 min at 2°C and 25°C, respectively. Among Gram-negative foodborne pathogens, *E. coli* O157:H7 is the most pressure resistant. This particular strain of *E. coli* showed a very high-pressure resistance than the other strains of same species (*E. coli*). In orange juice of low pH (4.5), 550 MPa pressure treatments for 5 min duration is required to reduce the population of *E. coli* O157:H7 by around 6 $\log_{10}$. Among food-poisoning bacteria, *Yersinia enterocolitica* is considered as one of the most pressure-sensitive bacteria. In one study, 275 MPa pressure treatment for 15 min in phosphate buffered saline reduced the population of this organism by around 5 $\log_{10}$.[47] For similar level of reduction, *S. typhimurium* requires 350 MPa, *L. monocytogenes* requires 375 MPa, *Salmonella enteritidis* requires 450 MPa, *E. coli* O157:H7 requires 700 MPa and *S. aureus* requires 700 Mpa in case of same duration of pressure treatment.

High-pressure treatment was found to reduce the level of total viable count (TVC) and total *Enterobacteriaceae* count in Indian white prawn and yellowfin tuna leading to increase in their shelf like on chilled storage on ice.[16,28] In case of yellowfin tuna, 100, 200, and 300 MPa pressure treatments for 5 min reduced the TVC by 0.41, 1.03, and 1.54 $\log_{10}$ cfu/g, respectively, with respect to control (5.65 $\log_{10}$ cfu/g) immediately after pressure treatment. The total *Enterobacteriaceae* count was reduced by 1.5 and 1.9 $\log_{10}$ cfu/g, respectively, after 200 and 300 MPa pressure treatments. And 100 MPa pressure treatment showed no effect on total *Enterobacteriaceae* count.[28] In case of Indian white prawn (*Fenneropenaeus indicus*), 100, 270, 435, and 600 MPa pressure treatments for 5 min reduced TVC of mesophilic bacteria by 0.3, 1.78, 2.05, and 4.02 $\log_{10}$ cfu/g, respectively, as compared to control (6.5 $\log_{10}$ cfu/g). This reduced level of TVC after pressure treatment maintained as compared to control during chilled storage on ice. The control (untreated) and 100 MPa pressure-treated reached rejection level before 7 days of storage whereas, 270 and 435 MPa pressure treated samples reached rejection level before 21st and 28th days of storage. The samples treated with 600 MPa pressure was still acceptable after 28th days of storage.[16] In case of total *Enterobacteriaceae*, the reduction in the form of $\log_{10}$ cfu/g found was 0.74, 1.41, 3.02, and 3.17 $\log_{10}$ cfu/g, respectively, as compared to control (5.23 $\log_{10}$ cfu/g) after 100, 270, 435, and 600 Mpa pressure treatment, respectively.[16] Ramirez-Suarez and Morrissey[50] studied the effect of HP treatment on shelf-life of albacore tuna and observed that 275 and 310 MPa pressure treatment increases the shelf life of albacore tuna by >22 and >93 days, while stored on 4°C and −22°C, respectively. The profound reduction of total *Enterobacteriaceae* count was observed in case of octopus muscle due to pressure treatment of 200, 300, and 400 MPa for 15 min.[26] Miyao et al.[39] investigated the effects of HPP on microorganisms in surimi paste. All of the microbes were destroyed at 300–400 MPa, fungi showed highest sensitivity to HPP, followed by Gram-negative and Gram-positive bacteria. They also identified pressure resistant bacteria, for example, *Moraxella* spp. (viable at 200 MPa), *Acinetobacter calcoaceticus* (viable at 300 MPa), *Streptococcus faecalis* (viable at 400 MPa), and *Corynebacterium* spp. (viable at 600 MPa). They also reported a long lag phase in the growth curve of pressure-treated bacteria as compared to nontreated bacteria (e.g., *S. faecalis* subjected to 400 MPa showed a lag phase extended by 20 h). In the studies with pressurized minced mackerel meat, Fuji et al.[13] reported that in pressure-treated samples, *Bacillus*, *Moraxella*, *Pseudomonas*, and *Flavobacterium* spp. were totally inactivated, whereas *Staphylococcus* spp. and *Micrococcus* spp. dominated during storage after pressurization.

The main limiting factor of HP processing is very high cost of HP equipment. From microbiology point of view, the important limiting factor is that the sterility of the food cannot be achieved using HP processing alone. The sublethally injured cells, which generally do not come in count, are very frequently produced by high-pressure treatment.[37] The potential hazards are sublethally injured bacteria needs to be addressed in details. These sublethally injured microbial cells sometimes contribute to overestimation of microbial inactivation in terms of pressure destruction kinetics.

HPP is also very much effective in prevention of microbial spoilage. Microbial spoilage of food is caused by microbial growth on food leading to formation of several metabolites including amines, sulfides, alcohols, aldehydes, ketones, and organic acids with unacceptable off-flavors.[17] Among fish spoilage bacterial community, *Pseudomonas* spp., *Shewanella* spp., *Photobacterium phosphoreum*, some species of *Vibrio,* lactic acid bacteria, *Carnobacterium* spp., *Brochothrix thermosphacta,* swarming *Bacillus* spp., and sulfite-reducing *Clostridia* are the most important. Among different groups of bacteria, psychrotrophs, and H_2S forming bacteria are the most important spoilage forming bacteria. One positive aspect of microbial spoilage with reference to HP treatment is that most of the spoilage-producing bacteria are Gram-negative, which are generally sensitive to high pressure. But till date, sufficient research work has not been carried out on effect of HPP on individual species of spoilage bacteria as done in case of pathogenic bacteria.

The data on pressure destruction kinetics of fish spoilage bacteria is mostly unavailable. But the inhibitory effect of HPP on spoilage bacteria is mostly evidenced by extension of shelf-life of the product due to HP treatment.

15.2.2 EFFECT ON VIRUSES

In case of viruses, the sensitivity of HP treatment varies greatly among different viruses and in most of the cases, viral envelop is the target of HP treatment. For example, calicivirus is relatively high pressure labile and usually gets inactivated at 275 MPa for 5 min, but poliovirus is very much resistant to pressure treatment and can withstand 600 MPa pressure treatment for 1 h.[62] Very often, due to high-pressure treatment, the viruses lost infectivity even though they are not killed. This is because of fact that high pressure damages the receptors (e.g., hemagglutinin in case of rotavirus) and because of that it cannot attach to the host cell.[49] But it has been observed that immunogenic properties of the viruses are not inactivated due to high-pressure treatment and[49] these characteristics may help in future in development of vaccines.

15.2.3 EFFECT ON YEAST AND MOLDS

In comparison to different bacterial species, the yeast and molds are generally much more pressure sensitive. Generally, most of the yeast and molds species are inactivated in around 400 MPa pressure treatment, but the fungal ascospores need a higher pressure to get inactivated.

Apart from bacteria, viruses, and fungi (yeast and molds) high-pressure treatment was also found to inactivate bacterial exotoxins. The combination of ultra high-pressure homogenization and HPP was found very much effective in inactivation of *S. aureus* enterotoxin.[31]

15.2.4 EFFECT ON BACTERIAL ENDOSPORES

The extreme resistance of bacterial spores to both thermal and nonthermal treatment possessed a great public health hazards. The bacterial endospores are highly resistant to HP treatment and endospores of some bacterial species can withstand even 1200 MPa pressure treatment.[57] Even combination of heating and pressure treatment is not sufficient to kill the spores. Farkas and Hoover[12] observed that pressure treatment at 827 MPa for 30 min at 75°C was not sufficient to destroy spores of *Clostridium botulinum*. Gao and Ju[14] reported that combined treatment of heat, nisin, and high pressure is very much effective against *C. botulinum* spores. L′opez-Pedemonte et al.[30] observed that a treatment with low pressure for long duration (60 MPa for 210 min) followed by high-pressure treatment for short period (400 MPa for 15 min) can significantly inactivate *Bacillus cereus* spores in cheese and addition of nisin in milk augmented the inactivation of the spores in cheese. But they also found that addition of lysozyme did not make much difference. Among the endospores of different pathogenic bacteria, the spores of *Clostridium* spp. are the most pressure resistant.[45] The destruction of bacterial spores can be achieved to some extent by mild heat treatment before HP processing. This is because of the fact that heat treatment induces germination of spores to vegetative form, which is susceptible to subsequent pressure treatment. Mills et al.[38] observed that heat treatment at 80°C for 10 min followed by pressure treatment 400 MPa for 30 min can reduce *Clostridium sporogenes* spores by around 3 $\log_{10}$ value. Unlike the vegetative cells of bacteria, low pH does not augment the inactivation of bacterial spores by high pressure. In one study, it was observed that inactivation of spores of *Bacillus subtilis* was more at neutral pH rather than low pH.[63] So, it is recommended to acidify the food after low pressure induced germination of spores. This is because of fact that germination of spores is generally favored at neutral pH and the vegetative cells are more pressure sensitive at low pH. The pressure treatment itself is also considered as an inducing factor in germination of bacterial endospores.[38,64] So, the repeated cycles of pressure treatment can also be considered as one of the methods of inactivation of bacterial endospores. This type of repeated cycles of pressure treatment is called as oscillatory HP treatment and it generally involves treatment with low pressure followed by treatment with high pressure. Low pressure induces spore to germinate. It was observed that a single pressure treatment of 600 MPa for 60min at 70°C reduced the spores of *Bacillus stearothermophilus* by around 4 $\log_{10}$, whereas oscillatory treatment of six 600 MPa treatment for 5 min each reduced the same count by around 8 $\log_{10}$.[18] The cyclic pressure treatment also aids in release of dipicolinic acid, which is typical compounds present in the core of bacterial endospores. In a study on *B. subtilis* endospores, it was observed that spores germinated through low-pressure treatment was more sensitive to subsequent pressure treatment than the spores germinated through high pressure treatment.[64] The low or moderate pressure (50–300 Mpa) triggers germination of spores by activating the nutrients receptors of the spore leading to nutrients triggered germination pathway, but very high pressure does not trigger the same receptors. That is why, the treatment with low pressures aids more in the germination process rather than very high pressure. It has also been observed that pressure resistance of bacterial spores is not related to heat resistance. For example, the spore of *B. stearothermophilus* is one of the most heat resistant spores ever known, but much less pressure resistant than the spores of *B. subtilis*.[41] It is almost understood that high pressure alone is not sufficient to inactivate bacterial spores and the cyclic pressure treatment as done in the case of oscillatory HP treatment is also not enough to eliminate bacterial spores totally because of the fact that in almost all cases, some fraction of spores does not germinate during the pressure treatment. In this case, the most promising approach may be combination of both pressure and heat and this process is known as pressure-assisted thermal sterilization.[4]

15.3 COMBINATION TREATMENT WITH HPP

The present research on high-pressure inactivation of microorganisms has focused on combination treatment of high pressure and other inhibitory substances. The combination of mild heating and high pressure showed very high activity against most of the foodborne organisms.[57] Different substances like nisin, lysozyme, EDTA, etc., were also found to render the microorganisms more pressure sensitive.[22] Nisin is a bacteriocin, which is secreted by *Lactococcus lactis* subsp. *lactis* and this is most effective against Gram-positive bacteria. But when nisin is used in combination with pressure treatment, then the combined effect was able to inactivate both Gram-positive and Gram-negative bacteria. Black et al.[68] observed that combination of nisin and pressure treatment was very much effective in reducing the number of *E. coli*, *Pseudomonas fluorescens*, *Listeria innocua*, and *Lactobacillus viridescens* in milk. The application of alternating current (50 Hz) increases the high-pressure assisted bacterial destruction in *E. coli*. The lethal effect of high pressure was increased when the organisms were exposed to alternating current before and after high-pressure treatment.[53] Nisin is only bacteriocin, which is approved for use in food. The combined treatment of HPP and nisin was found very much effective in destroying *L. monocytogenes* in Indian white shrimp (*Fenneropenaeus indicus*). It was observed that the nisin (400 IU/mL) with 350 and 250 MPa pressure treatment was able to reduce the *L. monocytogenes* population in white shrimp muscle by 5.819 and 2.912 $\log_{10}$ cfu/g, respectively, with respect to control (untreated). But when only pressure was applied without nisin treatment, then 250, 350, and

450 MPa pressure treatments reduced the population by 0.496, 2.532, and 4.164 $\log_{10}$ cfu/g, respectively, as compared to untreated control. On the other hand, only nisin treatment could reduce the population of same by 1.965 $\log_{10}$ value only (Fig. 15.2a,b).[11] Apart from heat, low pH, and nisin; the other inactivating factors, which can be combined with high pressures are lactoperoxidase, alternating current, gamma irradiation, and ultrasound.

15.4 APPLICATIONS IN AQUATIC PRODUCTS

Seafoods are highly perishable and usually spoil faster than other muscle foods. They are more vulnerable to postmortem texture deterioration than other meats.[55] Fish is characterized by the presence of odorless compounds called trimethylamine oxide (TMAO) which on spoilage is converted to trimethylamine by bacterial enzymes, and used as the assessment of quality. Generally, volatile bases are produced in fish muscle by autolytic enzymes, putrefactive microorganisms, or by chemical reactions. HPP can play a vital role in reducing the microbial load and thereby maintaining the quality of the product without bringing about drastic changes in the raw product.

HPP can be applied in a wide area of aquatic foods. The process inactivates vegetative microorganisms, reduces bacterial contamination, and can eliminate pathogens like *E. coli*, *Salmonella*, and *Listeria* and spoilage organisms without affecting its inherent qualities. HPP is used in shellfish processing for 100% removal of meat from the shells and for reducing the microbial risks during raw seafood consumption. The application of HPP in muscle foods is either for tenderization of the muscle or for extension of shelf-life. High pressure can be used to modify functional properties of the food material to develop new gel-based products with desirable sensory attributes and mouthfeel, while simultaneously enhancing safety of raw seafood. Since the processing is usually done at low or moderate temperatures, this does not affect the covalent bonds, but disrupts secondary and tertiary bonds and reduces the enzymatic activity and thereby minimize loss in flavor bearing components.[61] High-pressure treatment can be combined with salting and smoking to improve the quality of smoked fish. Applications for marination and impregnation of desired flavors and colors in fish fillets and steaks can also be effectively undertaken. Pressure-assisted thermal processing for development of shelf stable ready to eat products is another promising area of research. Pressure-assisted freezing and thawing helps in retaining the microstructure of and reducing drip loss in fish products.

15.4.1 SHELF-LIFE EXTENSION

HP processed fish products should always be stored at chilled temperatures since the normal temperature process does not ensure safety of the product. Fish being a low acid food requires chilled temperatures for minimizing the bacterial growth and for extension of shelf-life. Only high acid products, like fruit juices can be stored at ambient temperatures. Extended shelf-life of more than 22 days was observed in HP treated albacore tuna stored at 4°C.[50] Pressure level of 220 MPa and a 30 min holding time were optimal and most effective in prolonging the storage period of tuna muscle (up to 9 days), as well as in reducing the proteolysis activity, texture degradation, TVB-N, and histamine formation.[67] HPP maintained low microorganisms levels, changed the color of the muscle, and induced the formation of high molecular weight polypeptides most likely through disulfide bonding promoting texture improvement and thereby improved the shelf-life of minced albacore muscle for more than 22 days at 4°C and more than 93 days at −20°C when high pressure of 275 and 310 MPa were applied for 2, 4, and 6°s to minced albacore muscle.[27]

HP treated abalone has a life of more than 65 days of shelf-life irrespective of HP treatment than control which had a shelf life of 30 days. Red mullet stored at 4°C[43] showed a shelf life of 12 days for untreated red mullet and, 14 and 15 days for treated red mullet at 220 MPa for 5 min at 25°C and 330 MPa for 5 min at 3°C, respectively. High pressure treated sliced raw squids had a high reduction in the psychrophilic count and the trimethylamine and biogenic amine content also reduced considerably.[25] Studies on pressure treated mussel samples at 500 MPa or higher and stored under 2°C showed that, the extension of shelf-life was evolved up to 14 days, when psychotropic count of 10^6–10^{7}°cfu/mL was assumed as the level of spoilage.[46] He also stated that fresh and untreated samples were most acceptable, while pressure treated samples with 600 MPa were also acceptable after 2 weeks of storage based on the sensory analysis of the cooked mussels. Kaur et al.[29] reported a shelf-life extension of 15 days in high pressure treated tiger shrimp at 435 MPa compared with 5 days in untreated sample. The vacuum packed and pressure treatment of 200 and 400 MPa resulted in a shelf-life extension up to 21 and 35 days, respectively, when compared to 14 days for untreated shrimp samples.[35]

15.4.2 SHUCKING OF MEAT OF SHELLFISHES

High pressure-treated oysters are also being commercially produced and marketed in the USA. Besides microbial preservation and shelf-life extension, high-pressure treatment helps in shucking of oyster meat out of their shells which is favorable. HPP application at 400 MPa resulted in 5 log reduction of target microorganisms in oysters and enabled 41 days of storage at 2°C.[34] Cruz-Romero et al.[9] compared the physical and biochemical changes in oysters subjected to following high-pressure treatment at 260 MPa for 3 min or heat treatment (cool pasteurization at 50°C for 10 min or traditional pasteurization at 75°C for 8 min) on the shucking yield and reported that HP treated was best among the three. Studies revealed that the optimum shucking pressures that caused minimum changes to pacific oyster appearance are in the range of 240–275 MPa.[21]

15.4.3 DEVELOPMENT OF FISH GELS

Hydrostatic pressure has been used to induce the gelation of different kinds of surimi[19] from pollack, sardine, skipjack tuna, and squid. Pressure-induced gels from marine species were smoother and more elastic than those produced by heat and considered to be organoleptically superior. Several studies have been reported on the use of pressure technology to improve the gel-forming ability. Nagashima et al.[40] applied high pressure to squid meat paste in order to improve its thermal gelation ability and reported that the breaking strength value of the gel formed by the pressure of 400 MPa was twice as high as that of thermal induced gel. Sareevoravikul et al.[51] applied high pressure at levels of 300–3740 atm for 30 min to formulate gels from bluefish meat paste, and the properties of the resulting gels were compared with those of heat-induced gels. Gels formed by pressure were more translucent as compared with those formulated by heat. The salt extractable protein and protein digestibility studies indicated that pressure treatment formed gels had less protein denaturation and were more digestible. The results from proteolytic activity studies showed that the pressure range used in this study had less effect on the integrity of endogenous proteases than heat. High-pressure treatment (up to 1000 MPa for 20 min) applied to squid meat paste before heating was effective in inducing thermal gelation of squid meat hence improving its lower thermal gelation ability.[40]

15.4.4 EFFECT ON LIPID OXIDATION

High-pressure treatment resulted in higher lipid oxidation thiobarbituric acid (TBA) values (mg malonaldehyde/kg) for pressure-processed samples when compared to the control samples. Similar results have been found for cold smoked Salmon when subjected to 300 MPa pressures.[33] This is mainly due to the fact that during high pressure treatment there is a possibility for the cell structure to break, and make available inter-cellular lipid deposits for oxidation. Another reason for the accelerated lipid oxidation during lower pressure treatment may be due to the auto oxidation of fat which is promoted by the release of metal ions from the denatured heme protein.[59] Lipid oxidation was also reported by Yagiz[65] in trout dark muscle at pressure levels beyond 300 MPa, and by Chevalier et al.,[7] where levels of 200 MPa for 30 min duration increased lipid oxidation in turbot muscle.[59] Hultin,[24] suggested that the changes occurring in the cell membrane during high-pressure treatment may also make fish muscle less susceptible to oxidation even though the heme proteins are released. Pressure treatment can also bring about changes in the fish so that the antioxidant properties of astaxanthin present in the muscle is more available to retard further oxidation.[54] A decrease in the lipid oxidation has been observed in dark muscle of salmon treated at 300 MPa when compared to 150 MPa and cooked salmon.[66]

15.4.5 CHANGES IN COLOR

Pressure treated samples appeared like cooked fish, due to the loss of transparency. HP causes denaturation of the myofibrillar and sarcoplasmic proteins which leads to the loss of translucency of flesh affecting the fish color. The whitening effect observed in the flesh after high pressure treatment may also be due to the denaturation of the globin protein and release of the heme as observed in cod treated to 608 MPa and mackerel treated subjected to high pressure.[44] The initial orange-pink color of the flesh of salmon which is the most desirable indicator of quality also decreased after pressurization which may be due to lipid oxidation due to degradation of the highly unsaturated carotenoids such as astaxanthin, one of the major pigments in Salmon. Some authors have observed that fish color changed from raw to cooked appearance depending on the pressure level applied and this in turn is depended upon the fish species.[36,72] described that bluefish and sheep head generally became lighter with increasing pressure beyond 200 MPa. Atlantic salmon is probably one of the most sensitive fish to high pressure induced color change.[2,66]

15.4.6 CHANGES IN TEXTURE

Denaturation of sarcoplasmic and myofibrillar proteins occur during high-pressure treatment. The hardness may be due to the unfolding of the sarcoplasmic proteins and formation of new hydrogen bonded linkages.[3] The myosin from fish source may denature by pressure and form a gel-like texture.[5] HPP also affects contractile proteins especially actin which unfolds at higher. Proteolytic enzymes in fish get activated at relatively low pressures and inactivated at higher pressures leading to changes in fish texture thus improving the tenderness.[43] Post-mortem changes of fish texture are mainly caused by modifications of myofibrillar proteins, due to both proteases action and to variation of physical and chemical conditions. TMAO-ase converts TMAO into dimethylamine and formaldehyde. This formaldehyde reacts with the proteins in fish and forms cross-links resulting in a decrease of the solubility of the proteins and a decrease of the tenderness. This occurs in the fish muscle during frozen storage when concentration of oxygen is low. Matser et al.[36] found out that texture of cod-like fish species can be changed by high pressure as the authors observed an increase in hardness at 200 and 400 MPa since high pressure can influence the enzyme activity. High pressure causes modification of hydrogen and hydrophobic bonds that bring about changes in protein structure which in turn alters water bond, protease activity, myosin gel formation, and sarcoplasmic proteins.[3]

15.5 RESEARCH INITIATIVES OF THE ICAR–CENTRAL INSTITUTE OF FISHERIES TECHNOLOGY ON HIGH-PRESSURE PROCESSING OF FISH PRODUCTS

15.5.1 HIGH-PRESSURE PROCESSING OF PRAWNS

The process parameters for arriving at optimum pressure time and pressuring rate for processing headless Indian white prawn was studied by using response surface methodology. About 15 combinations were statically evaluated and further validated based on chemical and microbiological parameters and the best combination was 250 MPa, pressure 400 ramp rate and 6 min holding time at a temperature of 25°C. Chilled storage evaluation of the prawn revealed that the pressure treated samples had a shelf-life of 32 and 35 days at 6°C and 2°C of storage temperature respectively whereas the control was rejected after 12 days. The 50 MPa pressure treated prawn muscle had a firmer texture than untreated sample. In pressure-treated sample fibers looked tightened and round. This observation was attributed to the fact that the extracellular space decreases when pressure increases, in relation to the compaction of muscle and the possible protein gel network formation which leads to a slightly higher hardness value (Fig. 15.1).

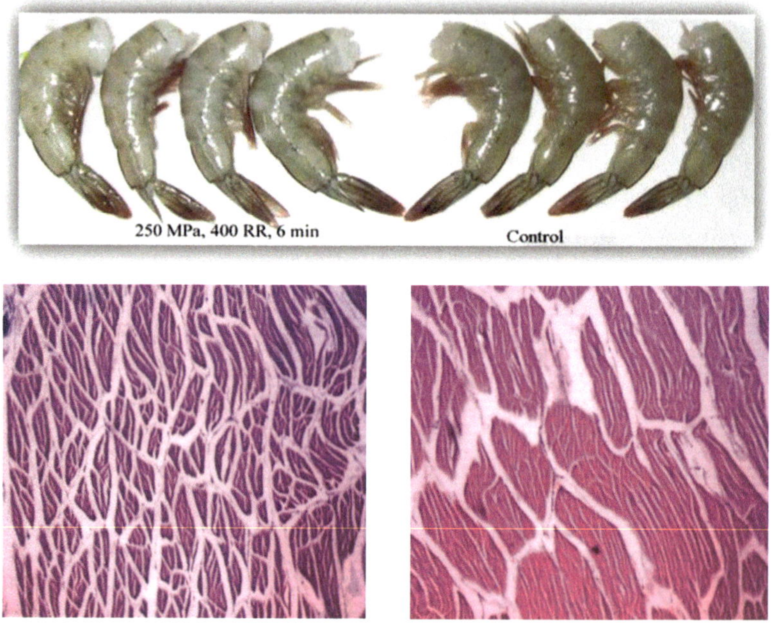

Control prawn (10 X) 250 MPa treated prawn

FIGURE 15.1 High pressure treated prawns and control.

15.5.1.1 READY TO FRY MARINATED PRAWN

Prawns were dry marinated with different flavors and masalas and subjected to high-pressure treatment of 200, 250, and 300 MPa. Then, 250 MPa showed better acceptability and had a shelf-life of 35 days (Fig. 15.2a,b).

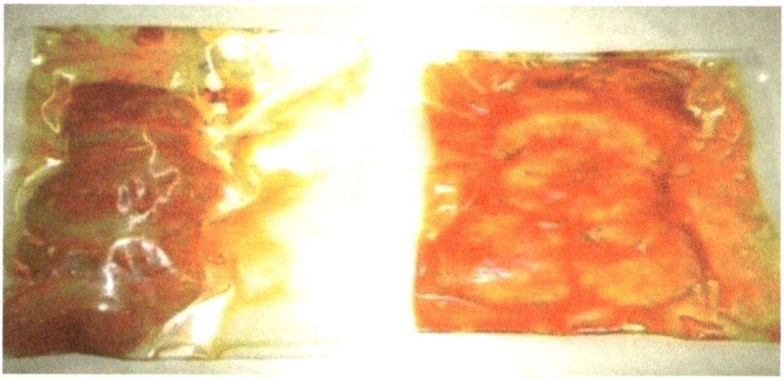

FIGURE 15.2 (a) Untreated marinated prawn. (b) Pressure-treated marinated prawn.

15.5.1.2 HIGH PRESSURE PROCESSED PRAWN CURRY

Ready to serve prawn curry was prepared and packed in ethyl vinyl alcohol (EVOH) pouches. The curry was then subjected to 100, 250, and 400 MPa pressure treatments with a holding time of 5 min. A control was also kept without pressure treatment. Samples were stored at 2 ± 1°C and analyzed for TVC and organoleptic characteristics. Control had an initial bacterial load of 2.2 $\log_{10}$°cfu/g and for 100, 250, and 400 MPa. It was 1.8, 1.33, and 0.8 $\log_{10}$°cfu/g, respectively. The prawn curry was acceptable up to 2 months microbiologically (Fig. 15.3).

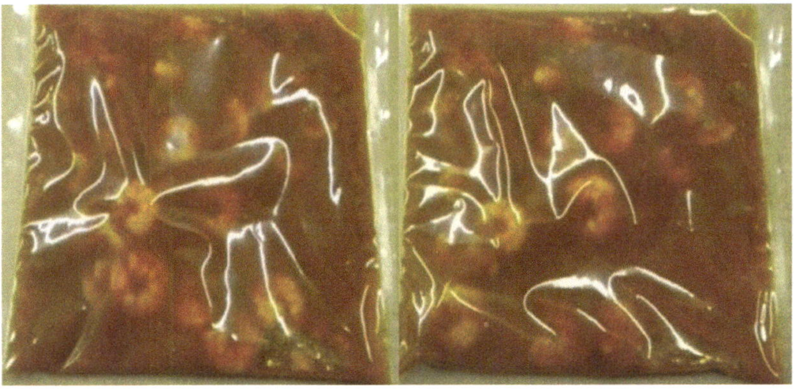

Control Pressure treated

FIGURE 15.3 High-pressure-processed prawn curry.

15.5.2 HIGH PRESSURE PROCESSING OF YELLOWFIN TUNA (Thunnus albacares)

Tuna chunks were packed under vacuum in EVOH multilayer films and subjected to three different pressures of 150, 200, and 250 MPa with a holding time of 5 min at a temperature of 25°C and subsequently stored at 2 ± 1°C. It was found that pH increases slightly during storage. TBA value got increased by HPP. TVB-N value got reduced by HPP. Free fatty acid content was less in treated samples compared to control and showed an increasing trend during storage. Texture (hardness) increased with pressure treatment. Color values (L^* and b^*) increased during storage, whereas a^* value showed slight decreasing trend during storage. TVC gave an increasing trend during storage (Fig. 15.4).

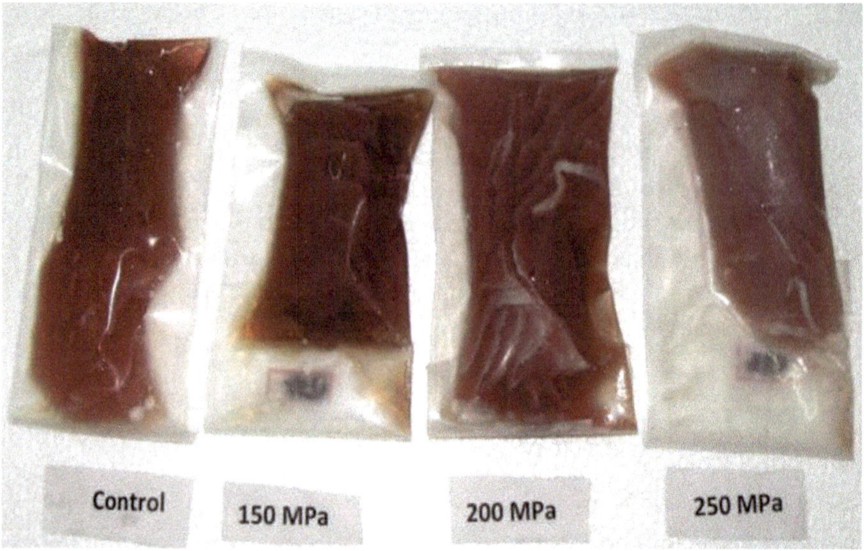

FIGURE 15.4 Tuna chunks.

15.5.2.1 EFFECT OF HIGH PRESSURE ON QUALITY OF YELLOWFIN TUNA (Thunnus albacares) CHUNKS IN EVOH FILMS DURING CHILL STORAGE

The effect of different high-pressure treatments on *K*-value, total plate count, *Enterobacteriaceae*, and organoleptic characteristics of yellowfin tuna chunks packed in EVOH films during chill storage (2 ± 1°C) was studied. An aliquot of 50 g of fresh tuna chunks were placed in EVOH multilayer film pouches and vacuum packed for the trials. Tuna chunks were subjected to 100, 200, and 300 MPa for 5 min at 25°C. Control was vacuum-packed tuna without pressure treatment. *K*-value, microbiological analysis, and sensory characteristics were evaluated at periodic intervals. The *K*-value of the samples was found to decrease with increase in pressure when compared to the control. However, *K*-value was found to increase in all the samples during storage. Higher-pressure treatment showed a decrease in the total plate count in the samples which increased during the storage. The *Enterobacteriaceae* decreased with increasing pressure and during storage. Control samples were sensorally acceptable up to 20 days of storage. During the storage period of 30 days, 200 MPa treated tuna chunks were found most acceptable based on above parameters.

15.5.2.2 QUALITY CHANGES IN HIGH PRESSURE TREATED MARINATED YELLOWFIN TUNA (Thunnus albacares) STEAKS DURING CHILL STORAGE

Effect of high pressure on marinated tuna steaks at 200 and 300 MPa pressure, 5 min hold time at 25–28°C to evaluate quality changes during chill storage (2 ± 1°C). Tuna steaks were marinated with 2% chili powder, 1.5% salt, and 0.1% turmeric and kept aside for 30 min. An aliquot of 100 g marinated tuna steaks were packed in EVOH pouches and pressure processed. Changes in biochemical parameters like pH, TBA value, trimethylamine nitrogen (TMA), total volatile base nitrogen (TVB-N), and instrumental hardness, and microbiological and sensory analysis during chill storage were evaluated periodically. Slight increase in pH in control and pressure treated samples was observed during chill storage. There was a significant difference in TBA values for 300 MPa when compared to 200 MPa and that of control. There was no significant difference in the TMA values during storage period. TVB-N values of treated samples were lower when compared to control and increased during storage. Hardness was higher in pressure treated sample and increased during storage. Microbiologically control 200 and 300 MPa treated samples had a shelf life of 22 days, 36 and 41 days, respectively. Sensory score also correlated with microbiological values. Based on the above parameters, 200 MPa treated samples were rated superior.

15.5.2.3 EFFECT OF HIGH PRESSURE ON MICRO FLORA ASSOCIATED WITH TUNA

Bacterial reduction pattern in high pressure processed tuna steaks was studied at 100, 200, 400, and 600 MPa with a holding time of 5 min. The aerobic plate count decreased from 1.6×10^6 cfu/g (control) and 1.2×10^2 cfu/g (600 MPa). The incremental log reduction observed for 100, 200, 400, and 600 MPa treatments were 0.38, 0.49, 1.41, and 4.12, respectively. The *Enterobacteriaceae* count reduced by 0.67 log cfu/g at high-pressure treatment of 200 MPa for 5 min. No count was observed from 300 MPa onwards.

15.5.2.4 EFFECT OF PRESSURE ON HISTAMINE FORMERS IN TUNA STEAKS

Tuna steaks were subjected to pressure treatment of 100, 200, 300, 400, 500, and 600 MPa and analyzed for changes in load of histamine forming bacteria. However, 2.4 log reduction of histamine forming bacteria was observed after exposure to 300 MPa. No histamine forming bacteria was observed on application at 400–600 MPa high pressure. *Morganella morganii* was inoculated to sterile tuna meat and exposed to high pressure of 200 and 300 MPa, holding time of 5 min at 28°C. In artificial medium, that is, BHI broth z (P) value of *M. morganii* culture ATCC25829 was estimated to be 66.52 MPa. *M. morganii* was inoculated to sterile tuna meat and subjected to high-pressure treatment of 200 and 300 MPa for 6, 24, and 48 h of interval. There was significant reduction in the load of *M. morganii* up to 6 h duration. The load of *M. morganii* in all HPP treated samples remained lower than in control. The load of *M. morganii* increased to 1.01×10^9 cfu/g in control, whereas it increased to 7.3×10^8 and 6.4×10^8 cfu/g in 200 and 300 MPa, respectively. Histamine content in control increased from 1 to 160 ppm in 48 h. It increased to 102 and 39.3 ppm in 200 and 300 MPa. At the end of 48 h there was comparable load of histamine forming bacteria in untreated and treated tuna meat. This result indicated the loss of histamine forming ability of *M. morganii* with HPP treatment. The threshold pressure required for inactivation of *Yesinia enterocolitis* in tuna steaks was observed to be 300 MPa. There was 0.8 log reduction of *Y. enterocolitis* exposed to 300 MPa for 5 min, but a sharp reduction of this pathogen was observed beyond 300 MPa. At 250 MPa the inactivation kinetic parameters of *Y. enterocolitis* in tuna meat were estimated to be as follows. *D* value 14.91 min. *K* is 0.15 min^{-1}. *Y. enterocolitis* being a psychrophilic pathogen, temperature assisted HP treatment was carried out. Tuna chunks were inoculated with *Y. enterocolitis* @ 9.8×10^5 cfu/g and subjecting to HP treatment at 250 MPa at 60°C for 3, 6, 9, and 12 min. With 3 min exposure at 60°C, 2.54 log reduction of this pathogen was noticed. The pathogen was completely eliminated in 6, 9, and 12 min holding time.

15.5.2.5 EVALUATION OF GEL STRENGTH OF FISH MINCE BY HIGH-PRESSURE TREATMENT

Studies on effect of low temperature on the gelling properties of fish mince from Japanese threadfin bream (*Nemipterus japonicas*) by high-pressure treatment was conducted based on a statistical design for determining the optimum gelling property. Fish mince was washed in chilled water 4°C in the ratio 1:4 (wt. of mince:volume of water) by constant stirring for 10 min and then the water drained by using a muslin cloth and excess water squeezed out. This process removes the water-soluble proteins and concentrates the myofibrillar protein. The washed mince was mixed with 2.5% salt in a silent cutter for 3 min then transferred into a sausage stuffer and stuffed into polypropylene casings of 5 cm diameter to a length of 3 inches. The edges of the casing are then clipped using a clipper and the same was then packed in covers made of EVOH films, vacuum sealed and subjected to high-pressure treatment. The following combination of pressure (200, 400, and 600 MPa), temperature (15–25°C, and 35°C) and holding time (10, 20, and 30 min) were tried out. The following parameters like pH, color (L^*, a^*, and b^* values), gel strength, folding test, and other instrumental texture parameters (hardness, cohesiveness, chewiness, and springiness), water holding capacity, TVB-N, and TVC were determined.

15.5.2.6 SHELF-LIFE EVALUATION OF TUNA SAUSAGE DURING CHILLED STORAGE

A study on the effect of high pressure on the biochemical, physical, and microbiological properties of tuna (*Thunnus albacares*) sausage was conducted. Tuna mince was washed twice in chilled water 4°C in the ratio 1:4 (wt. of mince:volume of water) by constant stirring for 10 min and then the water drained by using a muslin cloth and excess water squeezed out. All the ingredients (70% washed mince, 2.5% salt, 1.5% sugar, 0.2% polyphosphate, 9% corn starch, 0.1% guar gum, 5% hydrogenated fat, and 10% crushed ice) were mixed in a silent cutter for 3 min then transferred into a sausage stuffer and stuffed into polypropylene casings of 5 cm diameter to a length of 3 inches. The edges of the casing are then clipped using a clipper and the same was then packed in covers made of EVOH films, vacuum-sealed. Out of the four sets of samples, one was kept untreated, other was cooked, and the last two sets were subjected to high-pressure treatment. One set was cooked at 85–90°C for 30 min in a water bath and immediately cool it by putting in ice for 15 min. The following combination of pressure (500 and 600 MPa), temperature (24–29°C) and holding time (15 min) were tried out. All the sausages were chill stored at 2 ± 1°C. The following parameters like pH, TVB-N, gel strength, folding test, and other instrumental texture parameters (hardness, cohesiveness, chewiness, and springiness), total plate count, Brocothyrix count, Pseudomonas count, and Psychrophilic count were determined. The initial count of the raw sausage 4.08 log cfu/g was reduced to below 2.5 log cfu/g and the count in all sausages were below 3 log cfu/g throughout the storage. While the raw sample was rejected on 24th day, psychrophilic bacteria was below the detection limit immediately after high-pressure treatment and during 40 days of chill storage.

15.6 THE HIGH-PRESSURE EQUIPMENT

The high-pressure equipment consists of three modules, consisting of a console, a pressurizing unit, and a servicing unit. This equipment has capacity to pressurize up to 900 MPa. The packed product is pressurized in a thick vessel having a 2 L capacity. Thermocouples are provided to record the temperature at the top and bottom of the vessel. A percentage of 30 mono propylene glycol in water is used as the pressure transmitting liquid. The vessel can be operated at temperatures ranging from −20°C to 80°C. A chiller unit and heating facility are provided to control and maintain the temperature during the processing cycle.

15.7 CONCLUSIONS

Seafood is a highly perishable commodity and technologies like HPP are essential to increase the market value of some high-value fishes. HPP has a growing demand in the global market. A lot of studies are being done on HPP from the past decade. Further studies on the effects of this technology on the biochemical characteristics and microflora of shellfish are necessary. The effectiveness of high pressure on microbial and enzyme inactivation, while maintaining optimal product quality is a crucial factor for the commercialization of the technology. HP processing offers many advantages over conventional processing methods known to seafood. This is exemplified by the success of HP-processed oysters in USA by Motivatit Seafood, Goose Point Oysters, and Joey Oysters. However, as HP processing becomes more widely available, initial capital costs may be reduced, making technology accessible to more producers. In addition, the commercialization of the technology for other foods may provide encouragement for seafood processors, by allaying apprehension regarding the use of this novel technology and demonstrating consumer acceptance of HP-processed products. HPP of seafood is a promising method of preservation of food because high pressure inactivates the microbial population also at the same time it maintains the texture, sensory characters, and freshness. But, the huge initial investment due to high cost of high-pressure equipment is the limiting factor.

ACKNOWLEDGMENTS

The authors acknowledge the financial assistance provided by the National Agricultural Innovation Project (NAIP) (Grant No: NAIP/C4/C-30027/2008-09), Indian Council of Agricultural Research, for funding.

KEYWORDS

- **Gram-positive bacteria**
- **high pressure**
- **lactic acid bacteria**
- **bacterial endospores**
- **seafood**

REFERENCES

1. Alpas, H.; Kalchayanand, N.; Bozoglu, F.; Ray, B. Interactions of High Hydrostatic Pressure, Pressurization Temperature and pH on Death and Injury of Pressure-resistant and Pressure-sensitive Strains of Food-borne Pathogens. *Int. J. Food Microbiol.* **2000,** *60*, 33–42.
2. Amanatidou, A.; Schluter, O.; Lemkau, K.; Gorris, L. G. M.; Smid, E. J.; Knorr, D. Effect of Combined Application of High Pressure Treatment and Modified Atmospheres on the Shelf life of Fresh Atlantic Salmon. *Innovat. Food Sci. Emerg. Technol.* **2000,** *1* (2), 87–98.
3. Angsupanich, K.; Ledward, D. A. High Pressure Treatment Effects on Cod (*Gadus morhua*) Muscle. *J. Food Chem.* **1998,** *63* (1), 39–50.
4. Black, E. P.; Etlow, P.; Hocking, A. D.; Stewart, C. M.; Kelly, A. L.; Hoover, D. G. Response of Spores to High Pressure Processing. *Comp. Rev. Food Sci. Food Safety.* **2007,** *6*, 103–119.
5. Cheftel, J. C. Review: High-pressure, Microbial Inactivation and Food Preservation. *Food Sci. Technol. Int.* **1995,** *1*, 75–90.
6. Cheftel, J. C.; Culioli, J. Effects of High Pressure on Meat: A Review. *Meat Sci.* **1997,** *46* (3), 211–236.
7. Chevalier, D.; Le Bail, A.; Ghoul, M. Effects of High Pressure Treatment (100–200 MPa) at Low Temperature on Turbot (*Scophthalmus maximus*) Muscle. *Food Res. Int.* **2001,** *34*, 425–429.
8. Considine, K. M.; Kelly, A. L.; Fitzgerald, G. F.; Hill, C.; Sleator, R. D. High Pressure Processing: Effect on Microbial Food Safety and Food Quality. *FEMS Microbiol. Lett.* **2008,** *281*, 1–9.

9. Cruz-Romero, M.; Kelly, A. L.; Kerry, J. P. Effects of High-pressure and Heat Treatments on Physical and Biochemical Characteristics of Oysters (*Crassostrea gigas*). *Innovat. Food Sci. Emerg. Tech.* **2007,** *8*, 30–38.
10. Das, S.; Lalitha, K. V.; Ginson, J.; Bindu, J. Combination of Nisin and High Pressure Processing: An Effective Method of Inactivation of Listeria monocytogenes in White Shrimp. Presented at International Conference on Food Technology (INCOFTECH) at Indian Institute of Crop Processing Technology (IICPT), Thanjavur, India, January 4–5, 2013.
11. Das, S.; Lalitha, K. V.; Ginson, J.; Bindu, J. High Pressure Destruction Kinetics of Listeria monocytogenes in Indian White Prawn. Presented at International Conference on Innovation in Food Processing, Value Chain Management and Food Safety Held at National Institute of Food Technology Entrepreneurship Management (NIFTEM), Haryana, India, January 10–11, 2013.
12. Farkas, D. F.; Hoover, D. G. High Pressure Processing. Supplement-Kinetics of Microbial Inactivation for Alternative Food Processing Technologies. *J. Food Sci.* **2000,** *65*, 47–64.
13. Fuji, T.; Satomi, M.; Nakatsuka, G.; Yamaguchi, T.; Okuzumi, M. Changes in Freshness Indexes and Bacterial Flora during Storage of Pressurized Mackerel. *J. Food Hyg. Soc. Jpn.* **1994,** *35*, 195–200.
14. Gao, Y. L.; Ju, X. R. Exploring the Combined Effect of High Pressure and Moderate Heat with Nisin on Inactivation of *Clostridium botulinum* Spores. *J. Microbiol. Meth.* **2008,** *72*, 20–28.
15. Gervilla, R.; Sendra, E.; Ferragut, V.; Guamis, B. Sensitivity of *Staphylococcus aureus* and *Lactobacillus helveticus* in Ovine Milk Subjected to High Hydrostatic Pressure. *J. Dairy Sci.* **1999,** *82*, 1099–1107.
16. Ginson, J.; Kamalakanth, C. K.; Bindu, J.; Venkateswarelu, R.; Das, S.; Chauhan, O. P.; Gopal, T. S. K. Changes in K Value, Microbiological and Sensory Acceptability of High-pressure Processed Indian White Prawn (*Fenneropenaeus indicus*). *Food Bioprocess Technol.* **2013,** *6*, 1175–1180. DOI 10.1007/s11947-012-0780-2.
17. Gram, L.; Dalgaard, P. Fish Spoilage Bacteria-problems and Solutions. *Curr. Opin. Biotechnol.* **2002,** *13*, 262–266.
18. Hayakawa, I.; Kanno, T.; Yoshiyama, K.; Fujio, Y. Oscillatory Compared with Continuous High Pressure Sterilization of *Bacillus stearothermophilus* Spores. *J. Food Sci.* **1994,** *59*, 164– 167.
19. Hayashi, R. Use of High Pressure in Food. In *Engineering and Food;* Spiess, W. E. L., Schubert, H., Eds.; Elsevier: England, 1989; pp 815.
20. Hayman, M. M.; Anantheswaran, R. C.; Knabel, S. J. The Effects of Growth Temperature and Growth Phase on the Inactivation of *Listeria monocytogenes* in Whole Milk Subject to High Pressure Processing. *Int. J. Food Microbiol.* **2007,** *115*, 220–226.
21. He, H.; Adams, R. M.; Farkas, D. F.; Morrissey, M. T. Use of High Pressure Processing for Oyster Shucking and Shelf-life Extension. *J. Food Sci.* **2002,** *67*, 640–645.
22. Heuben, K. J. A.; Wuytack, E. Y.; Soontjes, C. C. F.; Michiels, C. W. High-pressure Transient Sensitization of *Eschericia coli* to Lysozyme and Nisin by Disruption of Outer-membrane Permeability. *J. Food Protect.* **1996,** *59*, 350–355.
23. Hite, B. H. The Effect of Pressure in the Preservation of Milk. *Bull. West Virginia Univ. Agric. Exp. Station* **1899,** *58*, 15–35.
24. Hultin, H. O. Oxidation of Lipids in Seafoods. In *Seafoods: Chemistry, Processing Technology and Quality;* Shahidi, F., Botta, J. R., Eds.; Chapman and Hall, Academy Press: London, 1994; pp 49–74.

25. Jingyu, G.; Hua, X.; Geun-Pyo, C.; Hyeon_Young, L.; Juhee, A. Application of High Pressure Processing for Extending Shelf Life of Sliced Raw Squid. *Food Sci. Biotechnol.* **2010,** *19* (4), 923–927.
26. Jose, L. H.; Montero, P.; Borderías, J.; Solas, M. High-pressure/temperature Treatment Effect on the Characteristics of Octopus (*Octopus vulgaris*) Arm Muscle. *Eur. Food Res. Technol.* **2001,** *213*, 22–29.
27. Juan, C.; Ramirez, S.; Morrissey, M. T. Effect of High Pressure Processing (HPP) on Shelf Life of Albacore Tuna (*Thunnus alalunga*) Minced Muscle. *Innovat. Food Sci. Emerg. Technol.* **2005,** *7*, 19–27.
28. Kamalakanth, C. K.; Ginson, J.; Bindu, J.; Venkateswarelu, R.; Das, S.; Chauhan, O. P.; Gopal, T. S. K. Effect of High Pressure on K-value, Microbial and Sensory Characteristics of Yellowfin Tuna (*Thunnus albacares*) Chunks in EVOH Films during Chilled Storage. *Innov. Food Sci. Emerg. Technol.* **2012,** *12*, 451–455.
29. Kaur, B.; Kaushik, N.; Rao, P. S.; Chauhan, O. P. Effect of High-Pressure Processing on Physical, Biochemical, and Microbiological Characteristics of Black Tiger Shrimp (*Penaeus monodon*). *Food Bioprocess Technol.* **2013,** *6* (6), 1390–1400.
30. L′opez-Pedemonte, T. J.; Roig-Sagues, A. X.; Trujilo, A. J.; Capellas, M.; Guamis, B. Inactivation of Spores of *Bacillus cereus* in Cheese by High Hydrostatic Pressure with Addition of Nisin and Lysozyme. *J. Dairy Sci.* **2003,** *86*, 3075–3081.
31. L′opez-Pedemonte, T.; Brine˜z, W. J.; Roig-Sagu′es, A. X.; Guamis, B. Fate of *Staphylococcus aureus* in Cheese Treated by Ultrahigh Pressure Homogenization and High Hydrostatic Pressure. *J. Dairy Sci.* **2006,** *89*, 4536–4544.
32. Lakshmanan, R.; Dalgaard, P. Effect of High Pressure Processing on *Listeria monocytogenes*, Spoilage Microflora and Multiple Compounds Quality Indices in Chilled Cold-smoked Salmon. *J. Appl. Microbiol.* **2004,** *96*, 398–408.
33. Lakshmanan, R.; Miskin, D.; Piggott, J. R. Quality of Vacuum Packed Cold-smoked Salmon during Refrigerated Storage as Affected by High-pressure Processing. *J. Sci. Food Agr.* **2005,** *85* (4), 655–661.
34. Lopez-Caballero, M. E.; Perez-Mateos, M.; Borderias, A. J.; Montero, P. Oyster Preservation by High Pressure Treatment. *J. Food Protect.* **2000,** *63*, 196–201.
35. Lopez-Caballero, M. E.; Perez-Mateos, M.; Borderías, A. J.; Montero, P. Extension of the Shelf Life of Prawns (*Penaeus japonicus*) by Vacuum Packaging and High-pressure Treatment. *J. Food Protect.* **2000,** *63* (10), 1381–1388.
36. Matser, A. M.; Stegeman,D.; Kals, J.; Bartels, P. V. Effects of High Pressure on Colour and Texture of Fish. *High Pressure Res.* **2000,** *19*, 109–115.
37. Metrick, C.; Hoover, D. G.; Farkas, D. F. Effects of High Hydrostatic Pressure on Heat Resistant and Heat Sensitive Strains of *Salmonella*. *J. Food Sci.* **1989,** *54*, 1547–1564.
38. Mills, G.; Earnshaw, R.; Patterson, M. F. Effects of High Hydrostatic Pressure on *Clostridium sporogenes* Spores. *Lett. Appl. Microbiol.* **1998,** *26*, 227–230.
39. Miyao, S.; Shindoh, T.; Miyamori, K.; Arita, T. Effects of High-pressure Processing on the Growth of Bacteria Derived from Surimi (Fish Paste). *J. Jpn. Soc. Food Sci. Technol.* **1993,** *40*, 478–484.
40. Nagashima, Y.; Ebina, H.; Tanaka, M.; Taguchi, T. Effect of High Hydrostatic Pressure on the Thermal Gelation of Squid Mantle Meat. *Food Res. Int.* **1993,** *26* (2), 119–123.
41. Nakayama, A.; Yano, Y.; Kobayashi, S.; Ishikawa, M.; Sakai, K. Comparison of Pressure Resistance of Spores of Six *Bacillus* Strains with Their Heat Resistances. *Appl. Environ. Microbiol.* **1996,** *62*, 3897–900.

42. Niven, G. W.; Miles, C. A.; Mackey, B. M. The Effects of Hydrostatic Pressure on Ribosome Conformation in *Escherichia coli*: An In Vivo Study Using Differential Scanning Calorimetry. *Microbiology* **1999,** *145*, 419–425.
43. Nuray, E.; Uretener, G.; Alpas, H. Effect of High Pressure (HP) on the Quality and Shelf Life of Red Mullet (*Mullus surmelutus*). *Innovat. Food Sci. Emerg. Technol. 11* (2), 259–264.
44. Ohshima, T.; Ushio, H.; Koizumi, C. High-pressure Processing of Fish and Fish Products. *Trends Food Sci. Technol.* **1993,** *4* (11), 370–375.
45. Patterson, M. High-pressure Treatment of Foods. *In The Encyclopedia of Food Microbiology*; Robinson, R. K., Batt, C. A., Patel, P. D., Eds.; Academic Press: New York, NY, 1999; pp 1059–1065.
46. Patterson, M. F.; Linton, M.; McCiements, J. M. J.; Farmer, L. J.; Johnston, D. E.; Dynes, C. Effect of High Hydrostatic Pressure on the Microbiological, Chemical and Sensory Quality of Fish, 1FT Annual Meeting, Chicago.
47. Patterson, M. F.; Quinn, M.; Simpson, R.; Gilmour, A. Sensitivity of Vegetative Pathogens to High Hydrostatic Pressure Treatment in Phosphate-buffered Saline Treatment and Foods. *J. Food Protect.* **1995,** *58*, 524–529.
48. Pauling, L. *College Chemistry: An Introductory Textbook of General Chemistry;* W. H. Freeman and company: San Francisco, CA, 1964.
49. Pontes, L.; Cordeiro, Y.; Giongo, V.; Villas-Boas, M.; Barreto, A.; Araújo, J. R.; Silva, J. L. Pressure-induced Formation of Inactive Triple Shelled Rotavirus Particles is Associated with Changes in the Spike Protein VP4. *J. Mol. Biol.* **2001,** *307*, 1171–1179.
50. Ramirez-Suarez, J. C.; Morrissey, M. T. Effect of High Pressure Processing (HHP) on Shelf Life of Albacore Tuna (*Thunnus alalunga*) Minced Muscle. *Innovat. Food Sci. Emerg. Technol.* **2006,** *7*, 19–27.
51. Sareevoravikul, R.; Simpson, B. K.; Ramaswamy, H. S. Comparative Properties of Bluefish Gels Formulated by High Pressure and Heat. *J. Aquatic Food Prod. Technol.* **1996,** *5*, 65–79.
52. Shigehisa, T.; Ohmori, T.; Saito, A.; Taji, S.; Hayashi, R. Effects of High Pressure on the Characteristics of Pork Slurries and Inactivation of Microorganisms Associated with Meat and Meat Products. *Int. J. Food Microbiol.* **1991,** 12, 207– 216.
53. Shimada, K. Effect of Combination Treatment with High Pressure and Alternating Current on the Total Damage of *Escherichia coli* Cells and *Bacillus subtilis* Spores. In *High Pressure and Biotechnology;* Balny, C., Hayashi, R., Heremans, K., Masson, P., Eds.; John Libbey and Co. Ltd.: London, UK, 1992; pp 49–51.
54. Shimidzu, N.; Goto M.; Miki, W. Carotenoids as Singlet Oxygen Quenchers in Marine Organisms. *Fisheries Sci.* **1996,** *62* (1), 134–137.
55. Simpson, B. K. High Pressure Processing of Fresh Sea Foods. *Adv. Exp. Med. Bio.* **1998,** *434*, 67–80.
56. Simpson, R. K.; Gilmour, A. The Effect of High Hydrostatic Pressure on the Activity of Intracellular Enzymes of *Listeria monocytogenes*. *Lett. Appl. Microbiol.* **1997,** *25*, 48–53.
57. Smelt, J. P. P. M. Recent Advances in the Microbiology of High Pressure Processing. *Trends Food Sci. Technol.* **1998,** *9*, 152–158.
58. Smelt, J. P. P. M.; Hellemons, J. C.; Wouters, P. C.; van Gerwen, S. J. C. Physiological and Mathematical Aspects in Setting Criteria for Decontamination of Foods by Physical Means. *Int. J. Food Microbiol.* **2002,** *78*, 57–77.

59. Tanaka, M.; Zhuo, X. Y.; Nagashima, Y.; Taguchi, T. Effect of High Pressure on the Lipid Oxydation in Sardine Meat. *Nippon Suisan Gakk.* **1991,** *57*, 957–963.
60. Thakur, B. R.; Nelson, P. E. High-pressure Processing and Preservation of Food. *Food Rev. Int.* **1998,** *14* (4), 427–447.
61. Torres, J. A.; Velazquez, G. Commercial Opportunities and Research Challenges in the High Pressure Processing of Foods. *J. Food Eng.* **2005,** *67*, 95–112.
62. Wilkinson, N.; Kurdziel, A. S.; Langton, S.; Needs, E.; Cook, N. Resistance of Poliovirus to Inactivation by Hydrostatic Pressures. *Innov. Food Sci. Emerg. Technol.* **2001,** *2*, 95–98.
63. Wuytack, E. Y.; Boven, S.; Michiels, C. W. Comparative Study of Pressure-induced Germination of *Bacillus subtilis* Spores at Low and High Pressures. *Appl. Environ. Microbiol.* **1998,** *64*, 3220–3224.
64. Wuytack, E. Y.; Michiels, C. W. A Study on the Effects of High Pressure and Heat on *Bacillus subtilis* at Low pH. *Int. J. Food Microbiol.* **2001,** *64*, 333–341.
65. Yagiz, Y.; Kristinsson, H. G.; Balaban, M. O.; Marshall, M. R. Effect of High Pressure Treatment on the Quality of Rainbow Trout (*Oncorhynchus mykiss*) and Mahi mahi (Colyphaena hippurus). *J. Food Sci.* **2007,** *72* (9), 509–515.
66. Yagiz, Y.; Kristinsson, H. G.; Balaban, M. O.; Welt, B. A.; Ralat, M.; Marshal, M. R. Effect of High Pressure Processing and Cooking Treatment on the Quality of Atlantic Salmon. *Food Chem.* **2009,** *116*, 828–835.
67. Zahra, Z. Department of Agricultural and Food Chemistry. M.Sc. Thesis, MacDonald Campus of McGill University, Montreal, Quebec, Canada, 2004.
68. Black, E. P.; Kelly, A. L.; Fitzgerald, G. F. The Combined Effect of High Pressure and Nisin on Inactivation of Microorganisms in Milk. *Innov. Food Sci. Emerg. Technol.* **2005,** *6*, 286–292.
69. Lucore, L. A.; Shellhammer, T. H.; Yousef, A. E. Inactivation of *Listeria monocytogenes* Scott A on Artificially Contaminated Frankfurters by High-Pressure Processing. *J. Food Prot.* **2000,** *63* (5):662–664.
70. Myers, K.; Montoya, D.; Cannon, J.; Dickson, J.; Sebranek, J. The Effect of High Hydrostatic Pressure, Sodium Nitrite and Salt Concentration on the Growth of *Listeria monocytogenes* on RTE Ham and Turkey. *Meat Sci.* **2013,** *93* (2):263–268.
71. Nagashima, Y.; Ebina, H.; Tanaka, M.; Taguchi, T. Effect of High Hydrostatic Pressure on the Thermal Gelation of Squid Mantle Meat. *Food Res. Int.* **1993,** *26*, 119–123.
72. Ashie, I. N. A.; Simpson, B. K. Application of High Hydrostatic Pressure to Control Enzyme Related Fresh Seafood Texture Deterioration. *Food Res. Int.* **1996,** *29*(5–6), 569–575.

INDEX

L

M

N

O

P

R

S